# The JCT 05 Standard Building Sub-Contract

*To Carol*

*Gracias para su paciencia y comprensión.*

# The JCT 05 Standard Building Sub-Contract

Peter Barnes

**Blackwell**
Publishing

Blackwell Publishing editorial offices:
Blackwell Publishing Ltd, 9600 Garsington Road, Oxford OX4 2DQ, UK
Tel: +44 (0)1865 776868
Blackwell Publishing Inc., 350 Main Street, Malden, MA 02148-5020, USA
Tel: +1 781 388 8250
Blackwell Publishing Asia Pty Ltd, 550 Swanston Street, Carlton, Victoria 3053, Australia
Tel: +61 (0)3 8359 1011

First published 2006 by Blackwell Publishing Ltd

ISBN-13: 978-1-4051-4048-5
ISBN-10: 1-4051-4048-8

Library of Congress Cataloging-in-Publication Data
Barnes, Peter.
The JCT 05 standard building sub-contract/Peter Barnes. – 1st ed.
   p.   cm.
Includes bibliographical references and index.
ISBN-13: 978-1-4051-4048-5 (hardback : alk. paper)
ISBN-10: 1-4051-4048-8 (hardback : alk. paper)   1. Construction contracts–Great Britain.
2. Construction industry–Subcontracting–Great Britain.   I. Title.
  KD1641.B375 2006
  343.41′07869–dc22
2006021262

A catalogue record for this title is available from the British Library

Set in 9.5/11.5 pt Palatino
by SNP Best-set Typesetter Ltd., Hong Kong
Printed and bound in Great Britain
by TJ International Ltd, Padstow, Cornwall

The publisher's policy is to use permanent paper from mills that operate a sustainable forestry policy, and which has been manufactured from pulp processed using acid-free and elementary chlorine-free practices. Furthermore, the publisher ensures that the text paper and cover board used have met acceptable environmental accreditation standards.

For further information on Blackwell Publishing, visit our website:
www.blackwellpublishing.com/construction

# Contents

# Preface

It is commonly recognized that the vast majority of construction work is carried out by sub-contractors, and yet there is a dearth of books that deal with contract law in the specific context of sub-contracting.

Long gone are the days of the naive sub-contractor. Sub-contractors in the twenty-first century are generally extremely sophisticated and are keen to establish their rights and obligations, and main contractors and others that deal with sub-contractors need to keep fully abreast of developments in contract law and general law that affect sub-contracts and sub-contractors.

Because of the apparent lack of up-to-date law books that deal with the issues of contracts in respect of the contractor/sub-contractor relationship, I had planned to produce a more general book on sub-contracts. However, just as I was on the point of commencing my planned book, the Joint Contracts Tribunal announced that it was bringing out a new 2005 suite of contracts and sub-contracts, including a Standard Building Sub-Contract (which, as an option, incorporates sub-contractor's design conditions).

This particular sub-contract was designed specifically to be used where the main contract is the Standard Building Contract (with quantities, without quantities, or with approximate quantities).

The Standard Building Contract is the successor of the JCT 98 Standard Form of Building Contract, the form of main contract that is still used on a very large percentage of all construction works carried out in England and Wales.

Traditionally, where the main contract has been the JCT 98 Standard Form, the sub-contract used has been the domestic sub-contract form DOM/1, and this tradition largely continued despite the fact that in 2002 the JCT produced its own Standard Form of Domestic Sub-Contract for use with all versions of the Standard Form of Building Contract (i.e. the Standard Form of Domestic Sub-Contract 2002 Edition (DSC 2002)).

However, my understanding is that it is not planned to publish further editions of the DOM/1 form, and that the DSC 2002 (which did not appear to be heavily used in any event) has now been made largely redundant by the publication in 2005 of the new JCT suite of contracts and sub-contracts.

Given the above factors it is almost inevitable that the JCT Standard Building Sub-Contract will become the regular sub-contract form adopted by contractors when sub-contracting works under the JCT Standard Building Contract.

Therefore, in respect of this book, it seemed eminently sensible to provide a detailed commentary on the Standard Building Sub-Contract, whilst at the same time providing a review of the latest position on the relevant law (particularly as it affects sub-contractors); and this is what I have attempted to do.

Whilst this book is principally aimed at sub-contractors and main contractors, it will also be of particular interest to other construction professionals and lawyers who need to have an understanding of the contractual relationship and the allocation of risk between main contractors and sub-contractors utilizing the Standard Building Sub-Contract.

Although this book can be used as an academic guide, my approach to the subjects raised has, naturally, been influenced by my direct involvement in the contractor/sub-contractor relationship for over 30 years. Through this experience, I have faced nearly every type of problem that can occur in a contractor/sub-contractor relationship and I have attempted to incorporate some of the knowledge that I have gained through that experience into the text of this book.

As anyone involved with the law knows, the law is continually evolving, and this factor may change some of the conclusions that I have reached. Also, of course, it is quite possible that main contractors will amend the text of the Standard Building Sub-Contract in an attempt to gain an advantage.

I obviously have no control over either of these factors, but I believe that the principles of the Standard Building Sub-Contract, as dealt with within this book, will provide a very useful guide as to the possible effects that either a change in the law, or amendments made to the text, will have on the parties' respective rights and obligations.

Peter Barnes
June 2006

# Definitions

In the JCT 2005 Standard Building Sub-Contract, certain defined terms, as listed under clauses 1.1 and 6.1 of SBCSub/C and SBCSub/D/C are included in capitalized form. This convention is not generally followed within this book (see further commentary on definitions on pages 55 and 56).

However, for the ease of general understanding, the definitions of certain defined terms are provided below.

| | |
|---|---|
| Article | An article in the sub-contract agreement (SBCSub/A or SBCSub/D/A). |
| Bills of quantities | Bills of quantities supplied by the contractor and included in the numbered documents within SBCSub/A or SBCSub/D/A. |
| Business day | Any day which is not a Saturday, a Sunday or a public holiday. |
| Completion date | The date for completion of the main contract works or of a section of the main contract works. |
| Contract bills | The contract bills as defined in the main contract. |
| Contractor's persons | The contractor's employees and agents and all other persons employed or engaged on or in connection with the main contract works or any part of them or any other person properly on the site in connection therewith excluding the architect/contract administrator, the quantity surveyor, the employer, the employer's persons, the sub-contractor, the sub-contractor's persons and any statutory undertaker. |
| Contractor's requirements | The documents relating to the sub-contractor's designed portion referred to in the fourth recital and included in the numbered documents. |
| Employer's persons | All persons employed, engaged or authorized by the employer, excluding the contractor, the contractor's persons, the architect/contract administrator, the quantity surveyor, the sub-contractor, the sub-contractor's persons and any statutory undertaker. |
| Excepted risks | Ionizing radiations or contamination by radioactivity from any nuclear fuel or from any nuclear waste from the combustion of nuclear fuel, radioactive toxic explosive or other hazardous properties of any explosive nuclear assembly or nuclear component thereof, pressure waves caused by aircraft or other aerial devices travelling at sonic or supersonic speeds. |

| | |
|---|---|
| Health and safety plan | Those parts of the health and safety plan for the main contract works applicable to the sub-contract works and annexed to the schedule of information, together with any developments of it by the principal contractor. |
| Main contract particulars | The contract particulars for the main contract annexed to the schedule of information. |
| Numbered documents | The documents listed in the sub-contract particulars (item 15, SBCSub/A; item 16, SBCSub/D/A) and annexed to the sub-contract agreement. |
| Planning supervisor | The person named as such in the schedule of information or any successor appointed by the employer. |
| Pre-agreed adjustment | An adjustment to value and/or time agreed between the contractor and sub-contractor before variation works are undertaken. |
| Principal contractor | The person named as such in the schedule of information or any successor appointed by the employer. |
| Provisional sum | A sum provided for work that the contractor may or may not decide to have carried out, or which cannot be accurately specified in the sub-contract documents. |
| Public holiday | Unless amended otherwise, Christmas Day, Good Friday or a day which under the Banking and Financial Dealings Act 1971 is a bank holiday. |
| Recitals | The recitals in SBCSub/A or SBCSub/D/A. |
| SCDP | Sub-contractor's designed portion. |
| SCDP analysis | The sub-contractor's price analysis of the SCDP. |
| SCDP documents | The contractor's requirements, the sub-contractor's proposals, the SCDP analysis and any further documents referred to in clause 2.6.1.1 of SBCSub/D/C. |
| SCDP works | That part of the sub-contract works comprised in the SCDP. |
| Schedule of information | The schedule of information in respect of the main contract in SBCSub/A or SBCSub/D/A. |
| Schedule of modifications | The schedule of modifications to the sub-contract conditions. |
| Section | One of the sections into which (where applicable) the main contract works have been divided, as referred to in the schedule of information. |
| Specified perils | Fire, lightning, explosion, storm, tempest, flood, escape of water from any water tank, apparatus or pipes, earthquake, aircraft and other aerial devices or articles dropped therefrom, riot and civil commotion, but excluding 'Excepted risks'. |

| | |
|---|---|
| Statutory requirements | Any statute, statutory instrument, regulation, rule or order made under any statute, directive, regulation or bye-law having the force of law which affects the sub-contract works or performance of any obligations under the sub-contract. |
| Statutory undertaker | Any local authority or statutory undertaker where executing work solely in pursuance of its statutory obligations. |
| Sub-contract documents | The documents referred to in article 1 of SBCSub/A or SBCSub/D/A. |
| Sub-contract particulars | The particulars as completed by the parties in SBCSub/A or SBCSub/D/A. |
| Sub-contractor's design | The drawings, details and specifications of materials, goods and workmanship and other related documents prepared by or for the sub-contractor in relation to the sub-contractor's designed portion. |
| Sub-contractor's persons | The sub-contractor's employees and agents and all other persons employed or engaged on or in connection with the sub-contract works or any part of them or any other person properly on the site in connection therewith excluding the architect/contract administrator, the quantity surveyor, the employer, employer's persons, the contractor, contractor's persons and any statutory undertaker. |
| Sub-contractor's proposals | The documents relating to the sub-contractor's designed portion referred to in the fifth recital and included in item 16 of the sub-contract particulars (SBCSub/D/A). |
| Terminal date | Whichever occurs first:<br><br>(a) the date of practical completion of the sub-contract works or, in respect of a section, of such works in the section, as determined in accordance with clause 2.20; or<br>(b) the date of termination of the sub-contractor's employment under the sub-contract, however arising. |
| Terrorism cover | Insurance provided by a joint names policy under insurance option A, B or C of the main contract conditions for physical loss or damage to work executed and site materials or to an existing structure and/or its contents caused by terrorism. |

# Chapter 1
# Background and Introduction

## *The purpose and use of the JCT Standard Building Sub-Contract*

The Joint Contracts Tribunal (JCT) was established in 1931 and for 75 years has produced standard forms of contracts, guidance notes and other standard documentation for use in the construction industry. The JCT became incorporated as a company limited by guarantee and commenced operation as such in May 1998.

Prior to May 1998, JCT forms were issued and updated by eleven constituent bodies (reduced to nine when the three local authority associations merged as the local government association). All new forms and amendments required the agreement of all nine bodies before they could be issued as JCT forms.

The nine bodies at that time were:

- the Association of Consulting Engineers
- the British Property Federation
- the Construction Confederation
- the Local Government Association
- the National Specialist Contractors' Council
- the Royal Institute of British Architects
- the Royal Institution of Chartered Surveyors
- the Scottish Building Contract Committee Limited
- the Specialist Engineering Contractors' Group

The Latham Report (*Constructing the Team*, 1994) suggested that only those bodies whose members would be a signatory to a form should have the right to approve and authorize its publication; and this suggestion was adopted by the JCT's constituent bodies (with the exception of the Specialist Engineering Contractors' Group).

Because of this, there are now only eight bodies that have the right to approve and authorize JCT forms – being those listed above, excluding only the Specialist Engineering Contractors' Group.

The listed bodies are intended to be reasonably representative of the interests across the construction industry, namely, the employers, the consultants, the contractors and the sub-contractors; and the JCT forms are naturally a reflection of these competing interests.

In respect of sub-contractors, although the National Specialist Contractors' Council is representative of a large proportion of sub-contractors, it could be said that the sub-contractors' representation has been reduced somewhat due to the fact that the Specialist Engineering Contractors' Group is no longer a body that has the right to approve and authorize JCT forms. The Specialist Engineering Contractors' Group estimates that the specialist engineering sector (which includes telecommunications, power and lighting, heating and ventilation, air conditioning, air movement and refrigeration, ductwork, plumbing, lifts and escalators) now accounts for over 25% of construction output, and states (on its web-page) that it represents over 60,000 companies, with a combined workforce of more than 300,000 people.

In 2005, the JCT published an entire new suite of contracts and sub-contracts, etc. and one of those documents was the JCT Standard Building Sub-Contract which is the subject matter of this book.

The JCT Standard Building Sub-Contract is only suitable for sub-contracts where the main contract is one of the three versions of the 2005 edition of the JCT Standard Building Contract, noted below:

- with quantities (SBC/Q);
- with approximate quantities (SBC/AQ);
- without quantities (SBC/XQ).

Each of these three versions of main contract may be used by both private and local authority employers, and each contains optional provisions for:

- a contractor's designed portion; and
- for the works to be divided into sections.

The JCT Standard Building Sub-Contract is *not* for use when the main contract is either:

- the 2005 edition of the JCT Intermediate Building Contract (denoted as 'IC') (the JCT published the Intermediate Sub-Contract Agreement and Conditions (ICSub/A and ICSub/C) for use with this version of main contract); or
- the 2005 edition of the JCT Design and Build Contract (denoted as 'DB') (the JCT published the Design and Build Sub-Contract Agreement and Conditions (DBSub/A and DBSub/C) for use with this version of the main contract).

For ease of general understanding, the JCT Standard Building Sub-Contract is similar to the JCT Standard Form of Domestic Sub-Contract 2002 edition (DSC 2002), which related to the JCT 98 Standard Form of Building Contract; and which was intended to replace the DOM/1 form of sub-contract.

The JCT Standard Building Sub-Contract is published in two versions:

- one version is denoted as SBCSub for use when the sub-contractor is *not* required to design any of the sub-contract works; and
- the other version is denoted as SBCSub/D which is for use when the sub-contractor is required to design a discrete part (or parts) of the main contract works (contractor's designed portion); and all, or part, of the sub-contract works.

The SBCSub/D is *not* to be confused with the DBSub.

As noted above, SBCSub/D is for use with the 2005 edition of the JCT Standard Building Contract *only* (even where there is a design element for the sub-contractor). SBCSub/D is, in effect, a modern day version of the well known DOM/1 form of sub-contract, with a design portion supplement.

On the other hand, DBSub is for use with the 2005 edition of the JCT Design and Build Contract *only*, and this new JCT sub-contract is in effect a modern day version of the DOM/2 form of sub-contract.

Both SBCSub and SBCSub/D can be used when the main contract works are to be carried out in sections; and each version comprises two documents:

- the agreement; and
- the conditions.

Where work is to be carried out in sections this must be detailed in the main contract. A section must have its own discrete commencement and completion date, and any liquidated damages applicable will relate to each individual section.

For the version without sub-contractor's design the agreement and the conditions are respectively:

- SBCSub/A; and
- SBCSub/C.

For the version with sub-contractor's design these are respectively:

- SBCSub/D/A; and
- SBCSub/D/C.

The JCT also published a guide to the JCT Standard Building Sub-Contract and gave this the designation of SBCSub/G.

This particular guide provides some explanation and clarification of certain clauses. However, it must be noted that SBCSub/G does not form part of the sub-contract between the contractor and the sub-contractor. Further, there is an overriding caveat on the back cover of SBCSub/G which says:

> 'All parties must rely exclusively upon their own skill and judgment or upon those of their advisers when using this document and neither Sweet & Maxwell nor its associated companies assume any liability to any user or any third party in connection with such use.'

The sub-contract agreements (SBCSub/A and SBCSub/D/A) incorporate by reference (under article 1 of SBCSub/A and SBCSub/D/A) the respective sub-contract conditions (SBCSub/C and SBCSub/D/C); therefore it is not necessary in formulating a sub-contract to include these conditions in the documents to be exchanged. Despite this, SBCSub/G recommends that, unless both parties regularly use the JCT Standard Building Sub-Contract, SBCSub/C or SBCSub/D/C (as appropriate) should be exchanged so that each party has a complete set of the documentation applicable to the sub-contract.

Within this book, when referring to JCT Standard Building Sub-Contract (2005 edition), the agreement, the conditions, the guide and the 2005 edition of the JCT Standard Building Contract, the following references will generally be used:

- JCT Standard Building Sub-Contract will be used when referring to the JCT Standard Building Sub-Contract (2005 edition).
- SBCSub/A and SBCSub/D/A will be used when referring to any matters that are *common to both versions* of the JCT Standard Building Sub-Contract Agreement.
- SBCSub/D/A will be used when referring to any matter that is *only* applicable to SBCSub/D version of the JCT Standard Building Sub-contract Agreement.
- SBCSub/C and SBCSub/D/C will be used when referring to any matters that are *common to both versions* of the JCT Standard Building Sub-Contract Conditions.
- SBCSub/D/C will be used when referring to any matter that is *only* applicable to the SBCSub/D version of the JCT Standard Building Sub-contract Conditions.
- SBCSub/G will be used when referring to the JCT Standard Building Sub-Contract Guide.
- The JCT Standard Building Contract will be used when referring to the 2005 edition of the JCT Standard Building Contract.

## Sub-contracting generally

As noted above, the JCT Standard Building Sub-Contract is for use with the JCT Standard Building Contract.

The relationship between these two JCT forms is:

- The employer *contracts* with the contractor using the JCT Standard Building Contract.
- The contractor *contracts* with each of his sub-contractors using the JCT Standard Building Sub-Contract, or, more accurately, the contractor *may* contract with each of his sub-contractors using the JCT Standard Building Sub-Contract, since, as will become evident later in this chapter, there is no obligation on a contractor to use the JCT Standard Building Sub-Contract when agreeing terms with its sub-contractors.

The key point is that, although the term 'sub-contract' is used in the latter case so that it can be distinguished from the main contract, and so that it is apparent that a sub-contractor is operating under the general 'umbrella' of a main contract, in both of the above cases a *contract* is formed between two parties only. It would be useful therefore to understand some basic principles of contract law.

Most aspects of the law on contract are set down in common law (i.e. case law). However, there are some notable exceptions where provision is made in statute (for example the Sale of Goods Act 1979, the Unfair Contract Terms Act 1977 and the Supply of Goods and Services Act 1982).

The basic principles of contract law provided below are, necessarily, dealt with in outline only in this book. If further in-depth analysis is required in respect of any of these, reference must be made to one of the standard works on the law of contract.

## What is a contract?

There are many definitions of a contract, but in simple terms it can be considered as being 'an agreement which gives rise to obligations which are enforced or recognized by law'.

Under English law, only the actual parties to a contract can acquire rights and liabilities under the contract. This is known as 'Privity of contract'.

In respect of a main contract situation, the practical consequences of the doctrine of privity of contract are two-fold:

- the main contractor carries responsibility for a sub-contractor's work, etc. so far as the employer is concerned; and
- the employer cannot take direct action in contract against the sub-contractor, unless there is a separate contract between the employer and the sub-contractor.

The effect that the Contracts (Rights of Third Parties) Act 1999 has on this position in respect of the JCT Standard Building Sub-Contract will be dealt with in Chapter 3.

## How is a contract formed?

The essence of any contract is *agreement*.

In deciding whether there has been an agreement, and what its terms are, the court looks for an *offer* to do or to forbear from doing something by one party, and an unconditional *acceptance* of that offer by the other party, turning the offer into a promise.

In addition, the law requires that a party suing on a promise must show that he has given *consideration* for the promise, unless the promise was given by *deed*.

Further, it must be the *intention of both parties to be legally bound* by the agreement, and the parties must have the *capacity* to make a contract, and any formalities required by law must be complied with.

Both the consideration and the objects of the contract *must not be illegal*.

If there is *fraud or misrepresentation* the contract may be *voidable*, while if there is a *mutual mistake* about some serious fundamental matter of fact this may have the effect of making the contract *void*.

Finally, there must be sufficient *certainty of terms*.

The other matter that needs to be considered is the significance, or otherwise, of 'Letters of intent'.

## Offer and acceptance (and 'invitations to treat')

In terms of offer and acceptance, the courts adopt what is sometimes known as the 'mirror image' rule of contract formation. That is to say, there must be a clear and unequivocal offer which is matched by an equally clear and unequivocal acceptance.

An offer is a statement by one party of a willingness to enter into a contract on stated terms, provided that these terms are, in turn, unequivocally accepted by the party or parties to whom the offer is addressed. There is generally no requirement that the offer be made in any particular form; it may be made orally, in writing or by conduct. Of course, if a dispute arose in the future then it would be beneficial for the offer to be in writing.

Care must be taken, however, in distinguishing between an offer and an invitation to treat. An invitation to treat is simply an expression of willingness to enter into negotiations which may lead to a contract at a later date. Therefore, a tender enquiry would normally be taken to be an invitation to treat, that is, it is not an offer, but an invitation to another party to make an offer.

Thus, when a contractor sends a tender enquiry to a sub-contractor, the tender enquiry would, in effect, be an invitation to the sub-contractor to provide an offer (or, in other words, submit a tender) to execute the required sub-contract works.

The question whether a statement is an offer or an invitation to treat depends primarily on the intention with which the statement was made. The actual wording of that statement is not necessarily conclusive.

Thus, a statement may be construed as being an invitation to treat even if it contains the word offer;[1] whilst on the other hand a statement may be an offer although it is expressed to be an acceptance,[2] or where it requests the person to whom it is addressed to make an offer.[3]

Of course, if the invitation to treat expressly provides that the issuing party is not to be bound merely by the receiving party's notification of acceptance, but only when he has signed the document in which the statement is contained, then it cannot be construed as being an offer.

Now, moving away from invitations to treat, an offer may be a sub-contractor's tender, *even if that tender does not comply with the tender enquiry requirements.*

This is a very important point, and is something that will be considered later in this chapter. But for now, it is enough that both the contractor and the sub-contractor are aware that a sub-contractor's tender does not have to match the contractor's tender enquiry letter to be a valid offer.

For example, a tender enquiry may ask a sub-contractor to carry out roofing works using Welsh slates. The sub-contractor may submit a tender on the basis of using Spanish slates (usually a much less expensive product). The fact that the tender enquiry (the invitation to treat) asked for Welsh slates does not detract in any way from the validity of the sub-contractor's tender as being an offer (even though an entirely different product was allowed for within the sub-contractor's tender to that required within the contractor's tender enquiry).

Another example may be where a tender enquiry asks that the sub-contractor carries out the required foundation works in five weeks; but the sub-contractor's tender (i.e. the offer) states that the sub-contractor will require ten weeks.

Clearly, there could be numerous other examples given, but the main point to note is that there is no need at all for the sub-contractor's tender to be a mirror image of the tender enquiry (the invitation to treat) for it to be a valid offer in its

---

[1] *Spencer* v. *Harding* (1870) LR 5 CP 561.
[2] *Bigg* v. *Boyd Gibbins Ltd* [1971] 1 WLR 913.
[3] *Harvis Investments Ltd* v. *Royal Trust Co. of Canada* (1789) 3 TR 148; *British Car Auctions Ltd* v. *Wright* [1972] 1 WLR 1519.

own right. The tender may have (and often does have) different terms and conditions to the initial tender enquiry document (or the invitation to treat).

In fact, for a sub-contractor's tender to fulfil the legal requirements of a valid offer it must simply be something which invites, and is intended by the person making the offer (i.e. the sub-contractor) to invite, acceptance.

In addition, an offer must be sufficiently definite to be capable of resulting in a contract if accepted; its terms and conditions must be clear and unequivocal, and it must be made with the intention that it is to become binding as soon as it is accepted by the person to whom it is addressed. In this context, a person includes a corporation because, in law, a corporation is a legal person; that is to say, a corporation is regarded by the law as a legal entity quite distinct from the person or persons who may for the time being be the member or members of the corporation.

With the above in mind, it must be noted by sub-contractors that the submission of a tender does not normally conclude a contract. Therefore, the preparation of a tender in response to a tender enquiry (which would, in the normal course of events, become an offer when submitted) may involve the sub-contractor in considerable expense, but the cost of tender preparation is not recoverable as a discrete cost, and the tenderer normally incurs this expense at its own risk. Obviously, the general cost of tender preparations is normally recovered within the head office overhead percentage that is added by sub-contractors on to their tenders, and the tender preparation costs so incorporated are therefore recovered by sub-contractors when their tenders are successful.

It is clear, therefore, that an offer is a statement by one party of a willingness to enter into a contract on stated terms, provided that these terms are, in turn, accepted by the party or parties to whom the offer is addressed.

As noted above, for agreement to be reached there must be a clear and unequivocal acceptance of a clear and unequivocal offer.

Therefore, if in the above example of the Welsh slates/Spanish slates, the contractor, upon receiving the sub-contractor's tender, accepted the sub-contractor's tender without qualification, then the contract, when formed, would be on the basis of the terms and conditions which formed part of the tender (i.e. based on using Spanish slates).

In such a situation, if a future dispute arose, the contractor would not be able to rely on the terms and conditions forming part of the tender enquiry (i.e. that Welsh slates were required) because those terms and conditions would not form part of the contract between the parties.

The general rule is that an acceptance has no effect until it is communicated (either in writing or orally) to the party making the offer. The main reason for this being that it could cause hardship to the party making the offer if he were bound without knowing that his offer had been accepted.

Of course, the terms of the offer may be such that the requirement of communication of acceptance is waived by the person making the offer, which is commonly the case in unilateral contracts.

A unilateral contract is a contract which results from an act made in response to an offer. As for example in its simplest terms, 'I will pay you five pounds if you wash my car.' No obligation to wash the car exists in law and the obligation to pay the five pounds only comes into being upon the performance of the invited act.

However, the person making the offer may not waive the requirement of communication in the sense of proclaiming that silence will constitute acceptance.[4]

The well known, but little used, 'postal rule' was considered in the *Adams* v. *Linsell*[5] case and only applies where postal acceptance is prescribed by the person making the offer, or is indicated by the manner or terms of the offer, or is the common sense mode of acceptance in the circumstances. The 'postal rule' is that postal acceptance takes effect when the letter of acceptance is posted and is in the control of the Post Office, not when the acceptance has been received by the person making the offer.

Another rule is that refusal of an offer puts an end to that offer.[6]

An offer can be revoked (i.e. withdrawn) at any time before it is accepted.[7] This rule even applies where an offer is stated to be open for a fixed time. Therefore, if a sub-contractor submitted a tender (or a quotation) and stated that it remained open for acceptance within 30 days, there would be nothing to prevent the sub-contractor from revoking that offer in a period of less than 30 days (unless the express terms of the sub-contract stated otherwise).

Further, if an offer is stated to be open for a fixed time, then it cannot (without the agreement of the party making the offer) be accepted after that time.

However, if no time is stated in the offer, then the offer is taken to lapse after a reasonable time. The word 'reasonable' is, of course, open to interpretation, but is based on the facts. Thus, in the case of *Quenerduaine* v. *Cole*,[8] it was found that if an offer was sent by telegram (at a time when telegrams were a common means of communication), then an acceptance by letter may well be held to be too late.

Alternatively, acceptance may be made by conduct[9] (e.g. allowing possession, making payments in line with the agreed terms, etc.) or by performance[10] (e.g. by carrying out and completing the sub-contract works, etc.), as appropriate.

If, however, instead of simply making an unqualified acceptance of the tender, the contractor submitted a 'qualified' acceptance (e.g. in the case of the Welsh slates/Spanish slates example noted above, the submitted price was acceptable but the terms and conditions would need to be in line with the tender enquiry, i.e. that Welsh slates would need to be used rather than Spanish slates as allowed within the tender) then this would not be an acceptance at all, because, as noted previously, an acceptance must clearly and unequivocally accept the offer.

If the acceptance does not clearly and unequivocally accept the offer, then it is simply a counter-offer.

A counter-offer has the same status as an offer in the formation of a contract, and consequently, a counter-offer must be clearly and unequivocally accepted before agreement has been reached.

Counter-offers should be distinguished from requests for information, which will not necessarily amount to a counter-offer[11] but, having said that, great care

---

[4] *Felthouse* v. *Bindley* (1862) aff'd (1863).
[5] *Adams* v. *Linsell* (1818) 1 B & Ald 681.
[6] *Hyde* v. *Wrench* (1840) 3 Beav 334.
[7] *Dickinson* v. *Dodds* (1876) 2 Ch D 463.
[8] *Quenerduaine* v. *Cole* (1883) 32 WR 185.
[9] *G. Percy Trentham* v. *Archital Luxfer Ltd* (1992) 63 BLR 44, CA.
[10] *Brogden* v. *Metropolitan Railway Co.* (1877) 2 App Cas 666.
[11] *Stevenson, Jaques & Co.* v. *McLean* (1880).

needs to be taken when requesting further information that this is not construed, in fact, as being a counter-offer.

In respect of construction works, in particular, there are frequently a whole series of counter-offers made before an acceptance is made. This may be because the parties are negotiating about the terms or because they are each trying to impose their own terms.

This latter situation is often called the 'battle of the forms', and variants of this battle of the forms are at the base of many disputes between contractors and sub-contractors.

### The battle of the forms

The expression 'battle of the forms' refers to an offer followed by a series of counter-offers where each party successively seeks to stipulate different terms, often based on their own standard printed terms. It is sometimes extremely difficult to determine whether or when a concluded contract came into existence, particularly where no formal contract is ever signed.

An illustration of this may be seen by reference to an actual case, for example, the *Chichester Joinery* v. *John Mowlem* case.[12]

In that case Mowlem was the main contractor for works at the Royal Holloway College, and Chichester Joinery was the prospective joinery sub-contractor. The exchange of offers/counter-offers was as follows.

- November 1984
  Chichester Joinery submitted a quotation accompanied by its standard terms and conditions.

- January 1985
  Mowlem sent Chichester Joinery a pro forma enquiry form; although the face of this document referred to conditions on its reverse, no conditions were printed on the original.

- January to March 1985
  At two subsequent meetings the parties discussed the proposed sub-contract terms.

- 14 March 1985
  Mowlem sent Chichester Joinery its purchase order, and this document stated:
  **'THE TERMS AND CONDITIONS OF PURCHASE NOS 1–18 INCLUSIVE AS SET OUT ON THE REVERSE HEREOF ARE EXPRESSLY DECLARED TO APPLY TO THE PURCHASE ORDER.**

    You are requested to sign the acceptance of order and return the same within seven days. Any delivery made and accepted will constitute an acceptance of this order.'

- 30 April 1985
  Chichester Joinery did not sign and return the acceptance of order but instead on 30 April 1985, before any delivery had been made, sent to John Mowlem a

---

[12] *Chichester Joinery* v. *John Mowlem* (1987) 42 BLR 100.

printed form headed **ACKNOWLEDGMENT OF ORDER**, which purported to accept Mowlem's order 'subject to the conditions overleaf'.

Also on the Acknowledgement of order form were the words '**A COPY OF THIS ACKNOWLEDGMENT IS ALSO A SPECIFICATION TO OUR WORK SO PLEASE READ THIS CAREFULLY**'.

On the back of the Acknowledgement of order form were the identical conditions to those on Chichester Joinery's original 1984 quotation.

At court it was found that the sub-contract was governed by Chichester Joinery's terms and conditions.

This judgment was made on the basis that Mowlem's purchase order dated 14 March 1985 had been a counter-offer to Chichester Joinery's original offer (i.e. its quotation submitted in November 1984), and had therefore extinguished that earlier offer.

However, Mowlem's counter-offer (i.e. its purchase order dated 14 March 1985) had in turn been extinguished by Chichester Joinery's own counter-offer (i.e. its acknowledgement of order dated 30 April 1985).

It was found that this final counter-offer had been accepted by Mowlem by conduct, namely by accepting the joinery later delivered to site by Chichester Joinery.

This case follows the basic principles of law set out earlier in this chapter; but the judge accepted that it was a very difficult case and stated that it highlighted the risks facing parties who sought to impose their own respective conditions rather than using some well established standard form of contract.

Based upon the above analysis of the *Chichester Joinery* v. *John Mowlem* case, it may be thought that the last document in place before acceptance by conduct will be deemed to be the offer that has been accepted. This is sometimes referred to as the 'Last shot' doctrine.

Although this doctrine very often applies, in certain cases, such as the *Butler* v. *Ex-Cell-O*[13] case, this doctrine was not followed, and it is now very often the case that the courts will look at the whole correspondence to try and glean from it the general intention of the parties.

In the *Butler* v. *Ex-Cell-O* case, Butler offered to supply a machine for a specified sum. The offer was said to be subject to certain terms and conditions, including a 'price escalation clause' which meant that the amount payable was to depend on the 'prices ruling upon date of delivery'.

In reply, Ex-Cell-O placed an order for the machine on a form setting out its own terms and conditions, which varied from the terms and conditions on the original offer and which excluded a price escalation clause. The order contained a tear-off slip which was to be signed by Butler and was to be returned to Ex-Cell-O, stating that Butler accepted the order.

Butler did not sign the slip, but returned it with a letter saying that it was 'entering' the order 'in accordance with' its (original) offer.

When a dispute arose, it was found that Butler's letter returning the order slip was an acceptance of Ex-Cell-O's terms and conditions, and that the reference by Butler to its (original) offer was (in the court's view) only made to identify the subject matter.

---

[13] *Butler Machine Tool Co. Ltd* v. *Ex-Cell-O Corporation (England) Ltd* [1979] 1 WLR 401.

Therefore, this is a case in point where the last shot doctrine was not followed.

However, if Butler's last-shot letter had spelt out the terms of its original offer again, rather than simply referring to the subject matter, then it may have been that the court would have found that (because of subsequent acceptance by conduct, when the machine was delivered) the parties had contracted on the terms of Butler's original offer.

Alternatively, another school of thought is that if Butler had not returned the tear-off slip from the order at all, then Butler's original offer would have been applicable.

The above analysis shows just how difficult it can be in attempting to determine when a contract has been formed, particularly where a battle of the forms situation arises.

So, perhaps the lesson to be learnt is to ensure that the sub-contract terms are agreed and recorded in writing to avoid the need to rely on a court interpreting the intention of the parties from an analysis of the battle of the forms.

## Agreement

As noted previously, the essence of any contract is agreement.

The test for the existence of an agreement is objective rather than subjective. In other words, the existence is tested on the facts, rather than on what may have been perceived to be the intention of the parties.

The principal justification for the adoption of this test is the need to promote certainty.

An agreement is reached either:

- when a statement of agreement is signed (e.g. signing the articles of agreement of the SBCSub/A or SBCSub/D/A): or
- when one party makes an unambiguous offer capable of being accepted, and the other party accepts it unequivocally.

## Consideration

Except where a contract is executed as a deed (see further commentary later within this chapter), an agreement made with the intention of being legally bound will be unenforceable where the parties have not provided consideration.

The classic definition of consideration was expressed in *Currie* v. *Misa*[14] in the following terms:

'A valuable consideration, in the sense of the law, may consist either in some right, interest, profit or benefit accruing to the one party, or some forbearance, detriment, loss or responsibility given, suffered or undertaken by the other.'

---

[14] *Currie* v. *Misa* (1875) LR 10 Ex 153.

The basic feature of the doctrine of consideration is the idea of reciprocity; 'something of value in the eye of the law'[15] must be given for a promise in order to make it enforceable as a contract.

In the ordinary building sub-contract situation the consideration given by the contractor is the price paid or the promise to pay, and the consideration given by the sub-contractor is the carrying out of the works or promise to carry them out.

The rules which make up the doctrine of consideration may be divided into three categories.

(1) The first rule is that *consideration must be sufficient but it need not be adequate.*

This means that the courts will not enforce a promise unless something of value is given in return for the promise, that is what is meant by saying that consideration must be *'sufficient'*, but, other than in the case of duress and undue influence, the courts are not concerned whether the consideration is adequate. Therefore, if a Rolls Royce worth £100,000 is sold for £100 that is sufficient consideration, even though it is quite clearly inadequate.

The consideration must contribute to the bargain, but as long as the court can recognize it as something of value then even a relatively trivial value will suffice.

Of course, if someone is forced into signing a contract, it will not bind them later, since it was made under duress. Examples of duress include a threat to physically harm someone or their close family, a threat of dishonour and also economic duress. A contract made under duress will be voidable, not void. 'Voidable' means that if the party who was coerced into the agreement chooses to go ahead with it, it will become binding.

Similarly, if undue influence is used to unfairly pressurize or influence someone in such a way that they do not make a free choice, then a contract made as a result will also be voidable. Undue influence falls into two categories, presumed undue influence and actual undue influence. Presumed undue influence may arise in relationships between, for example, solicitor and client, doctor and patient, etc. In cases of presumed undue influence, the burden of proof is with the party that may be presumed to have exerted undue influence to show that no such undue influence was exercised. In the case of actual undue influence, undue influence must be proved by the party seeking to establish its existence.

(2) The second rule is that *past consideration is not good consideration.*

The consideration for a promise must be given in return for a promise. Therefore, if a promise is made to reward for acts which have already been performed, the general rule is that the promise cannot be enforced because the consideration that has been provided is past (i.e. the consideration came before the promise).

This rule can best be illustrated by reference to the (rather sad) case of *Re McArdle*.[16] In that case a widow decorated a house which had been left to her

---

[15] *Thomas* v. *Thomas* (1842) 2 QB 851.
[16] Re McArdle [1951] Ch 669.

children. The children later wrote a note promising to pay her £488 for the work that she had already completed. When the children refused to pay, the widow took her children to court. However, the court found that the children did not have to make payment and were not bound by their note, as the work had already been completed before the promise to pay occurred (i.e. it was past consideration).

However, if consideration satisfies the following three criteria the courts will recognize it as good consideration:

(a)  where the act was done at the request of the person making the promise;
(b)  where the parties knew that payment would be made;
(c)  where payment is legally recoverable.

(3)  The third rule is that the *consideration must move from the party to whom the promise has been made.*

A person to whom a promise is made can only enforce the promise if he himself provides consideration for the promise. If the consideration is provided by a third party, the promisee cannot enforce the promise.

This rule is closely linked to the common law doctrine of privity of contract, which, as noted earlier in this chapter, means (under the general rule) that contracts cannot be enforced either by or against third parties.

Whilst the common law position remains unchanged, the Contracts (Rights of Third Parties) Act 1999, which came into force on 11 November 1999, gives a third party the right to enforce a term of a contract, if there is an express provision that he should, or if he is named in a contract which purports to confer a benefit on him. However, parties to a contract can, and often do, expressly exclude the operation of this Act.

## Executed as a deed

As noted above, consideration is not required in the case where a contract is executed as a deed. Such contracts were formerly known as being entered 'under seal', rather than 'under hand', which applies to simple contracts in writing. Contracts, other than those made by deed, are called simple contracts, whether made orally or in writing.

The need for company seals has now been abolished (by way of the Law of Property (Miscellaneous Provisions) Act 1989), and has been replaced by the simpler requirement for signature in the presence of a witness.

The other significant feature of contracts executed as a deed is that the limitation period for an action to be commenced is twelve years, whereas for simple contracts, the limitation period is six years (the question of limitation periods is further dealt with in Chapter 2).

## Intention to create legal relations

For an agreement to be recognizable in court, the parties must have intended it to be legally binding. This is called the intention to create legal relations.

The courts will assume a commercial agreement to be binding unless the parties expressly state otherwise. Thus in the case of *Rose & Frank* v. *J.R. Compton*,[17] an agreement between two businesses contained the words 'This arrangement is not entered into as a formal or legal agreement', and the courts found that the agreement was not legally enforceable.

Family arrangements are usually presumed not to be legally binding[18] and agreements between spouses are usually not regulated by the common law.[19]

## Capacity

Adults of sound mind have full contractual capacity. Minors (i.e. persons under the age of 18), the mentally incapacitated and companies have limited contractual capacity.

In the case of minors and the mentally incapacitated, contract law seeks to protect such persons from the consequences of their own inexperience or inability. Despite this, a minor is bound by a contract to supply him with 'necessaries' where the contract as a whole is for the benefit of the minor.

A company is, in the eyes of the law, a legal person that is separate and distinct from its shareholders. The capacity of a company is limited by the objects for which the company is set up and which are contained in the company's memorandum of association. If the company acts beyond its objects then it has acted *ultra vires*, that is to say, it has acted beyond its capacity.

In the case of *Ashbury Railway Carriage* v. *Riche*[20] it was held that in the case where, in respect of a contract, a company acted *ultra vires* the contract was deemed void. A contract that is 'void' is a contract that has no legal effect (i.e. a nullity) as distinct from a contract that is voidable which is simply a contract that is capable of being set aside (or rescinded) by one of the parties.

A strict application of the *ultra vires* rule could cause hardship to an innocent party who entered into a contract with a company, unaware of the *ultra vires* nature of the contract. However, significant protection is provided to such innocent parties by way of statute in section 35(1) of the Companies Act 1985 (as amended by the Companies Act 1989) which states that:

'The validity of an act done by a company shall not be called into question on the grounds of lack of capacity by reason of the fact that it is beyond the objects of the company stated in the memorandum of association.'

The intention behind this section of the Companies Act is to abolish the *ultra vires* rule as regards innocent parties who deal in all good faith with the company, while retaining it for internal purposes concerning the relationship between the shareholders and the company.

---

[17] *Rose & Frank Co.* v. *J.R. Compton & Bros Ltd* [1923] 2 KB 261.
[18] *Jones* v. *Padavatton* [1969] 1 WLR 325.
[19] *Balfour* v. *Balfour* [1919] 2 KB 571.
[20] *Ashbury Railway Carriage and Iron Co.* v. *Riche* (1875) LR 7 HL 653.

## Must not be illegal

As a general rule, the courts will not enforce a contract which is illegal or which is otherwise contrary to public policy. Nor, as a general rule, will they permit the recovery of benefits conferred under such a contract (although they will permit enforcement of the contract by the innocent party where the innocent party had no knowledge of the illegality).

A contract is illegal if its formation is expressly or impliedly prohibited by statute.

A contract may be illegal at common law on the ground that it is contrary to public policy. Contracts which are contrary to public policy include contracts which are contrary to good morals, contracts which are prejudicial to family life, contracts to commit a crime or a civil wrong, contracts which are prejudicial to the administration of justice, contracts prejudicial to public relations and contracts in unreasonable restraint of trade.

## Fraud or misrepresentation

A misrepresentation may be defined as an unambiguous, false statement of fact which is addressed to a party and which misleads that party and which materially induces that party to enter into a contract.

A representation is a statement of fact relating to a contract made by one party which does not become a term of the contract. If it is untrue, whether made fraudulently or innocently, it is a misrepresentation and the general effect is to render the contract voidable. As a general rule, simple silence is not an example of misrepresentation.[21]

The statement in question must be one of existing fact, and the following four categories of statement have been held not to constitute statements of existing fact and therefore cannot amount to actionable misrepresentation.

(1) 'Mere puff'. These are statements that are so vague that they have no effect in law or equity. For example, it has been found[22] to describe land as 'fertile and improvable' is mere sales puff which gives no ground for relief.

(2) A statement of opinion or belief which proves to be unfounded is not a false statement of fact. Assertions that an anchorage was safe[23] and that a piece of land had the capacity to support 2000 sheep[24] have been held to be merely statements of opinion or belief, because in both cases the party making the statement had (as the other party knew) no personal knowledge of the facts upon which it was based, and it was understood that he could only state his belief. On the other hand, if the party has or professes to have some special knowledge or skill as to the matter stated, the statement is likely to be treated as one of fact.[25]

---

[21] *Keates v. Cadogan* (1851) 10 CB 591.
[22] *Dimmock v. Hallett* (1866) LR 2 Ch App 21.
[23] *Anderson v. Pacific Fire & Marine Insurance Co.* (1872) LR 7 CP 65.
[24] *Bissett v. Wilkinson* (1927) AC 177.
[25] *Esso Petroleum Ltd v. Mardon* [1976] QB 801.

(3) A representation of English law is not a statement of existing fact, since everyone is presumed to know the law (although it should be noted that a representation of foreign law is regarded as one of fact).

(4) A statement of intention is not a statement of fact, and also a promise is not a statement of fact. A person who fails to carry out his stated intention does not thereby make a misrepresentation,[26] but a person who misrepresents his present intention does make a false statement of fact because the state of his intention is a matter of fact.

The statement must be addressed to the party misled. This may either be by way of direct communication to that party, or it may be communicated to that party by a third party acting under the instruction of the party making the representation.

Finally, the statement must be a material inducement to make a party enter into the contract. The statement must be material; that is, it must be a representation that would affect the judgment of a reasonable man in deciding whether to enter into the contract on those terms (although a person who has been fraudulent cannot succeed with an argument that the representation was immaterial); and the statement must also actually induce a party to enter into the contract.

There are four different types of misrepresentation.

(1) Fraudulent misrepresentation
Fraudulent misrepresentation is made when it is proved that a false representation has been made knowingly, or without belief in its truth, or recklessly or, carelessly, whether it is true or false.

(2) Negligent misrepresentation (actionable at common law)
Negligent misrepresentation is actionable at common law where there is a 'Hedley Byrne'[27] relationship (i.e. some special relationship of proximity) between the parties. The existence of such a relationship depends upon a number of factors, including the knowledge of the party making the representation), the purpose for which the statement was made and the reasonableness of the reliance by the party receiving the representation.

(3) Negligent misrepresentation (arising under section 2(1) of the Misrepresentation Act 1967)
Section 2(1) of the Misrepresentation Act 1967 states that where a misrepresentation has been made by one contracting party to another, the party making the representation is liable to the other for damages unless he can prove that he had reasonable grounds to believe and did believe up to the time that the contract was made that his statement was true.

(4) Innocent misrepresentation
An innocent misrepresentation is a misrepresentation which is neither fraudulent, nor negligent.

---

[26] *Wales v. Wadham* [1977] 1 WLR.
[27] *Hedley Byrne & Co. v. Heller & Partners* [1963] 3 WLR 101, HL.

The principal remedies for misrepresentation are rescission (i.e. revoking the contract) and damages.

Rescission is in principle available for all types of misrepresentation. The effect of rescission is to put the parties as far as possible into the position in which they would have been if the contract had not been concluded.

Damages can be claimed for all types of misrepresentation. However, in respect of innocent misrepresentation only, the court has discretion to award damages in lieu of rescission under section 2(2) of the Misrepresentation Act 1967.

The ability of a party to exclude liability for misrepresentation is controlled by section 3 of the Misrepresentation Act 1967, which subjects any term which purports to exclude or restrict liability or a remedy for misrepresentation to a reasonableness test.

## Mistake

In *Bell* v. *Lever Bros*,[28] Lord Atkin said: 'If mistake operates at all, it operates so as to negative or in some cases to nullify consent.' In other words, in certain circumstances, a mistake as to the form or content of an agreement may result in that agreement being void.

Mistake does, however, operate within very narrow confines. A mistake can be mutual, unilateral or common.

A mutual mistake occurs when both parties make a mistake; a unilateral mistake occurs where only one party makes a mistake; and a common mistake occurs when both parties make the same mistake.

Where both parties make the same common mistake, the existence of an agreement is undisputed, but one party may say that the mistake has deprived the contract of its effectiveness. Where the mistake relates to the existence of the subject matter the contract is void, as in the case of sale of goods which have perished at the time of sale, or which never existed. However, less fundamental mistakes, particularly those regarding the quality of the goods, may not make a contract void or voidable.

The other alternative is where both parties have different intentions, whether they are both mistaken (a mutual mistake) or whether the mistake is unilateral. In this situation, the contract is void only if the mistake prevents one party from appreciating the fundamental character of what he is offering or accepting.

A mistake which affects only motives, as when one party thinks he had a much better bargain than was the case, cannot affect the validity of the contract.

If the parties had different intentions, the court will, wherever possible, ascertain the true meaning of the contract. If this cannot be done the contract is void.

In addition to common law remedies for mistake, other relief may be available in equity.

Principally, if both parties, or even one party, intended something different from that which the documents record, the courts have the power, in certain and narrowly confined circumstances, to rectify or alter the document so as to give effect to the true agreement.

---

[28] *Bell* v. *Lever Bros Ltd* [1932] AC 161.

## Void or voidable

A contract that is 'void' is a contract that has no legal effect (i.e. a nullity) as distinct from a contract that is voidable, which is simply a contract that is capable of being set aside (or rescinded) by one of the parties.

## Certainty of terms

Even if there is clearly agreement through offer and acceptance, a contract may fail to come into existence because of uncertainty as to what has been agreed.[29]

Although it has been found that 'the parties are to be regarded as masters of their contractual fate in determining what terms are essential'[30] and 'it is for the parties to decide whether they wish to be bound and, if so, by what terms, whether important or unimportant',[31] it is generally considered that agreement as to parties, price, time and description of works is normally the minimum necessary to make the contract commercially workable.

Silence by the parties as to either price or time may not alone prevent a contract coming into existence, for if the other essential terms are agreed a reasonable charge or time for completion may be implied by the Supply of Goods and Services Act 1982.

For example, in the *Drake & Scull* v. *Higgs & Hill*[32] case, there was no contract but the court observed that if an agreement as to dayworks were the only matter which prevented there being a binding contract, the gap could be made good by the implication of a term that a reasonable rate would be paid.

## *Letters of intent*

A letter of intent arises out of a situation where parties are in the course of negotiating a contract and they express in writing their intention to enter into a contract in the future.

Although there is often much debate regarding this matter, it is in fact a misconception that letters of intent are or can be contracts. A true letter of intent does no more than set out the party's intention to enter into a contract at a later date.

Despite this, there have been circumstances when the courts have found that what one of the parties intended to be a letter of intent was actually a contract in its own right.

In reality, as identified in the *Monk* v. *Norwich Union*[33] case, the question of the legal effect of letters of intent has only three possible outcomes. These are:

● that a simple contract may be formed under which each party assumes reciprocal obligations to the other; or

---

[29] *Scammell* v. *Ouston* (1941) 1 All ER 14.
[30] *Pagnan* v. *Feed Products* [1987] 2 Lloyds Rep 601, CA.
[31] *Mitsui Babcock Energy Ltd* v. *John Brown Engineering Ltd* (1996) CILL 1189.
[32] *Drake & Scull Engineering Ltd* v. *Higgs & Hill Northern Ltd* (1994) 11 Const LJ.
[33] *Monk Construction Ltd* v. *Norwich Union Life Assurance Society* (1992) 62 BLR 107.

- that there may be an 'if' contract, under which Party A requests performance from Party B and, if Party B performs, he will receive remuneration (this is sometimes referred to as a 'quasi-contract'); or
- there is no contract formed at all.

## How can a letter of intent be an ordinary contract?

The main consideration is whether the terms set out in the letter of intent are sufficiently certain for a contract to be formed, and to what extent the matters that still need to be resolved are contained within the letter of intent.

The intention of the parties from the surrounding circumstances is also relevant. In the *British Steel* v. *Cleveland Bridge*[34] case, Mr Justice Goff said:

'the question whether . . . any contract has come into existence must depend on a true construction of the relevant communications which have passed between the parties and the effect (if any) of their actions pursuant to those communications.'

Even if some matters remain outstanding but all essential terms have been agreed, a contract may still be formed. Authority for this is found in the *Mitsui* v. *John Brown*[35] case, where it was stated:

'My review of the authorities leads me to the conclusion that there is no reason in principle why two parties should not enter into a binding agreement, even though they have agreed that some proposed terms should be the subject of further discussions and later agreement.'

From the above, it would appear that for a letter of intent to be enforceable in law as a contract the following elements must be present:

- the parties must have demonstrated their intention to contract, either from the content of the letter of intent or from their conduct;
- the parties must have agreed the essential terms (e.g. parties, description of works, price and time) with sufficient certainty so as to allow the contract to become commercially workable;
- acceptance of the offer must be clear either by words or by conduct.

It should be noted that there are many instances where a letter of intent has a capped price or a cap on time. In such a situation, the courts have generally enforced spending limits in letters of intent strictly (see *Monk Construction v. Norwich Union*,[36] where the letter of intent authorized mobilization and ordering of materials up to a maximum expenditure of £100,000 and the court held that Monk Construction could not recover proven costs for works performed in excess of that sum).

---

[34] *British Steel Corporation* v. *Cleveland Bridge* [1984] 1 All ER 504.
[35] *Mitsui Babcock Energy Ltd* v. *John Brown Engineering Ltd* (1996) CILL 1189.
[36] *Monk Construction Ltd* v. *Norwich Union Life Assurance Society* (1992) 62 BLR 107.

## Standard forms of contract and sub-contract

There are obvious benefits in using standard forms of contract and sub-contract.

These include:

- there is no need to produce (and incur the legal costs of producing) ad hoc contracts and sub-contracts for every project;
- there is a degree of certainty regarding the interpretation of the clauses of the contract (particularly those standard forms that have been in existence for some time and where some of the more important clauses may have been tested in the courts);
- the parties know (with reasonable certainty) the consequences of various possible courses of action.

Standard forms of contract (perhaps less so of standard forms of sub-contract) can be traced back to the nineteenth century (if not earlier).

However, it appears to be a fairly recent phenomenon, largely emanating from the Latham Report (*Constructing the Team*, 1994), that standard forms and sub-contracts are now seen as having a dual purpose, these being:

- to set out the rights and obligations of the parties;
- to ensure that the risk of the project is allocated to the party that can best manage that risk.

In this regard there have been great steps taken, particularly in respect of main contracts, but it is questionable whether that same progress has been made in sub-contracts.

Concern regarding the practice of main contractors drafting their own forms of sub-contract, often containing one-sided provisions, has been expressed for many years. In 1964, the Banwell Report (*The Placing and Management of Contracts for Building and Civil Engineering Work*, 1964) recommended that a single standard form of contract for the entire construction industry was both desirable and practicable, and that standardization of sub-contract conditions should follow. That objective was, however, never achieved.

It is still prevalent for main contractors to produce their own sub-contract forms, or, more commonly, issue their own amendments to the standard sub-contract forms.

In respect of the JCT Standard Building Contract, before a contractor enters into a sub-contract he is obliged to obtain the written consent of the architect/contract administrator, as required by clause 3.7.1.

Clause 3.7.1 of the JCT Standard Building Contract states:

'The Contractor shall not without the written consent of the Architect/Contract Administrator sub-let the whole or any part of the Works. Such consent shall not be unreasonably delayed or withheld but the Contractor shall remain wholly responsible for carrying out and completing the Works in all respects in accordance with clause 2.1 notwithstanding any such sub-letting.'

It should be noted that clause 3.7.3 of the JCT Standard Building Contract states that the execution of part of the works by a statutory undertaker (acting in that capacity) must not be a sub-contractor within the terms of the contract.

Clause 3.7.2 of the JCT Standard Building Contract (which deals with the sub-letting of design) states:

> 'Where there is a Contractor's Designed Portion, the Contractor shall not without the written consent of the Employer sub-let the design for it. Such consent shall not be unreasonably delayed or withheld but shall not in any way affect the obligations of the Contractor under clauses 2.2 and 2.19 or any other provision of this Contract.'

The significant point to note is that under clause 3.7.1 written consent of the architect/contract administrator is required, whereas under clause 3.7.2 the written consent needs to be obtained from the employer (i.e. the contractor's employer).

However, apart from the provisions of clause 3.9 of the JCT Standard Building Contract (detailed below), the actual terms on which a contractor contracts with a sub-contractor are a matter for agreement between the contractor and the sub-contractor.

Clause 3.9 states:

> 'It shall be a condition of any sub-letting to which clause 3.7 or 3.8 applies that:
>
> 3.9.1 the employment of the sub-contractor under the sub-contract shall terminate immediately upon the termination (for any reason) of the Contractor's employment under this Contract;
>
> 3.9.2 the sub-contract shall provide:
>
> 3.9.2.1 that no unfixed materials and goods delivered to, placed on or adjacent to the Works by the sub-contractor and intended for them shall be removed, except for use on the Works, unless the Contractor has consented in writing to such removal (such consent not to be unreasonably delayed or withheld) and that:
>
> 3.9.2.1.1 where, in accordance with clause 4.10 and 4.16 of these Conditions, the value of any such material or goods has been included in any Interim Certificate under which the amount properly due to the Contractor has been paid to him, those materials or goods shall be and become the property of the Employer and the sub-contractor shall not deny that they are and have become the property of the Employer;
>
> 3.9.2.1.2 if the Contractor pays the sub-contractor for any such materials or goods before their value is included in any Interim Certificate, such materials or goods shall upon such payment by the Contractor be and become the property of the Contractor;
>
> 3.9.2.2 for the grant by the sub-contractor of the rights of access to workshops or other premises referred to in clause 3.1 of these Conditions;
>
> 3.9.2.3 that if by the final date for payment stated in the sub-contract the Contractor fails properly to pay any amount, or any part of it, due to the sub-contractor, the Contractor shall in addition to the properly paid pay simple interest at the Interest Rate for the period until such payment is made; such payment of interest to be on and subject to terms equivalent to those of clauses 4.13.6 and 4.15.6 of these Conditions;

3.9.2.4 where applicable, for the execution and delivery by the sub-contractor, in each case within 14 days of receipt of a written request by the Contractor, of such collateral warranties as comply with the Contract Documents;

3.9.2.5 that neither of the provisions referred to in clauses 3.9.2.1.1 and 3.9.2.1.2 shall operate so as to affect any vesting in the Contractor of any property in any Listed item required for the purposes of clause 4.17.2.1 of these Conditions;

3.9.3 the Contractor shall not give such consent as is referred to in clause 3.9.2.1 without the prior consent of the Architect/Contract Administrator under clause 2.24 of these Conditions.'

Therefore, although SBCSub/G states that when reaching an agreement, the contractor will need to ensure, amongst other things, that if an act or default of the sub-contractor causes the contractor to be in breach of the main contract he has redress under the sub-contract against the sub-contractor; and adds that the sub-contractor will need provisions in the sub-contract which provide, amongst other things, for:

- regular payment for work properly carried out;
- the recovery of any loss or expense caused to him by any act or default of the contractor or of the employer;
- the adjustment of the agreed period or periods for carrying out the sub-contract works where such adjustment is needed due to the occurrence of events outside the sub-contractor's control.

Also, despite the fact that SBCSub/G states that to assist contractors and sub-contractors in agreeing a sub-contract which meets the needs of the parties and the requirements of the main contract, JCT published the Standard Building Sub-Contract for use with the Standard Building Contract, adding that the requirements of clause 3.9.1 and 3.9.2 together with those in paragraphs A.3 and B.4 of the fluctuations options (schedule 7) of the Standard Building Contract, are met by the JCT Standard Building Sub-Contracts (SBCSub and SBCSub/D).

Because it is clear that, apart from the provisions of clause 3.9 noted above, the terms on which a contractor contracts with a sub-contractor are a matter for agreement between the contractor and the sub-contractor, it remains to be seen if the Standard Building Sub-Contract will be used by main contractors without amendment.

From past experience, where the DOM/1 sub-contract form, for example, has regularly been heavily amended by contractors, it seems unlikely that the Standard Building Sub-Contract will not be similarly amended by contractors to better protect their position.

# Chapter 2
# SBCSub/A and SBCSub/D/A

## *The JCT Standard Building Sub-Contract Agreement (SBCSub/A or SBCSub/D/A) – introduction*

People frequently refer to the terms and conditions of a contract.

Using this terminology, SBCSub/A or SBCSub/D/A (as appropriate) when completed set out the terms of the sub-contract, whereas SBCSub/C or SBCSub/D/C (as appropriate) set out the conditions (noting, of course, that clause 1.3 of SBCSub/C and SBCSub/D/C make it clear that the sub-contract is to be read as a whole).

SBCSub/A or SBCSub/D/A, as a blank document, does not (and cannot) set out the terms of the sub-contract; it is for the contractor and the sub-contractor to agree upon those terms and for those terms to be entered into SBCSub/A or SBCSub/D/A.

Whatever is inserted into the executed version of SBCSub/A or SBCSub/D/A will be the express terms agreed between the parties.

The law will understand these express terms to be freely agreed between the parties, and any argument put forward at a later date that the express terms were not freely agreed is unlikely to meet with any success.

Also, the terms will be the only express terms applicable to the sub-contract. Whatever may have been discussed and/or written about before the execution of the sub-contract will not form part of the express terms of the sub-contract unless the matter so discussed and/or written about is specifically listed within SBCSub/A or SBCSub/D/A (under the sub-contract particulars).

Therefore, it is important for both the contractor and the sub-contractor that everything that should be referred to and listed within the sub-contract particulars *is* referred to and listed within the sub-contract particulars.

The corollary to the above is, of course, that anything that should not be referred to and listed within the sub-contract particulars should *not* be referred to and listed within the sub-contract particulars.

If a matter was discussed and/or written about prior to the execution of the sub-contract – for example that the sub-contractor was to clear its waste materials and rubbish to skips to be provided by the contractor – but this matter was not listed within the sub-contract particulars, then, if this matter became an issue at a later date, the sub-contractor would not have the contractual right to insist that the contractor provide skips for the sub-contractor's waste materials and rubbish.

Conversely, if a matter was not discussed and/or written about prior to the execution of the sub-contract – for example that the sub-contractor would be charged for the use of mess rooms – but this matter was listed within the sub-contract particulars, then, if this matter became an issue at a later date, the contractor would have the contractual right to charge the sub-contractor when the sub-contractor used the mess rooms.

The parties need to be aware that the meaning of a contract will be ascertained by a court from the meaning of the words actually used in that written contract, and a court may not hear evidence from the parties as to what they may have actually meant.

This 'rule' is sometimes known as the 'Parole evidence rule'. Although there are some exceptions to this rule, it is generally futile (if a dispute arises) for a party to rely on any negotiations leading up to a written contract (if those negotiations are not recorded in some way within the written contract), in an attempt to change in some way the plain wording of the contract. This point can represent a trap for contractors and sub-contractors who may wish to refer to the sub-contract negotiations in an attempt to show that the alleged intention of the parties was different to the actual wording of the sub-contract. (The fairly remote possibility of void or voidable contracts because of mistake or misrepresentation is dealt with in Chapter 1.)

It is extraordinary the number of times that contractors and sub-contractors take great care over agreeing terms that they both feel comfortable with, only to fail to properly record those agreements in the sub-contract. Experience shows that when disputes arise, people tend to have very selective memories; therefore if a term has been agreed, it must be recorded in the sub-contract if a party wishes to rely upon that term in future.

## The structure of SBCSub/A and SBCSub/D/A

SBCSub/A and SBCSub/D/A both consist of:

- The date that the agreement is made (normally the date that the second party executes the agreement).
- The names and addresses of the parties (i.e. the contractor and the sub-contractor). Normally, the party's company number should be inserted, and their registered office address should be used. If the contractor or the sub-contractor is a company incorporated outside England and Wales, particulars of its place of incorporation should be inserted immediately before the company number. If the contractor or the sub-contractor are neither a company incorporated under the Companies Act nor a company registered under the laws of another country, the references to a company number and to a registered office address should be deleted.
- Details (by identification only) of the sub-contract works, and the main contract works of which they are a part.

For SBCSub/A, this is then followed by:

- first to fourth recitals;
- articles 1 to 6;
- sub-contract particulars items 1 to 15;
- a summary of interim and final payment provisions;

- attestation forms;
- a schedule of information;
- notes (to annexures to the agreement and to the schedule of information);
- supplementary particulars (containing information needed for formula adjustment, if applicable).

For SBCSub/D/A this is followed by:

- first to eighth recitals;
- articles 1 to 6;
- sub-contract particulars items 1 to 16;
- a summary of interim and final payment provisions;
- attestation forms;
- a schedule of information;
- notes (to annexures to the agreement and to the schedule of information);
- supplementary particulars (containing information needed for formula adjustment, if applicable).

## The recitals

The recitals set out certain facts as a background to the parties entering into the sub-contract.

The recitals are:

| SBCSub/D/A recital | Equivalent recital in SBCSub/A |
|---|---|
| First | First |
| Second | Second |
| Third | Not applicable |
| Fourth | Not applicable |
| Fifth | Not applicable |
| Sixth | Not applicable |
| Seventh | Third |
| Eighth | Fourth |

### The first recital of SBCSub/D/A (The first recital of SBCSub/A)

This recital refers to the main contract which is described in item 1 of the schedule of information annexed to the SBCSub/D/A or SBCSub/A. The said schedule of information should be completed by the contractor. The schedule of information is dealt with in detail later in this chapter.

### The second recital of SBCSub/D/A (The second recital of SBCSub/A)

This recital simply states that the contractor wishes the sub-contractor to execute the sub-contract works described in the numbered documents. (Refer to item 15 of the sub-contract particulars, SBCSub/A, and item 16 of the sub-contract particulars, SBCSub/D/A, dealt with later in this chapter.)

## The third recital of SBCSub/D/A

This recital is to be completed by the parties (i.e. the contractor and the sub-contractor) so as to show the part of the sub-contract works (which may, of course, be all of the sub-contract works) in which the sub-contractor is responsible for both the design and the construction. The entry made may be descriptive, or it may refer to the document(s) that more fully describe the relevant part of the sub-contract works.

The entry made in this recital is known as the sub-contractor's designed portion.

## The fourth recital of SBCSub/D/A

This recital confirms that the contractor has provided documents showing and describing or otherwise stating the requirements of the contractor for the design and construction of the sub-contractor's designed portion. These documents, which would probably include the 'Employer's requirements' and which may include the 'Contractor's proposals', should be listed under item 16 of the sub-contract particulars (SBCSub/D/A) as numbered documents.

These documents are known collectively as being the 'Contractor's requirements'.

The documents which comprise the contractor's requirements should be annexed to SBCSub/D/A and (for the avoidance of any future dispute regarding the documents referred to) should be signed or initialled by or on behalf of each party.

## The fifth recital of SBCSub/D/A

This recital confirms that in response to the contractor's requirements (as identified in the fourth recital of SBCSub/D/A) the sub-contractor has supplied to the contractor:

- documents showing and describing the proposals of the sub-contractor for the design and construction of the sub-contractor's designed portion. These documents are known collectively as the 'Sub-contractor's proposals';
- an analysis of the portion of the sub-contract sum (or sub-contract tender sum, if appropriate) relating to the sub-contractor's designed portion. This analysis is called the sub-contractor's designed portion analysis.

The documents which comprise the sub-contractor's proposals and the sub-contractor's designed portion analysis should be listed under item 16 of the sub-contract particulars (SBCSub/D/A) as numbered documents, and should be annexed to SBCSub/D/A.

For the avoidance of any future dispute regarding the documents referred to, the said documents should be signed or initialled by or on behalf of each party (i.e. the contractor and the sub-contractor).

## The sixth recital of SBCSub/D/A

This recital confirms that the contractor has examined the sub-contractor's proposals and, subject to the sub-contract conditions, is satisfied that they appear to meet the contractor's requirements.

The entire matter of design liability will be dealt with in more detail in Chapter 4. However, at this point it is important to note that the sixth recital does not require the contractor to check the sub-contractor's proposals or to confirm that they satisfy the contractor's requirements, but merely that 'they appear to meet' them.

Therefore, for example, if the contractor's requirements called for a certain lux level of lighting in a room, it is not necessary for the contractor to check that the lighting arrangement proposed by the sub-contractor will achieve the lux level required, as long as it appears that it will.

If, when installed, the lighting arrangement proposed by the sub-contractor fails to meet the contractor's requirements, the sub-contractor could be required to alter the lighting arrangement (at no cost to the contractor) so that it does achieve the lux level required. The word 'could' is used in the above sentence because the contractor may accept the lower lux level (that has been achieved by the sub-contractor) and therefore no change to the lighting arrangement would be necessary.

## The seventh recital of SBCSub/D/A (the third recital of SBCSub/A)

This recital relates to the provision of a priced schedule of activities (the priced activity schedule).

This recital should be deleted if either:

- article 3B is to apply; or
- a priced activity schedule is not provided.

Article 3B is for use where the sub-contract works are to be completely remeasured and valued on a remeasurement basis.

In the activity schedule, each activity should be priced so that the sum of those prices equals the sub-contract sum excluding:

- provisional sums; and
- the value of work for which approximate quantities are included in any bills of quantities.

## The eighth recital of SBCSub/D/A (the fourth recital of SBCSub/A)

This recital states that the sub-contractor has been given a copy of, or a reasonable opportunity to inspect, any other documents and information relating to the provisions of the main contract insofar as they relate to the sub-contract works and are listed in item 1.9 of the schedule of information (referred to generally in respect of the first recital of SBCSub/D/A and SBCSub/A above, and in more detail later in this chapter).

It is essential that both the contractor and the sub-contractor ensure that this requirement is properly fulfilled, and that item 1.9 of the schedule of information correctly lists *only* the other documents and information which relate to the sub-contract works and which the sub-contractor has been given a copy of or a reasonable opportunity to inspect.

The use of the words 'reasonable opportunity' above leaves the possibility of a future dispute as to whether the opportunity provided was reasonable or not. It is submitted that if such a dispute did arise, the onus would be on the contractor to give evidence that the opportunity given was indeed reasonable in the circumstances. Of course, if the sub-contractor did not consider that a reasonable opportunity to inspect the documents and information listed under item 1.9 had been given, then he should resist the inclusion of such a list of documents and information until copies have been provided to him or until he has been given a reasonable opportunity to inspect the documents and information.

## The articles

The articles set out certain basic elements of the sub-contract (i.e. the sub-contract documents; the sub-contract conditions; the sub-contract sum (or the sub-contract tender sum); the final sub-contract sum; and the dispute resolution procedures).

### Article 1: sub-contract

This article lists the documents comprising the sub-contract documents.
These documents are:

- SBCSub/A (or SBCSub/D/A, as appropriate).
- The sub-contract particulars (as annexed to SBCSub/A or SBCSub/D/A). (The sub-contract particulars are dealt with later in this chapter.)
- The schedule of information and the documents referred to and annexed to the schedule of information. (The schedule of information is dealt with later within this chapter.)
- The Standard Building Sub-Contract conditions (SBCSub/C or SBCSub/D/C as appropriate, incorporating the JCT amendments[1] stated in the sub-contract particulars (item 1), and any schedule of modifications[2] referred to in the sub-contract particulars (item 1) and listed (and annexed) as a numbered document in the sub-contract particulars (item 15 for SBCSub/A or item 16 for SBCSub/D/A).

---

[1] For further commentary on the JCT amendments, see notes regarding item 1 of the sub-contract particulars later in this chapter.
[2] For further commentary on the schedule of modifications, see notes regarding item 1 of the sub-contract particulars, and regarding item 15 of the sub-contract particulars (for SBCSub/A) or item 16 of the sub-contractor particulars (for SBCSub/D/A) later in this chapter.

- The numbered documents annexed to the SBCSub/A or SBCSub/D/A. The numbered documents are listed under item 15 of the sub-contract particulars (for SBCSub/A) and item 16 of the sub-contract particulars (for SBCSub/D/A).

## Article 2: sub-contractor's obligations

This article states the basic obligation of the sub-contractor which is to carry out and complete the sub-contract works in accordance with the sub-contract.

In SBCSub/D/A, the sub-contractor's obligations include the requirement to complete the design for the sub-contractor's designed portion:

- in compliance with regulation 13 of the CDM Regulations;[3] and
- in accordance with such directions as the contractor may give for the integration of the design for the sub-contractor's designed portion with the design for the main contract works as a whole.

The sub-contractor's obligations are more fully described in clauses 2.1 to 2.5 of SBCSub/C and SBCSub/D/C (refer to Chapters 4 and 5).

## Article 3A: sub-contract sum and final sub-contract sum, and article 3B: sub-contract tender sum and final sub-contract sum

Either article 3A or Article 3B is completed as appropriate. In both cases a sum of money in both words and figures needs to be inserted.

The article that is not completed is to be deleted.

Article 3A is used where a sub-contract sum is applicable. A sub-contract sum is applicable where the works are valued on the adjustment basis (i.e. that is where the sub-contract sum only changes if SBCSub/C or SBCSub/D/C so provides, for example when variations are carried out). For further information on this point, see the commentary in respect of clauses 4.1 and 4.3 of SBCSub/C and SBCSub/D/C in Chapter 8.

Article 3B is used where a sub-contract tender sum is applicable. A sub-contract tender sum is applicable where the works are to be completely remeasured (the remeasurement basis). For further commentary on this point, see the commentary in respect of clauses 4.2 and 4.4 of SBCSub/C and SBCSub/D/C in Chapter 8.

## Article 4: adjudication

This article simply confirms that if any dispute or difference arises under the sub-contract either party may refer it to adjudication in accordance with clause 8.2 of

---

[3] The CDM Regulations are the Construction (Design and Management) Regulations 1994. In certain (fairly uncommon) circumstances, the CDM Regulations may not apply to a project, and the sub-contractor's obligation to comply with regulation 13 of the CDM Regulations would not apply. For further commentary on this point and on regulation 13 of the CDM Regulations see commentary on item 2 of the schedule of information later in this chapter, and clause 2.2.3 of SBCSub/D/C in Chapter 4.

SBCSub/C or SBCSub/D/C. For further commentary on adjudication, refer to Chapter 13.

## Article 5: arbitration

This article relates to the position where arbitration is to be used to finally resolve any dispute or difference between the parties.

When the sub-contract is executed, under item 2 of the sub-contract particulars (SBCSub/A or SBCSub/D/A) article 5 should be marked either to apply (i.e. delete the words 'do not apply') or not to apply (i.e. delete the word 'apply').

If neither entry is deleted, article 5 and clauses 8.3 to 8.8 will not apply, and disputes and differences will be determined by legal proceedings. In other words, the default position is that legal proceedings will apply.

If disputes and differences are to be determined by arbitration and not by legal proceedings, item 2 of the sub-contract particulars (SBCSub/A or SBCSub/D/A) makes it clear that item 2 must be marked to show that article 5 and clauses 8.3 to 8.8 apply. Therefore a positive act must be taken if it is required that arbitration is to be used to finally determine disputes and differences.

For further commentary on arbitration, refer to Chapter 13.

## Article 6: legal proceedings

This article relates to the position where legal proceedings are to be used to finally resolve any dispute or difference between the parties.

When article 6 applies, the English courts shall have jurisdiction over any dispute or difference between the parties which arises out of or in connection with the sub-contract.

If the parties wish any dispute or difference to be determined by the courts of another jurisdiction (i.e. not the English courts) the appropriate amendment should be made to article 6 (refer also to commentary on clause 1.10 of SBCSub/C and SBCSub/D/C in Chapter 3).

When the sub-contract is executed, under item 2 of the sub-contract particulars (SBCSub/A or SBCSub/D/A) article 5 should be marked either to apply (i.e. delete the words 'do not apply') or not to apply (i.e. delete the word 'apply').

If neither entry is deleted, article 5 and clauses 8.3 to 8.8 will not apply, and disputes and differences will be determined by legal proceedings (i.e. in line with article 6). The default position is therefore that legal proceedings will apply.

For further commentary on legal proceedings, refer to Chapter 13.

## *The sub-contract particulars*

The sub-contract particulars set out certain basic terms of the sub-contract.

Wherever practicable, a default position has been provided in respect of the sub-contract particulars, and, in such cases, an entry only need be made where it is intended that the default position is not to apply.

Where an entry is made, and this entry requires a continuation sheet, then such continuation sheets should be identified as such, should be signed or initialled by or on behalf of each party, and should be annexed to SBCSub/A or SBCSub/D/A as appropriate.

The sub-contract particulars are:

| SBCSub/A sub-contract particulars item number | SBCSub/D/A sub-contract particulars item number | Sub-contract particulars item description | SBCSub/A or SBCSub/D/A article that relates to the sub-contract particulars item in question | SBCSub/C or SBCSub/D/C clause (or schedule to the SBCSub/C or SBCSub/D/C) that relates to the sub-contract particulars item in question |
|---|---|---|---|---|
| 1 | 1 | Conditions | Article 1 | |
| 2 | 2 | Arbitration | Article 5 | |
| 3 | 3 | Sub-contract base date | | Clause 1.1 |
| 4 | 4 | Electronic communications | | Clause 1.8 |
| 5 | 5 | Programme | | Clause 2.3 |
| 6 | 6 | Attendance | | Clause 3.16 |
| 7 | 7 | Listed items | | Clause 4.14 |
| 8 | 8 | Retention percentage | | Clause 4.15 |
| 9 | 9 | Sub-contractors retention bond | | Clause 4.16 |
| 10 | 10 | Fluctuations | | Clauses 4.17 and 4.18 and Schedule 4 |
| 11 | 11 | Dayworks | | Clause 5.9 |
| 12 | 12 | Insurance – personal injury and property | | Clause 6.5 |
| 13 | 13 | Incorporation of the sub-contract works into the main contract works | | Clause 6.7.8 |
| – | 14 | Sub-contractors designed portion (SCDP) professional indemnity insurance | | Clause 6.10 (SBCSub/D/C *only*) |
| 14 | 15 | Settlement of disputes | | Clauses 8.2 and 8.4.1 |
| 15 | 16 | Numbered documents | | |

## SBCSub/A sub-contract particulars item number 1 (SBCSub/D/A sub-contract particulars item number 1) – conditions

This refers to article 1 of SBCSub/A or SBCSub/D/A.

The conditions applicable are SBCSub/C or SBCSub/D/C.

The JCT amendments to SBCSub/C or SBCSub/D/C are either those specifically noted, or if no JCT amendments are specifically noted then the default position is that those JCT amendments current at the sub-contract base date will apply. The sub-contract base date is the date stated under item 3 of the sub-contract particulars (dealt with later in this chapter).

JCT amendments are amendments to SBCSub/C or SBCSub/D/C that are issued by the JCT.

In addition, there is the facility for a schedule of modifications to the SBCSub/C or SBCSub/D/C.

If a schedule of modifications is not provided then the reference to a schedule of modifications should be deleted from article 1.

If a schedule of modifications is provided then it should be listed (and annexed) as a numbered document in the sub-contract particulars (item 15 for SBCSub/A or item 16 for SBCSub/D/A).

The schedule of modifications can considerably alter the entire basis of the SBCSub/C or SBCSub/D/C. A schedule of modifications is usually drawn up either to reflect amendments to the main contract, or in an attempt to give an advantage to the contractor (and a disadvantage to the sub-contractor) in respect of particular clauses in the SBCSub/C or SBCSub/D/C. Therefore, a sub-contractor must take careful note of the implications of the proposed schedule of modifications before agreeing to accept the schedule.

Paragraph 15 of the SBCSub/G makes the comment that:

'As the Standard Building Sub-Contract has been agreed by representatives of contractors and sub-contractors, it is hoped that users will find it provides a fair and equitable basis and that it will not be amended by the contractor or by the sub-contractor other than to reflect any amendments in the main contract itself.'

However, the reality is that contractors are likely to produce their own schedule of modifications which they will attempt to incorporate as a sub-contract document.

Examples of the modifications to SBCSub/C or SBCSub/D/C that contractors may attempt to introduce are:

- an increase in the length of time for payment to be made (ref: clause 4.9.3 of SBCSub/C and SBCSub/D/C);
- a reduction in the period of time before the final date for payment that a withholding notice needs to be given (ref. clause 4.10.3 of SBCSub/C and SBCSub/D/C);
- a provision that health and safety file information must be provided before a release of retention is made;
- a provision that one of the parties will be liable for the costs of any adjudication action, irrespective of the result of that adjudication action.

Of course, there may be many other modifications that a contractor may propose, and each of the proposed modifications could have a serious impact upon the sub-contractor's contractual position.

## SBCSub/A sub-contract particulars item number 2 (SBCSub/D/A sub-contract particulars item number 2) – arbitration

This refers to article 5 of SBCSub/A or SBCSub/D/A.

Under this item, the parties agree whether arbitration will or will not apply.

If the parties agree that disputes and differences are to be determined by arbitration then it must be stated that article 5 of SBCSub/A or SBCSub/D/A and clauses 8.3 to 8.8 of SBCSub/C or SBCSub/D/C apply.

If it is decided that arbitration is not to apply then legal proceedings will apply.

If it is not decided by the parties that arbitration is, or is not to apply, then the default position is that legal proceedings will apply.

For further commentary on arbitration and on legal proceedings, refer to Chapter 13.

## SBCSub/A sub-contract particulars item number 3 (SBCSub/D/A sub-contract particulars item number 3) – sub-contract base date

This refers to clause 1.1 of SBCSub/C and SBCSub/D/C.

The sub-contract base date is to be inserted under this sub-contract particulars item.

The sub-contract base date will not necessarily be the same as the main contract base date; it is usually a date that is close to the sub-contractor's tender or commencement date.

The sub-contract base date is not just some arbitrary or irrelevant date; it is relevant, amongst other things, to:

- Clause 2.12.2.1 of SBCSub/D/C and SBCSub/D/C – divergences from statutory requirements. For commentary on clause 2.12.2.1 refer to Chapter 4.
- Schedule 4 of SBCSub/C and SBCSub/D/C – which relates to fluctuations provisions. For commentary on schedule 4 refer to Chapter 8.
- The applicable standard method of measurement to be used – refer to clause 1.1 of SBCSub/C and SBCSub/D/C. For commentary on clause 1.1 refer to Chapter 3.
- The applicable definitions of the prime cost of daywork to be used – refer to clause 5.9 of SBCSub/C and SBCSub/D/C. For commentary on clause 5.9 refer to Chapter 10.

It should be noted that the main contract base date and not the sub-contract base date is relied upon in respect of:

- the relevant sub-contract event in clause 2.19.14 of SBCSub/C and SBCSub/D/C (for commentary on clause 2.19.14 refer to Chapter 5); and
- the applicable CIMAR arbitration rules referred to in clause 8.3 of SBCSub/C and SBCSub/D/C (for commentary on clause 8.3 refer to Chapter 13).

## SBCSub/A sub-contract particulars item number 4 (SBCSub/D/A sub-contract particulars item number 4) – electronic communications

This refers to clause 1.8 of SBCSub/C and SBCSub/D/C.

The particular communications that may be made electronically (i.e. by e-mail, etc.) and the format in which those electronic communications are to be made are to be set out under sub-contract particulars item number 4, or are to be set out in a document referenced to item 4 and annexed to SBCSub/A or SBCSub/D/A.

If no communications are identified under sub-contract particulars item number 4, then the default position is that all communications are to be in writing, unless subsequently agreed otherwise.

It should be noted that certain notices (e.g. default notices under section 7 of SBCSub/C and SBCSub/D/C) must be given by actual, special or recorded delivery; therefore, in such a situation, electronic communications would not be appropriate.

## SBCSub/A sub-contract particulars item number 5 (SBCSub/D/A sub-contract particulars item number 5) – programme

This generally refers to clause 2.3 of SBCSub/C and SBCSub/D/C (refer to Chapter 5).

### Item 5.1 of SBCSub/D/A

Under sub-contract particulars item 5.1 of SBCSub/D/A, a time period in weeks is to be inserted for the preparation of all necessary sub-contractor's drawings, etc. (coordination, installation, shop or builder's work or other drawings as appropriate) from receipt of the instruction to proceed with such preparation *and* from receipt of all other relevant drawings and specifications, etc., to the submission to the contractor of the drawings etc. for comment.

It should be noted that the time period stated is from receipt of the instruction to proceed with such preparation and from receipt of all other relevant drawings and specifications, etc. Although not stated, the implication is that the time period would only commence after both of the above criteria have been met. The onus is therefore on the contractor to ensure that the required instruction and the relevant drawings and specifications are provided in time to suit the programme requirements.

### Item 5.2 of SBCSub/D/A

Under sub-contract particulars item 5.2 of SBCSub/D/A, a time period in weeks is to be inserted for the contractor's initial comments upon the drawings, etc. (provided by the sub-contractor in line with item 5.1 above) from when the drawings, etc. have been received by the contractor to when the drawings, etc. are returned to the sub-contractor.

Although not stated, the implication is that the time period would apply in respect of individual drawings, etc. rather than apply to the date of receipt by the contractor of a set of drawings.

The periods entered under items 5.1 and 5.2 need to take into account clause 2.6.2 of SBCSub/D/C which requires the sub-contractor to provide to the contractor design documents and other information to enable the contractor to observe and perform his obligations under clause 2.9.3 of the main contract and schedule 5 (the contractor's design submission procedure). Clause 2.6.2 of SBCSub/D/C makes it clear that the sub-contractor must not commence any work in respect of a design document or other information which is subject to the contractor's design submission procedure before that procedure has been complied with. The contractor's design submission procedure is dealt with in detail in Chapter 4.

### *Item 5.3 of SBCSub/D/A (Item 5.1 of SBCSub/A)*

Under sub-contract particulars item 5.3 of SBCSub/D/A (item 5.1 of SBCSub/A), a time period in weeks is to be inserted for the procurement of materials, fabrication (where appropriate) and delivery to site prior to commencing work on site or work on site in each section.

The period inserted here would need to take into account any extended manufacturing/delivery times for any non-standard materials.

If non-standard materials are needed for one section of the sub-contract works but not for another, it is conceivable that different periods may be inserted for different sections (although it should be noted that the SBCSub/A and the SBCSub/D/A makes no specific provision for this possibility).

### *Items 5.1 to 5.3 of SBCSub/D/A*

It is important to note that the periods indicated under items 5.1, 5.2 and 5.3 of SBCSub/D/A are consecutive periods. Therefore, the total period from the date of receipt of the information by the sub-contractor under item 5.1 (of SBCSub/D/A) to the delivery to site of the materials and goods under item 5.3 (of SBCSub/D/A) is the sum of the periods entered against items 5.1, 5.2 and 5.3 of SBCSub/D/A individually. This summated period is inserted as a total period for items 5.1 to 5.3 in the paragraph which is inserted between item 5.3 and 5.4 of SBCSub/D/A.

Therefore, if a period of three weeks was inserted against item 5.1; a period of two weeks was inserted against item 5.2; and a period of six weeks was inserted against item 5.3, then the total period for items 5.1 to 5.3 of SBCSub/D/A (to be inserted in the paragraph between items 5.3 and 5.4 of SBCSub/D/A) would be eleven weeks.

### *Item 5.4 of SBCSub/D/A (item 5.2 of SBCSub/A)*

Under sub-contract particulars item 5.4 of SBCSub/D/A (item 5.2 of SBCSub/A), a time period in weeks is to be inserted for the period of notice required to commence works on site to enable a start to be made to the sub-contract works, or to each section (if applicable). It is conceivable that different periods may be inserted for different sections (although it should be noted that SBCSub/A and SBCSub/D/A makes no specific provision for this possibility).

This time period (or periods) inserted under item 5.4 of SBCSub/D/A (item 5.2 of SBCSub/A) is not stated as being consecutive to the periods for items 5.1 to 5.3 of SBCSub/D/A (or item 5.1 of SBCSub/A) and therefore must be taken as being concurrent with that period.

Therefore, if a period of two weeks was inserted under item 5.4 of SBCSub/D/A, then using the example above of eleven weeks for the summated time period for items 5.1 to 5.3 of SBCSub/D/A, the said notice would need to be given no later than two weeks before the expiry of that eleven-week period, if the commencement of works on site were to commence at the end of that eleven-week period.

This position is made clear from item 5.5 of SBCSub/D/A (and item 5.3 of SBCSub/A) which notes that the period (or periods) required for carrying out the sub-contract works on site commences after delivery of materials (i.e. the summated period of items 5.1 to 5.3 of SBCSub/D/A in the above example) *and* after the expiry of the period of notice to commence works (as item 5.4 of SBCSub/D/A in the above example). The position is also confirmed under clause 2.3 of SBCSub/C that the on-site construction periods are 'subject to receipt by the sub-contractor of notice to commence works'. A similar provision is included under clause 2.3 of SBCSub/D/C.

Therefore, again using the above example, if the notice under item 5.4 of SBCSub/D/A was not given until ten weeks of the eleven weeks summated time period for items 5.1 to 5.3 of SBCSub/D/A had expired, then, even if the materials were delivered to site after the eleven-week period, because of the *late* notice given under item 5.4 of SBCSub/D/A, and because of the two-week notice period required under item 5.4 of SBCSub/D/A, the sub-contractor could conceivably not commence works until the expiry of a twelve-week period (i.e. ten weeks when notice was given plus two weeks notice period required). It should be noted of course that there is nothing to prevent a sub-contractor from commencing prior to the stated notice period, and this may well occur frequently. However, the sub-contractor is not contractually obliged to commence prior to the stated notice period.

Depending on the circumstances of the case, the unnecessary *delay* of one week, caused by the late issue of the notice required by clause 5.4 of SBCSub/D/A, could have a major impact upon the programme for the main contract works. It is therefore essential that the contractor provides the notice required by item 5.4 of SBCSub/D/A at a time to suit the required start date on site for the whole or a section of the sub-contract works.

The notice required under sub-contract particulars item 5.4 of SBCSub/D/A (item 5.2 of SBCSub/A) relates only to the period of notice required to commence works on site to enable a start to be made to the sub-contract works or to each section. The notice is not relevant to the commencement of design work (refer to clause 2.3 of SBCSub/D/C which states 'subject, as respects construction but not design work, to receipt by the Sub-Contractor of notice to commence works'). Also, when the sub-contractor makes several visits to carry out its sub-contract works (refer to sub-contract particulars item 5.6 of SBCSub/D/A (item 5.4 of SBCSub/A) below) the notice is not required before each visit.

This point is emphasized by clause 2.3 of SBCSub/C (and through very similar wording in clause 2.3 of SBCSub/D/C), which states:

'The Sub-Contract works shall be carried out and completed in accordance with the programme details stated in the Sub-Contract Particulars (item 5) and reasonably in accordance with the progress of the Main Contract works or each relevant Section of them, subject to receipt by the Sub-Contractor of notice to commence work in accordance with those particulars and subject to clauses 2.16 to 2.19.'

The sub-contractor's obligation is carry out the sub-contract works reasonably in accordance with the progress of the main contract works, and it is therefore submitted that, after the commencement of the sub-contract works on site, the onus is on the sub-contractor to remain aware of the progress of the main contract works to ensure that he carries out the sub-contract works reasonably in accordance with that progress. This requirement would be consistent with the sub-contractor's obligation (under clause 2.17.1 of SBCSub/C and SBCSub/D/C, dealt with in Chapter 5) to give notice that the commencement, progress or completion of the sub-contract works is being or is likely to be delayed.

Of course, as noted under paragraph 50 of SBCSub/G, it is good practice for the contractor to inform and update sub-contractors regularly on the progress of the main contract works, particularly prior to the commencement on site of the sub-contract works.

An actual start date for each section does not need to be given, but, for guidance, the sub-contractor is to be advised of the earliest starting date and the latest starting date for the sub-contract works to be carried out on site under item 3.1 of the schedule of information which forms part of SBCSub/A and SBCSub/D/A. The earliest starting date and the latest starting date are obtained from the main contract programme information.

### Item 5.5 of SBCSub/D/A (item 5.3 of SBCSub/A)

Under this item the period required for the carrying out of the sub-contract works on site after the delivery of materials (i.e. the summated period of items 5.1 to 5.3 of SBCSub/D/A or the period of item 5.1 of SBCSub/A) *and* after the expiry of the period of notice to commence works (as item 5.4 of SBCSub/D/A or item 5.2 of SBCSub/A) is to be inserted.

If sectional completion does not apply, then only one period (in weeks) needs to be inserted.

If sectional completion does apply, then for each section a period (in weeks) needs to be inserted. If there is insufficient room on SBCSub/A or SBCSub/D/A forms, further sheets should be used and should be annexed to SBCSub/A or SBCSub/D/A as appropriate.

The period or periods stated are the total on-site working periods, and are not calendar week periods.

Therefore, if a sub-contractor was required to carry out the following works then the period to be entered would be twenty-one weeks (i.e. ten weeks (first fix) plus six weeks (second fix) plus three weeks (final fix) plus two weeks (test and commission)) not the thirty calendar weeks from the beginning of the first fix to the completion of the test and commission:

| | |
|---|---|
| First fix | ten weeks |
| Leave site | six weeks |
| Second fix | six weeks |
| Leave site | three weeks |
| Final fix | three weeks |
| Test and commission | two weeks |
| Total period | thirty weeks |

When agreeing to the above period, the sub-contractor needs to take into consideration the information provided under item 3.2 and 3.3 of the schedule of information which forms part of SBCSub/A and SBCSub/D/A.

Item 3.2 of the schedule of information notifies the sub-contractor of the days that the site will be closed due to holidays, which may include, for example, that the site would not be open on Saturdays, Sundays and public holidays.

Item 3.3 of the schedule of information notifies the hours each day that the site will be open for the sub-contractor to carry out the sub-contract works.

Items 3.2 and 3.3 noted above may have a major impact upon the sub-contract period required by the sub-contractor.

For example, if, under item 3.2, the sub-contractor was advised that the site would not be open on Saturdays or Sundays, and under item 3.3 was advised that the site would only be open from 8 am to 1 pm on all other days, this would almost certainly have a major impact on the sub-contract period that would be required, compared to the position where the site was to be closed on Sundays only, and was to be open on all other days from 8 am to 5 pm.

Despite the above, on most construction projects, because of the close inter-relationship between the trades, defined activities do not fall into neat pigeon holes.

If a contractor agrees a period of four weeks with a sub-contractor for a particular activity, the contractor would normally expect approximately 25% of the activity to be complete after one week, 50% after two weeks, and so on, thereby allowing following trades to dovetail in with the progress of the sub-contractor's work activity. Unfortunately, this expectation of the contractor is usually not expressed, and is unlikely to be implied in the sub-contract. Without any express term to the contrary, the sub-contractor is permitted 'to plan and perform the work as he pleases, provided that he finishes it by the time fixed in the sub-contract'.[4]

Therefore, in the above case, the sub-contractor could complete only 10% of the activity in question after three (of the four) weeks and still (arguably) be compliant with the terms of the sub-contract. (The only case that the contractor may be able to make was that the sub-contractor was not proceeding regularly and diligently with its work, but this may be a very difficult argument to pursue.) Of course, if each sub-contractor performed in such a way, it would be extremely difficult for the contractor to coordinate the various trades.

### Item 5.6 of SBCSub/D/A (item 5.4 of SBCSub/A)

Under this item further details or arrangements that may qualify or clarify the period required for the carrying out of the sub-contract works on site (or a section) are provided. If there is insufficient room on SBCSub/A or SBCSub/D/A form, further sheets should be used and should be annexed to SBCSub/A or SBCSub/D/A as appropriate.

Therefore, using the above example, if a period of twenty-one weeks had been inserted under item 5.5 of SBCSub/D/A (or under item 5.3 of SBCSub/A) for the completion of the sub-contract works (or a section) then the further details or arrangements that may qualify or clarify that period would be:

---

[4] *Pigott Foundations Ltd* v. *Shepherd Construction Ltd* (1993) 67 BLR 53.

| | |
|---|---|
| First fix | ten weeks |
| Second fix | six weeks |
| Final fix | three weeks |
| Test and commission | two weeks |
| Total working period | twenty-one weeks |

In addition, in consideration of the difficulties outlined above in respect of the lack of proportional progress of an individual activity within an agreed time period, it may be desirable to insert some further milestone requirements within the individual time periods so as to restrict the sub-contractor's freedom to 'plan and perform the work as he pleases'.

Of course, as noted above, the contractor has some protection in this regard under clause 2.3 of SBCSub/C and SBCSub/D/C (dealt with in more detail in Chapter 5), which requires the sub-contractor to carry out and complete its work 'reasonably in accordance with the progress of the Main Contract works or each relevant Section of them'. However, it is always safer to rely on some specific requirement, rather than a more generalized obligation. This is especially so because the generalized obligation is relied upon in a similar clause in the DOM/1 form of sub-contract and was considered by Mr Justice Gilliland as *not* requiring:

> 'The sub-contractor to plan his sub-contract work so as to fit in with either any scheme of work of the main contractor or to finish any part of the sub-contract works by a particular date so as to enable the main contractor to proceed with other parts of the work.'[5]

Of course, the contractor can ask a sub-contractor to produce a programme for the sub-contract works for the contractor's approval, but this often creates more problems than it solves. First, the contractor would need to correlate all of the individual sub-contractors' programmes to ensure that they were all properly coordinated, which in itself would be extremely difficult given that it is not uncommon for sub-contracts for different trades on one project to be placed several months apart. Second, the more detailed the programme is, the more difficult it is for the contractor to meet the rigid constraints set, and the easier it is for the contractor to default in this regard.

## SBCSub/A sub-contract particulars item number 6 (SBCSub/D/A sub-contract particulars item number 6) – attendance

This refers generally to clause 3.16 of SBCSub/C and SBCSub/D/C (dealt with in Chapter 7).

### Item 6.1: items of attendance which the contractor will provide to the sub-contractor free of charge

Items of attendance which the contractor must provide to the sub-contractor free of charge are to be identified under this sub-contract particulars item.

---

[5] *Pigott Foundations Ltd* v. *Shepherd Construction Ltd* (1993) 67 BLR 53.

A basic list of attendance items is provided, and this list is to be amended (as appropriate) and added to (if appropriate). If there is insufficient room on SBCSub/A or SBCSub/D/A forms, further sheets should be used and should be annexed to SBCSub/A or SBCSub/D/A as appropriate.

It is important that this list is carefully considered and correctly amended because all attendances (other than those finally listed) are to be provided by the sub-contractor. The matter of attendances is considered in further detail under clause 3.16 of SBCSub/C and SBCSub/D/C, in Chapter 7.

### Item 6.2: joint fire code – additional items

The main contract particulars which are to be annexed to the schedule of information (under item 1.6) in SBCSub/A and SBC/D/A, will show under clause 6.13 whether or not the joint fire code applies and, if it does, whether or not the requirements in respect of large projects apply. If the joint fire code does apply, it will be the joint fire code that was current as at the main contract base date (not the sub-contract base date).

The joint fire code states, under paragraph 1, that a large project is one with an original value of £20 million and above. However, the joint fire code also adds that there may be exceptional circumstances where due to the high fire risk nature of a project that threshold may be reduced.

The joint fire code, which is actually called the *Joint Code of Practice on the Protection from Fire of Construction Sites and Buildings Undergoing Renovation*, is published by the Construction Confederation and the Fire Protection Association, and has the support of the Association of British Insurers, the Chief and Assistant Chief Fire Officers' Association and the London Fire Brigade.

The objective of the joint fire code is stated as being:

'The prevention of fires on construction sites. The majority of fires can be prevented by designing out risks, taking simple precautions and by adopting safe working practices. All parties involved must work together to ensure that adequate detection and prevention measures are incorporated during design and contract planning stages; and that the work on site is undertaken to the highest standard of fire safety thereby affording the maximum level of protection to the building and its occupants.'

The joint fire code is divided into the following parts:

(1) Objective of the code
(2) Compliance with the code
(3) Introduction
(4) Definitions used in this code
(5) Design phase
(6) Construction phase
(7) Emergency procedures
(8) Fire protection
(9) Temporary covering materials
(10) Portable fire extinguishers
(11) Site security against arson
(12) Temporary buildings and temporary accommodation

(13) Site storage of flammable liquids and LPG
(14) Electricity and gas supplies
(15) Hot work
(16) Waste materials
(17) Plant
(18) Materials storage
(19) Smoking

Where the joint fire code applies the sub-contractor is expected to comply with it; in addition sub-contract particulars item 6.2 of SBCSub/A and SBCSub/D/A states that the contractor must provide the following attendances:

- hand bells, whistles, klaxons and manually operated sounders;
- security guards;
- fire signage e.g. location of fire access routes and escape routes and positions of dry riser inlets and fire extinguishers;
- fire doors and fire stopping to lift shafts, service ducts and voids;
- water supplies for fire fighting;
- portable fire extinguishers;
- maintenance and inspection of fire fighting equipment.

In addition to the above, where clause 6.13 of the main contract particulars states that the requirements in respect of large projects apply, then the contractor must provide the service of an appropriate number of fire marshals.

### Item 6.3: location of sub-contractor's temporary buildings

Item 6.3 does not apply (and should be deleted) if:

- the sub-contractor is not to provide temporary buildings at all; or
- if the sub-contractor is to provide temporary buildings but these are to be located more than six metres from the building under construction.

If, on a project where the joint fire code applies, the sub-contractor is to provide temporary buildings either within the building under construction and/or within six metres of the building under construction, then the sub-contractor is to ensure that the temporary buildings are constructed to comply with paragraphs 12.4 and 12.8 of the joint fire code.

This means that the temporary buildings should be designed and constructed so as to comply with test specifications laid down in LPS (Loss Prevention Standard) 1195 *Specification for Testing of Temporary Buildings for Use on Construction Sites*, published by the Building Research Establishment Limited, or alternatively that they should meet the criteria set out under paragraphs 12.4 to 12.14 of the joint fire code. These criteria include for the surface spread of flames and fire resistance requirements, etc. of the materials used in the construction of the temporary buildings, and also require the sub-contractor to fit the temporary buildings with fire detection systems.

If the project is a large project under the joint fire code, then, in line with item 6.3 of the sub-contract particulars in SBCSub/A and SBCSub/D/A, the contractor is to connect the above noted fire detection systems to a central station (alarm receiving centre).

Footnote 12 on page 8 of SBCSub/A and footnote 14 on page 9 of SBCSub/D/A note that this latter requirement only applies if the contractor is responsible for the connection.

Given that where the joint fire code applies, item 6.3 of the sub-contract particulars in SBCSub/A and SBCSub/D/A makes it clear that the contractor is to connect the above noted fire detection systems to a central station, then the reference that this only applies if the contractor is responsible for the connection presumably relates to paragraph 12.8 of the joint fire code which notes that the linking of the fire detection systems to a central station is unnecessary if 24-hour site security is provided.

### Item 6.4: clearance of rubbish resulting from the carrying out of the sub-contract works

If all rubbish resulting from the carrying out of the sub-contract works is to be disposed of off-site (by the sub-contractor) then no entry needs to be made. If no entry is made, then the default position is that all rubbish resulting from the carrying out of the sub-contract works is to be disposed of off-site by the sub-contractor.

If there are specific requirements for the manner in which rubbish is to be disposed of, then details should be inserted under item 6.4.

These details could be, for example, that the sub-contractor is to dispose of his rubbish to a central skip to be provided by the contractor; or may be that certain waste (for example used fluorescent tubes which contain PCB (polychlorinated biphenyl)) must be disposed of in a particular manner.

## SBCSub/A sub-contract particulars item number 7 (SBCSub/D/A sub-contract particulars item number 7) – listed items

This refers generally to clause 4.14 of SBCSub/C and SBCSub/D/C dealt with in Chapter 8.

The contractor and the sub-contractor should agree as to whether any payments are to be made for goods and materials before their delivery to site.

Where it is agreed, the materials and goods in question should be listed under sub-contract particulars item number 7.

If a bond is required in respect of these items, this should be stated and the maximum liability of the surety should be stated.

The matter of listed items is dealt with in more detail in Chapter 8, under the part headed 'Clause 4.13.3 of SBCSub/C and SBCSub/D/C (unfixed materials and goods off-site) (listed items)'.

## SBCSub/A sub-contract particulars item number 8 (SBCSub/D/A sub-contract particulars item number 8) – retention percentage

This refers to clause 4.15 of SBCSub/C and SBCSub/D/C dealt with in Chapter 8.

An entry only needs to be made here if the retention percentage is to be different to the default percentage of 3%.The retention percentage can be at any level that the parties agree, and it does not have to be the same percentage as applies to the main contract.

The question of retention is dealt with in more detail in Chapter 8.

## SBCSub/A sub-contract particulars item number 9 (SBCSub/D/A sub-contract particulars item number 9) – sub-contractor's retention bond

This refers to clause 4.16 of SBCSub/C and SBCSub/D/C dealt with in Chapter 8.

An entry only needs to be made here if a retention bond is required. The retention bond is contained under schedule 3 part 2 to SBCSub/C and SBCSub/D/C.

The information to be provided under item 9 of the sub-contract particulars is:

- the maximum aggregate sum for the purposes of clause 2 of the retention bond; and
- the expiry date for the purposes of clause 6.3 of the retention bond.

A retention bond may be provided by a sub-contractor irrespective of whether or not a retention bond is provided by the main contractor under the terms of the main contract.

The issue of retention and the retention bond is dealt with in more detail in Chapter 8.

## SBCSub/A sub-contract particulars item number 10 (SBCSub/D/A sub-contract particulars item number 10) – fluctuations

This refers to clauses 4.17 and 4.18 and schedule 4 of SBCSub/C and SBCSub/D/C dealt with in Chapter 8.

Under this item a fluctuation option must be chosen. The fluctuation options are option A, option B or option C.

The only restriction regarding the choice made is that footnote 13 on page 9 of SBCSub/A (footnote 15 on page 10 of SBCSub/D/A) states that where option A or B applies in the main contract it is a requirement of that contract (see paragraphs A.3 and B.4 of Schedule 7 of the Standard Building Contract) that the same option applies to any sub-contract.

Where option C applies to the main contract, the parties are free to agree that either options A, B or C apply to the sub-contract.

Where option A or B applies, a percentage addition to the fluctuation payments or allowances applicable is to be inserted. This percentage will relate to option A, paragraph A.12 or option B, paragraph B.13.

Where option B applies, a list of basic transport charges as referred to in paragraph B.1.5.1 of schedule 4 of SBCSub/C and SBCSub/D/C is to be provided.

This list may either be provided on the SBCSub/A or the SBCSub/D/A or it may be contained on a separate document that is simply referred to in the appropriate part of SBCSub/A or SBCSub/D/A.

If a separate list is provided, it should be included as a numbered document under sub-contract particulars item 15 (of SBCSub/A) or item 16 (of SBCSub/D/A).

The list provided under the sub-contract can be a different list to the equivalent list provided under paragraph B.1.5.1 of schedule 7 of the main contract.

The issue of fluctuations generally is dealt with in more detail in Chapter 8 of this book.

## SBCSub/A sub-contract particulars item number 11 (SBCSub/D/A sub-contract particulars item number 11) – dayworks

This refers to clause 5.9 of SBCSub/C and SBCSub/D/C dealt with in Chapter 10.

In respect of dayworks, either a schedule of daywork rates or prices (for labour, materials and plant) can be included as a numbered document under sub-contract particulars item 15 (of SBCSub/A) or item 16 (of SBCSub/D/A), which can then be used to value daywork items, or daywork is to be calculated in accordance with the definition or definitions identified under sub-contract particulars item 11, together with the percentage entered for each section of the prime cost.

The three definitions referred to (one or more of which may be chosen) are:

RICS/CC        the definition agreed by the Royal Institution of Chartered Surveyors with the Construction Confederation.

RICS/ECA      the definition agreed by the Royal Institution of Chartered Surveyors with the Electrical Contractors Association.

RICS/HVCA    the definition agreed by the Royal Institution of Chartered Surveyors with the Heating and Ventilating Contractors Association.

The three categories of prime cost are labour, materials and plant.

In respect of labour (which encompasses tradesmen, craftsmen and non-skilled labour) one option is to use one or more of the three definitions noted above to obtain the labour daywork base rate at the time that the daywork was carried out, and then apply the appropriate percentage addition for labour as inserted against the appropriate definition (when using one of the three definitions to obtain the labour daywork base rate, then the wage fixing body or bodies whose labour rates are to be applied when charging for daywork are to be indicated under sub-contract particulars item 11 (SBCSub/A and SBCSub/D/A). Alternatively, use the all-in rate inserted for labour in lieu of the cost of labour calculated in accordance with one of the definitions.

In respect of materials, the only option is to use one or more of the three definitions noted above to determine the materials daywork base rate at the time that the daywork was carried out, and then apply the appropriate percentage addition to the prime cost of materials as inserted against the appropriate definition.

In respect of plant, the only option is to use one or more of the three definitions noted above to obtain the plant daywork base rate at the time that the daywork was carried out, and then apply the appropriate percentage addition to the schedule of plant hire rates (if appropriate) as inserted against the appropriate definition.

It should be particularly noted that the percentage addition to the daywork rates, generally, is to be included in the sub-contractor's tender; and this percentage addition would normally be expected to cover for items such as:

- head office charges;
- site staff including site supervision;
- time lost due to inclement weather;
- the additional cost of bonuses and all other incentive payments in excess of any guaranteed minimum union wage rates;
- fares and travel allowance;
- sick pay or insurance in respect thereof;
- third party and employer's liability insurance;
- liability in respect of redundancy payments to employees;
- employers' National Insurance contributions;
- use of erected scaffolding, staging, trestles or the like;
- any variation to basic rates required by the sub-contractor in cases where the building contract provides for the specified schedule of basic plant charges (this can, on occasion be a very significant factor because the basic plant charges schedule being relied upon may pre-date the date when the dayworks are being carried out by several years).

General commentary in respect of dayworks is provided in Chapter 10.

## SBCSub/A sub-contract particulars item number 12 (SBCSub/D/A sub-contract particulars item number 12) – insurance, personal injury and property

This refers to clause 6.5 of SBCSub/C and SBCSub/D/C dealt with in Chapter 11.

The insurance cover to be entered under this item is the minimum level required in respect of death or personal injury (other than to employees covered under the sub-contractor's statutory employers' liability policy) and in respect of damage to property (excluding specified perils damage to the main contract works and site materials) up to the terminal date.

The terminal date is defined under clause 6.1 of SBCSub/C and SBCSub/D/C as being:

'(a) the date of practical completion of the Sub-Contract works or, in respect of a Section, of such works in the Section, as determined in accordance with clause 2.20; or (b) the date of termination of the Sub-Contractor's employment under the Sub-Contract, however arising; whichever occurs first.'

The level of cover required should not normally exceed the level stated in the main contract particulars. However, the level could be less by agreement, if the sub-contractor was unable to provide the level of cover required, and if the contractor's insurance covered for any shortfall on cover.

General commentary in respect of insurance provisions is provided in Chapter 11.

## SBCSub/A sub-contract particulars item number 13 (SBCSub/D/A sub-contract particulars item number 13) – incorporation of the sub-contract works into the main contract works

This refers to Clause 6.7.8 of SBCSub/C and SBCSub/D/C dealt with in Chapter 11.

The parties are to list under this sub-contract particulars item those elements of the sub-contract works that the contractor is prepared to regard as fully, finally and properly incorporated into the main contract works prior to practical completion of the sub-contract works or section as applicable; and also to detail the extent to which each of the listed elements needs to be carried out to achieve the status of being fully, finally and properly incorporated into the main contract works. If there is insufficient space on form SBCSub/A or SBCSub/D/A to make the required entries, then a separate sheet may be used which should then be annexed to SBCSub/A or SBCSub/D/A as appropriate.

In the past there have frequently been differences of opinion between contractors and sub-contractors as to when elements of the sub-contract works have been fully, finally and properly incorporated into the main contract works prior to practical completion of the sub-contract works. The purpose of the entries to sub-contract particulars item 13 is to attempt to reduce any differences of opinion between the contractor and the sub-contractor regarding this issue by setting out some clear requirements to be met before the status of fully, finally and properly incorporated into the main contract works can be achieved.

Clearly, for the entries under this sub-contract particulars item to have any effect, both the elements of the sub-contract works, and the extent to which each of the listed elements needs to be carried out to achieve the status of being fully, finally and properly incorporated into the main contract works, need to be precisely set out. Therefore, great care needs to be taken regarding the entries made since, if general terms are used, it will be difficult to determine if the full extent of the element has been installed and/or that the full extent of the required criteria has been met. In fact, inaccurate or imprecise entries under this sub-contract particulars item may create more problems than they solve.

Clause 6.7.2 of SBCSub/C and SBCSub/D/C notes that:

'Where, during the progress of the Sub-Contract works sub-contract materials or goods have been fully, finally and properly incorporated into the Main Contract works before the Terminal Date, then, in respect of loss or damage to any of the Sub-Contract works so incorporated that is caused by the occurrence of a peril other than a Specified Peril, the Sub-Contractor shall only be responsible for the cost of restoration of such work lost or damaged and removal and disposal of any debris in accordance with clause 6.7.3 to the extent that such loss or damage is caused by the negligence, breach of statutory duty, omission or default of the Sub-Contractor or of any of the Sub-Contractor's Persons.'

It is therefore of great importance for the sub-contractor to be able to determine when certain elements of its works have been fully, finally and properly incorporated into the main contract works, since after that time the sub-contractor's responsibility for the cost of restoration of any such work lost or damaged and the removal and disposal of any debris in accordance with clause 6.7.3 is greatly reduced, and in fact only applies where the loss or damage is caused by the negligence, breach of statutory duty, omission or default of the sub-contractor or of any of the sub-contractor's persons.

Clause 6.7.8 of SBCSub/C and SBCSub/D/C notes that for the purposes of clause 6.7.2 *only*, materials and goods forming part of the sub-contract works shall be deemed to have been fully, finally and properly incorporated into the main contract works when, in each case, they have been completed by the sub-contractor to the extent indicated or referred to in the sub-contract particulars item 13. Emphasis on the word *only* has been made to underline the point that the satisfaction of the criteria set out under sub-contract particulars item 13 means that the materials and goods in question must be deemed to have been fully, finally and properly incorporated into the main contract works for the purposes of clause 6.7.2 only, and not for any other purpose.

## SBCSub/D/A sub-contract particulars item number 14 (there is no equivalent sub-contract particulars item in SBCSub/A) – sub-contractor's designed portion (SCDP) professional indemnity insurance

This refers to clause 6.10 of SBCSub/D/C dealt with in Chapter 11.

The entry here relates to the level of cover required in respect of professional indemnity insurance in respect of clause 6.10 of SBCSub/D/C.

*It should be particularly noted that if no amount of cover is inserted, clause 6.10 will not apply, regardless of whether or not the other parts of this item are completed.*

The amount of cover required either:

- relates to claims or series of claims arising out of one event; or
- it is the aggregate amount for any one period of insurance. SBCSub/D/A sub-contract particulars item number 14 makes it clear that, unless stated otherwise, a period of insurance for these purposes is one year.

One or other of the above options is to be deleted.

If a level of cover that is different to the full indemnity cover noted above is required for pollution/contamination claims, this level of cover should be entered under item 14 of the sub-contract particulars for SBCSub/D/A. If no level of cover is stated, then the required level of cover will be the full amount of the indemnity cover stated above.

The period (in terms of years) that the professional indemnity insurance needs to be maintained from the date of practical completion of the sub-contract works, needs to be entered. The choices are six years, twelve years, or some other period in years. If no entry is made, the expiry date will be taken as being six years from the date of practical completion of the main contract works (not from the practical completion of the sub-contract works).

As a general guide, if the sub-contract is executed under hand a professional indemnity insurance period of six years after the date of practical completion of the sub-contract works would be adequate (because the limitation period for instituting proceedings is six years where a contract is executed under hand), but if the sub-contract is executed as a deed then a professional indemnity insurance period of twelve years after the date of practical completion of the sub-contract works would be required (because the limitation period for instituting proceedings is extended to twelve years where a contract is executed as a deed).

Further commentary regarding limitation periods is given below, and general commentary in respect of insurance provisions is provided in Chapter 11.

## SBCSub/A sub-contract particulars item number 14 (SBCSub/D/A sub-contract particulars item number 15) – settlement of disputes

This refers to clauses 8.2 and 8.4.1 of SBCSub/C and SBCSub/D/C dealt with in Chapter 13.

In respect of adjudication, either an adjudicator may be named, or a nominator of an adjudicator may be selected.

If an adjudicator is named then that named adjudicator should be first approached when a party requires a reference to adjudication to be made.

Even if an adjudicator is named, it is recommended that a nominator of an adjudicator is selected in case the named adjudicator is unwilling or unable to act for some reason.

The choice of nominators of an adjudicator is the President or a Vice-President or Chairman or a Vice-Chairman of:

- the Royal Institute of British Architects
- the Royal Institution of Chartered Surveyors
- the Construction Confederation
- the National Specialist Contractors Council
- the Chartered Institute of Arbitrators

If no choice of nominator is made, and assuming that there is no named adjudicator, or the named adjudicator has made it apparent that he is unable or unwilling to act, the party requiring the reference to adjudication may select any of the nominating bodies listed above.

In respect of arbitration, there is no facility to name an arbitrator.

However, there is the option to choose one of three appointers of an arbitrator.

The choice of appointers of an arbitrator is the President or a Vice-President of:

- the Royal Institute of British Architects
- the Royal Institution of Chartered Surveyors
- the Chartered Institute of Arbitrators

If no choice is made of appointer, the default position is that the appointer will be the President or a Vice-President of the Royal Institution of Chartered Surveyors.

Further commentary regarding adjudication and arbitration is provided in Chapter 13 of this book.

## SBCSub/A sub-contract particulars item number 15 (SBCSub/D/A sub-contract particulars item number 16) – numbered documents

Any numbered documents (and, as appropriate, any annexures to any numbered documents) should be listed under SBCSub/A sub-contract particulars item number 15 (SBCSub/D/A sub-contract particular item number 16). These documents should be numbered sequentially, should be initialled or signed by each party, and should be annexed to SBCSub/A or SBCSub/D/A as appropriate.

The numbered documents should comprise all the documents which are to be the sub-contract documents other than for:

- SBCSub/A or SBCSub/D/A;
- SBCSub/C or SBCSub/D/C;
- the schedule of information and its listed annexures.

Examples of the numbered documents (where applicable) include:

- any documents that show or describe the sub-contract works in the form of:

  — drawings;
  — specifications;
  — bills of quantities.

- any documents that set out any conditions relating to the execution of the sub-contract works stipulated by the employer or the contractor (which are not already detailed in the schedule of information or in its listed annexures), including any:

  — special conditions on the employment of labour;
  — limitation of working hours;
  — limitation of access;
  — limitation on use of hoisting equipment.

- the schedule of rates;
- the schedule of modifications to the SBCSub/C or SBCSub/D/C;
- any special condition;
- the priced activity schedule;
- the schedule of daywork rates;
- any separate schedule of attendance items where item 6 in the sub-contract particulars is not adequate;
- a basic transport charges list where item 10 in the sub-contract particulars is not adequate.

## Summary of interim and final payment provisions

A summary of the interim and final payment provisions is provided within both SBCSub/A and SBCSub/D/A.

Details of the payment provisions under the Standard Building Sub-Contract are provided in Chapter 8.

## Attestation forms

Attestation simply refers to certifying the validity of the sub-contract agreement, and this validity is confirmed by the parties executing the sub-contract agreement either under hand or as a deed.

## What difference does it make if the sub-contract is executed under hand or as a deed?

If a sub-contract is executed under hand the limitation period for commencing proceedings due to a breach of the contract is six years, whereas if it is executed as a deed, the limitation period is extended to twelve years.

The question of the limitation period often becomes an issue when latent defects become apparent. A 'latent defect' is a defect that for some reason or another is concealed or lies dormant, as distinct from a 'patent defect' which is open and obviously exists.

Normally, a contractor would wish to have the same limitation period for any recourse against a sub-contractor as there exists against the contractor from the employer. Therefore, if the main contract was executed as a deed (as is usually the case) the contractor would normally require the sub-contract to be executed as a deed. However, if the main contract was executed under hand, the contractor would normally accept that the sub-contract should be executed under hand.

The principle behind limitation periods is to provide a cut-off date after which no claim can be brought (i.e. the claim becomes time barred). The reason behind this is that the law regards it as undesirable that liability could exist without time limit. Limitation periods are set down for both contract and tort in the Limitation Act 1980.

The question of limitation periods in tort is dealt with in Chapter 4.

In a contractor's action in contract against a sub-contractor in respect of defects, the position may depend upon whether the sub-contract contains an express clause whereby the sub-contractor indemnifies the contractor against breaches of contract. If it does, as it does in the case of the Standard Building Sub-Contract (under clause 2.5.1 of SBCSub/C and SBCSub/D/C) the contractor's action upon the indemnity clause may not be statute barred until the expiry of the limitation period (i.e. six or twelve years, as appropriate) between the employer and the contractor under the main contract.[6]

The Latent Damage Act 1986 does not affect the existing law on limitation in contract cases[7] where it has been found that section 14A of the Limitation Act could not be applied to actions in contract.

However, under section 32 of the Limitation Act 1980, the limitation period can be extended if the contractor (or a sub-contractor) has deliberately concealed the defect.

Section 32(1) provides:

'(1) Subject to subsections (3) and (4A) below, where in any case of any action for which a period of limitation is prescribed by this Act, either:

(a) the action is based on the fraud of the defendant; or
(b) any fact relevant to the plaintiff's right of action has been deliberately concealed from him by the defendant; or
(c) the action is for relief from the consequences of a mistake;

the period of limitation shall not begin to run until the plaintiff has discovered the fraud, concealment or mistake (as the case may be) or could with reasonable diligence have discovered it.'

[6] *County and District Properties* v. *Jenner* (1976) 3 BLR 38.
[7] See *Iron Trades Mutual Insurance Co. Ltd* v. *J.K. Buckenham Ltd* [1990] 1 All ER 808.

It should be noted that deliberate concealment has, in recent years, been construed widely in building contract cases and may apply even where the employer (or the contractor, as appropriate) had the benefit of agents overseeing the works.[8]

However, simply proceeding with the works does not necessarily give rise to a deliberate concealment; to establish deliberate concealment under section 32 of the Limitation Act 1980, the claimant must show that the defendant:

(1) has taken active steps to conceal a breach of duty after he has become aware of it; or
(2) is guilty of deliberate wrongdoing and has concealed or has failed to disclose such wrongdoing in circumstances where it is unlikely to be discovered for some time.

## Execution under hand

When SBCSub/A or SBCSub/D/A is executed under hand, each party or his authorized representative should sign SBCSub/A or SBCSub/D/A in the appropriate place on the page headed 'Execution under hand'. Each of those signatures should be witnessed, and each witness should sign the form, and provide his/her name and address.

In cases where the forms of attestation set out in SBCSub/A or SBCSub/D/A are not appropriate (for example, in the case of certain housing associations and partnerships) then the appropriate forms may be inserted by the parties in lieu of the forms provided.

## Execution as a deed

When SBCSub/A or SBCSub/D/A is executed as a deed, the parties may execute the agreement either as a company (or other body corporate) or as an individual, as appropriate. In all cases the contractor's name and the sub-contractor's name should be inserted in the appropriate place on the page headed 'Execution as a deed'.

When executing as an individual, the contractor and/or the sub-contractor should insert their name, and should then sign SBCSub/A or SBCSub/D/A in the appropriate place on the page headed 'Execution as a deed'. Either or both of those signatures (as applicable) should be witnessed, and the witness should sign the form, and provide his/her name and address.

When executing as a company (or other body corporate) two options are available to the parties:

(1) The contract may be executed by the signature of a director and the company secretary, or by the signature of two directors. There must be two signatures. The fact that the company secretary may be a director also does not mean that only one signature needs to be provided.

This alternative is available to companies incorporated under the Companies Act whether or not they have a common seal but is not available to local

---

[8] *Lewisham Borough Council* v. *Leslie & Co.* (1979) 12 BLR 22, CA.

authorities or to certain other bodies corporate (for example, those incorporated by Act of Parliament or by Charter) which fall outside of the provisions of the Companies Act 1985 (refer to section 718 of the Companies Act 1985).

(2) The contract may be executed by affixing the company's common seal in the presence of a director and the company secretary, or in the presence of two directors or other duly authorized officers.

As noted above, in cases where the forms of attestation set out in SBCSub/A or SBCSub/D/A are not appropriate (for example, in the case of certain housing associations and partnerships) then the appropriate forms may be inserted by the parties in lieu of the forms provided.

Where a foreign company is involved, it can execute deeds (under the Companies Act 1989, as applied by the Foreign Companies (Execution of Documents) Regulations 1994 and the 1995 amendments to those regulations) either:

(1) by affixing its common seal (if it has one) or any other manner of execution permitted under the laws of its place of incorporation; or

(2) by expressing the document to be executed under the signature of persons authorized to sign on its behalf in accordance with its domestic law.

However, because of the complications that may arise when foreign companies are involved, it is often best to obtain professional advice regarding the proposed execution method.

## Schedule of information

Item 1 of the schedule of information provides the following information (in respect of the main contract):

- The name of the employer.
- The name of the architect or the contract administrator.
- The name of the planning supervisor (if a planning supervisor is named in the main contract).
- Confirmation that the principal contractor is the contractor, or alternatively if it is not the contractor, the name of the principal contractor.
- Confirmation of which version of the Standard Building Contract applies (i.e. either with quantities, or with approximate quantities, or without quantities).
- Confirmation of the JCT amendments to the Standard Building Contract (if any) that apply.
- Details of any schedule of modifications (i.e. not JCT amendments but employer amendments) to the Standard Building Contract. If there is any such schedule of modifications, this should be annexed to SBCSub/D/A or SBCSub/A.
- A copy of the main contract particulars (which should be annexed to SBCSub/D/A or SBCSub/A, and which should obviously be the copy forming part of the main contract between the employer and the contractor).
- Any changes to the main contract particulars. For example, any change to the completion date for the main contract works or of any section(s).

- Confirmation as to whether or not the main contract works are divided into sections, and, if applicable, details of those sections (if the information is not already provided in the main contract particulars).
- Finally, under item 1.9 of the schedule of information, any other documents and information relating to the provisions of the main contract of which the sub-contractor has had a copy or a reasonable opportunity to inspect.

Item 2 of the schedule of information notes that a copy of the health and safety plan, or more accurately, those parts of the health and safety plan for the main contract that are applicable to the sub-contract works (together with any developments to the health and safety plan by the principal contractor notified to the sub-contractor before or during the progress of the sub-contract works) should be annexed to the schedule of information except where:

- the Construction (Design and Management) Regulations 1994 (the CDM Regulations) do not apply to the main contract works;[9] or
- only regulation 7 (which relates to the requirement to notify a project to the Health and Safety Executive) and/or regulation 13 (which relates to the requirements on designers) of the CDM Regulations apply.

Item 3 of the schedule of information sets out certain programme information, and the issue of programmes is dealt with under sub-contract particulars items 5.4 and 5.5 of SBCSub/A (sub-contract particulars items 5.2 and 5.3 of SBCSub/D/A).

It is important that the schedule of information is accurately provided by the contractor and noted by the sub-contractor particularly because clause 2.5.1 of SBCSub/C and SBCSub/D/C states:

'Insofar as the Contractor's obligations under the Main Contract, as identified in or by the Schedule of information, relate and apply to the Sub-Contract works (or any part of them), the Sub-Contractor shall observe, perform and comply with those obligations, including (without limitation) those under clauses 2.10 (Levels and setting out), 2.21 (Fees or charges legally demandable), 2.22 and 2.23 (Royalties and patent rights) and 3.22 and 3.23 (Antiquities) of the Main Contract Conditions, and shall indemnify and hold harmless the Contractor against and from:

2.5.1.1 any breach, non-observance or non-performance by the Sub-Contractor or his employees or agents of any of the provisions of the Main Contract; and

---

[9] The Health and Safety Executive's Guidance leaflet entitled *Having Construction Work Done?* clarifies that the CDM Regulations apply to all demolition and structural dismantling work, except where it is undertaken by a domestic client.

The CDM Regulations also apply to most construction projects. There are, however, a number of situations where the CDM Regulations do not apply. These include:

- some small-scale projects which are exempt from some aspects of the CDM Regulations;
- construction work for domestic clients (although there are always duties and the designer and the contractor should notify the Health and Safety Executive, where appropriate);
- construction work carried out inside offices and shops, or similar premises, that does not interrupt the normal activities in the premises and is not separated from those activities;
- the maintenance or removal of insulation on pipes, boilers or other parts of heating or water systems.

2.5.1.2 any act or omission of the Sub-Contractor or his employees or agents which involves the Contractor in any liability to the Employer under the provisions of the Main Contract.'

Any documents annexed to the schedule of information should be signed or initialled by or on behalf of each party.

## Notes (to annexures to the agreement and to the schedule of information)

This part of the SBCSub/A and SBCSub/D/A simply provides notes to the annexures to the agreement and to the schedule of information. These matters are all dealt with under the appropriate parts of this chapter.

## Supplementary particulars (containing information needed for formula adjustment)

These particulars only need to be given when fluctuation option C (formula adjustment) applies.

When fluctuation option C does apply, the relevant insertions should be made by the contractor or the sub-contractor as appropriate.

The issue of fluctuations generally is dealt with in more detail in Chapter 8.

# Chapter 3
# Definitions and Interpretations

THE JCT STANDARD BUILDING SUB-CONTRACT CONDITIONS (SBCSUB/C OR SBCSUB/D/C): SECTION 1 (INCLUDING CLAUSES 1.1 TO 1.10 OF SBCSUB/C AND SBCSUB/D/C)

## General

The SBCSub/C comprises eight sections and four schedules, as follows:

Section 1    Definitions and Interpretation
Section 2    Carrying out the Sub-Contract Works
Section 3    Control of the Sub-Contract Works
Section 4    Payment
Section 5    Valuation of Work and Variations
Section 6    Injury, Damage and Insurance
Section 7    Termination
Section 8    Settlement of Disputes

Schedule 1   Sub-Contract Code of Practice
Schedule 2   Schedule 2 Quotation
Schedule 3   Forms of Bonds
Schedule 4   Fluctuation Options

The SBCSub/D/C comprises all of the above sections and schedules, but has, in addition, one further schedule, namely:

Schedule 5   Contractor's Design Submission Procedure

Each of the above sections and schedules is dealt with in the following chapters.

## Section 1 – definitions and interpretation

### Definitions

Clause 1.1 of SBCSub/C and SBCSub/D/C states that:

'Unless the context otherwise requires or this Sub-Contract specifically provides otherwise, the following words and phrases, where they appear in capitalized form, in the Sub-Contract Agreement or these conditions, shall have the meanings stated or referred to below.'

This is a very important clause in the sub-contract because, when dealing with the sub-contract, any defined terms must be reported in the capitalized form rather than in a lower case form, so that there is no doubt about what is actually being referred to.

Therefore, when corresponding about, for example, the sub-contract works the phrase the 'Sub-Contract Works' should be used rather than the more generalized phrase 'the sub-contract works', to make it clear that the sub-contract works *as defined in the sub-contract* is being specifically referred to.

*Despite the importance of this clause, this book does not generally follow this convention.*

Therefore, generally, a word that is defined in the sub-contract is included in the text of this book in lower case, for example 'Variation' is simply included as variation.

The only exception to the above convention is where it is considered that this may cause unnecessary ambiguity to the reader.

With the above point in mind, the reader is strongly advised to refer to the Definitions list at the beginning of this book where some of the more important defined terms within the sub-contract are listed. This list of defined terms should be referred to, as appropriate, when reading the various chapters of this book.

## Interpretation

### Reference to clauses, etc.

Clause 1.2 of SBCSub/C and SBCSub/D/C simply clarifies, that, unless otherwise stated, a reference in the sub-contract agreement SBCSub/A or SBCSub/D/A, or in the sub-contract conditions SBCSub/C or SBCSub/D/C to a clause or schedule, is to that clause in, or that schedule to, the appropriate sub-contract agreement or conditions.

Also, unless the context otherwise requires, a reference in a schedule to a paragraph is to that paragraph of that schedule.

### Sub-contract to be read as a whole

Clause 1.3 of SBCSub/C and SBCSub/D/C states that the sub-contract is to be read as a whole, but then sets out the priority of documents in the event that there are any inconsistencies.

*This, again, is a very important clause and must be used to resolve disputes about which document should be relied upon in the event of there being any inconsistency between the documents.*

The priority is set out as follows:

- The SBCSub/A takes priority if there is any discrepancy between SBCSub/A and SBCSub/C; and SBCSub/D/A takes priority if there is any discrepancy between SBCSub/D/A and SBCSub/D/C.

- The sub-contract documents (i.e. the documents referred to in article 1) but excluding the numbered documents (i.e. the documents listed in the sub-contract particulars – item 15 of SBCSub/A or item 16 of SBCSub/D/A) take priority if there is any discrepancy between the sub-contract documents (excluding the numbered documents) and the numbered documents (excluding the schedule of modifications which is the schedule of amendments to the sub-contract conditions as detailed under article 1 and the sub-contract particulars item 1).

- The sub-contract documents take priority if there is any discrepancy between the sub-contract documents and the main contract conditions.

- Nothing in any descriptive schedule or similar document issued in connection with and for use in carrying out the sub-contract works will impose any obligation beyond those imposed by the sub-contract documents.

- Nothing in the sub-contract documents will be construed as imposing any liability on the sub-contractor in respect of any act, omission or default on the part of the employer, the employer's persons, the contractor or the contractor's persons.

### *Headings, references to persons, legislation, etc.*

Clause 1.4 of SBCSub/C and SBCSub/D/C clarifies that, unless the context requires otherwise, in SBCSub/A, SBCSub/D/A, SBCSub/C and SBCSub/D/C:

- the headings in the sub-contract are included for convenience only and must not affect the interpretation of the sub-contract;
- the singular includes the plural and vice versa;
- a gender includes any other gender;
- a reference to a 'person' includes any individual, firm, partnership, company and any other body corporate;
- a reference to a statute, statutory instrument or other subordinate legislation ('legislation') is to such legislation that is amended and in force from time to time, including any legislation which re-enacts or consolidates it, with or without modification.

### *Reckoning periods of days*

Clause 1.5 of SBCSub/C and SBCSub/D/C makes clear that where an act is required to be done within a specified period of days after or from a specified date, the period must begin immediately after that date. Where the period would include a day which is a public holiday that day must be excluded.

Therefore, if, for example, an interim payment due date was on a Thursday before a public holiday Monday, then the notice under clause 4.10.2 of SBCSub/C or SBCSub/D/C (which is required to be given not later than five days after the date on which an interim payment becomes due) would need to be given no later than the Wednesday of the following week (i.e. the five days being Friday, Saturday, Sunday, Tuesday and Wednesday, the Monday being excluded because it is a public holiday).

### *Contracts (Rights of Third Parties) Act 1999*

As noted in Chapter 1, under the common law doctrine of privity of contract the general rule is that contracts cannot be enforced either by or against third parties.

However, whilst the common law position remains unchanged, the Contracts (Rights of Third Parties) Act 1999, which came into force on 11 November 1999, gives a third party the right to enforce a term of a contract, if there is an express provision that he should, or if he is named in a contract which purports to confer a benefit on him.

Clause 1.6 of SBCSub/C and SBCSub/D/C makes it clear that notwithstanding any other provision of the sub-contract, nothing in the sub-contract confers or is intended to confer any right to enforce any of its terms on any person who is not a party to it.

Therefore, the Standard Building Sub-Contract does not allow for any third party rights.

### Giving of service of notices and other documents

Clause 1.7 of SBCSub/C and SBCSub/D/C sets out the default position in respect of the serving of notices and other documents if there are no other specific provisions set out in the conditions. This default position is:

- any notice or other document may be given or served by any effective means (which presumably means delivery by hand, by courier, etc., but not by e-mail, etc. unless specifically stated (see clause 1.8 of SBCSub/C and SBCSub/D/C)) and shall be deemed to be duly given or served if addressed and given by actual delivery or sent by pre-paid post to the party to be served at the address stated in the sub-contract agreement or such other address as may from time to time be agreed;

- if there is then no agreed address for the notice to be served then the notice or other document must be effectively served if given by actual delivery or sent by pre-paid post to the party's last known principal business address, or if a body corporate, its registered or principal office.

It is vitally important that certain notices are properly served (e.g. withholding notice from the final payment – clause 4.12.3 of SBCSub/C and SBCSub/D/C; default notice – clause 7.4.1 of SBCSub/C and SBCSub/D/C), and great care needs to be taken in such situations that the requirements of clause 1.7 of SBCSub/C or SBCSub/D/C are met.

### Electronic communications

Clause 1.8 of SBCSub/C and SBCSub/D/C appears, on face value, to relate to communications other than notices, etc. However, the clause then states that such communications may be made in accordance with any procedures stated or identified in the sub-contract particulars (item 4) or otherwise agreed in writing by the parties.

Sub-contract particulars item 4 requires the parties to state the communications that may be made electronically, and the format that those communications are to take. It is therefore submitted that the parties could agree (and certainly the phrase 'or otherwise agreed in writing by the Parties' would seem wide enough to permit) that notices could be served by electronic communication.

If the parties were to agree to this, the *Bernuth Lines* v. *High Seas Shipping*[1] case highlighted the problems that this could cause.

In that case, solicitors for High Seas e-mailed Bernuth Lines at the address shown on its website and in the Lloyd's Maritime Directory, setting out details of High Seas' claim.

The solicitors received no response, but they proceeded to issue confirmation of the appointment of an arbitrator and other arbitration case management details to the same e-mail address. Eventually, the arbitrator also sent his award to that same e-mail address, and it was only after Bernuth Lines received a hard copy of the award that it became apparent to them that the previous e-mail communication had been sent to an e-mail address that was only used for cargo bookings, and had been ignored as spam e-mails.

Bernuth Lines tried, unsuccessfully, to get the arbitrator's award set aside. The court found against Bernuth Lines because of an express term in the contract which said that any means of service (of adjudication and arbitration notices, etc.) will suffice 'provided that it is a recognized means of communication, and effective to deliver the document to the party to whom it is sent at his address'. In this case, Bernuth Lines had only provided one e-mail address on its website and the same e-mail address was used in the Lloyd's Maritime Directory – if other e-mail addresses had been given, it may not have been found that the notices etc. had been served 'effectively'.

This shows the importance, if e-mail communication is agreed for the service of notices (in particular), of making sure that any change to the e-mail address is notified to the other party to the sub-contract.

### Effect of the final payment notice

The effect of the final payment notice is dealt with by way of Clause 1.9 of SBCSub/C and SBCSub/D/C.

Clause 1.9.1 of SBCSub/C or SBCSub/D/C states that (other than in the case of fraud) the notice of the final payment given in accordance with clause 4.12.2 of SBCSub/C or SBCSub/D/C (the final payment notice) shall have effect in any proceedings (whether by adjudication, arbitration or legal proceedings) under or arising out of the sub-contract, as:

- Conclusive evidence that where and to the extent that any of the particular qualities of any materials or goods or any particular standard of an item of workmanship was described expressly to be for the approval of the architect/contract administrator in any of the sub-contract documents or in any instruction issued by the architect/contract administrator under the main contract that affects the sub-contract works, the particular quality or standard was to the reasonable satisfaction of the architect/contract administrator.

Clause 1.9.1 of SBCSub/C and SBCSub/D/C states that, despite the above, the final payment notice is not conclusive evidence that the materials or goods or workmanship noted above (or indeed that any other materials or goods or workmanship) comply with any other requirement or term of the sub-contract. This

---

[1] *Bernuth Lines Ltd v. High Seas Shipping Ltd* [2005] EWHC 3020 (Comm).

entire matter is dealt with in more detail in Chapter 4 under the heading 'Materials, goods and workmanship'.

- Conclusive evidence that any necessary effect has been given to all of the terms of the sub-contract which require that an amount is to be taken into account in the calculation of the final sub-contract sum.

  The issue of the final sub-contract sum is dealt with in detail in Chapter 8 under the headings 'Final sub-contract sum – adjustment basis' and 'Final sub-contract sum – remeasurement basis'.

  Clause 1.9.2 allows effectively for a 'slip rule' in that the final sub-contract sum can be subsequently adjusted where there has been an accidental inclusion or exclusion of any work, materials, goods or figure in any computation or any arithmetical error in any computation.

  Unfortunately, no guidance is given as to how far this slip rule can extend, but it is submitted that in the context of the significance of clause 1.9 the slip rule would be construed narrowly and that any accidental inclusion or exclusion would need to be self-evident to an independent third party.

- Conclusive evidence that all and only such extensions of time, if any, as are due under clause 2.18 of SBCSub/C or SBCSub/D/C have been given.

  Matters relating to extensions of time are dealt with under Chapter 5.

- Conclusive evidence that the reimbursement of direct loss and expense, if any, due to the sub-contractor pursuant to clause 4.19 of SBCSub/C or SBCSub/D/C is in final settlement of all and any claims which the sub-contractor has or may have arising out of the occurrence of any of the relevant sub-contract matters, whether such claim be for breach of contract, duty of care, statutory duty or otherwise.

  Matters relating to direct loss and expense are dealt with in Chapter 9.

  Although it appears that the intention of this clause is to ensure that all claims are allowed for, it is debatable whether this would include for common law damages, as these are not payable pursuant to clause 4.19 of SBCSub/C or SBCSub/D/C.

- Conclusive evidence that the reimbursement of direct loss and expense, if any, due to the contractor pursuant to clause 4.21 of SBCSub/C or SBCSub/D/C is in final settlement of all and any claims which the contractor has or may have arising out of the matters referred to in clause 4.21, whether such claim be for breach of contract, duty of care, statutory duty or otherwise.

  Matters relating to the contractor's reimbursement are dealt with in Chapter 9.

  Again, although it appears that the intention of this clause is to ensure that all claims are allowed for, it is doubtful that this would include for common law damages, as these are not payable pursuant to clause 4.21 of SBCSub/C or SBCSub/D/C.

It is self-evident that the effect of the final payment notice is far-reaching for both parties. It means that if it remains unchallenged, the final payment notice cannot be 'opened-up' in respect of the quality of any materials or goods or of the standard of an item of workmanship which had expressly been for the approval of the architect/contract administrator, the valuation of the sub-contract works, the sub-contractor's entitlement to extensions of time, the sub-contractor's entitlement to direct loss and expense, or the contractor's reimbursement entitlement.

Clause 4.12.2 of SBCSub/C and SBCSub/D/C makes it clear that for the final payment notice to be effective it must be in writing and must be sent to the sub-contractor by special or recorded delivery.

If the sub-contractor disagrees with the final payment notice he should commence adjudication, arbitration or other proceedings within ten days of receipt of the notice if he wishes to avoid the notice being deemed conclusive evidence as detailed above. It should be noted that if the proceedings only relate to one issue (for example, the sub-contractor's entitlement to loss and expense) then the final payment notice will still be taken as being conclusive evidence of all other matters (refer to clause 1.9.3 of SBCSub/C and SBCSub/D/C).

In the situation where either party has already commenced proceedings before the final payment notice is issued, unless the notice is to be taken as being conclusive evidence of all matters outlined above, either or both parties will have to take a further step in such proceedings within 12 months from or after the date the final payment notice was given (as clause 1.9.2.2 of SBCSub/C or SBCSub/D/C).

At the conclusion of such proceedings, the final payment notice will be taken as being conclusive evidence of the matters outlined above, subject only to the terms of any decision, award or judgment in or settlement of those proceedings (refer to clause 1.9.2.1 of SBCSub/C or SBCSub/D/C).

Clause 1.9.4 of SBCSub/C and SBCSub/D/C adds that if an adjudicator gives his decision on a dispute or difference after the date of the final payment notice and either party then wishes to have the dispute determined by arbitration or legal proceedings, such proceedings must be commenced within 28 days of the date on which the adjudicator gives his decision if the decision is not to be final in respect of the matters referred to in clause 1.9.1 of SBCSub/C and SBCSub/D/C as outlined above.

### Applicable law

Clause 1.10 of SBCSub/C and SBCSub/D/C states that the default position is that the sub-contract must be governed by and construed in accordance with the law of England.

The footnote to clause 1.10 notes that if the parties do not wish the law of England to apply, then the parties should make the appropriate amendments.

# Chapter 4
# Sub-Contractor's Obligations

SBCSUB/C AND SBCSUB/D/C: SECTION 2 – CARRYING OUT THE SUB-CONTRACT WORKS (CLAUSES 2.1, 2.2, 2.4 TO 2.15 INCLUSIVE AND SCHEDULE – 5 CONTRACTOR'S DESIGN SUBMISSION PROCEDURE)

## *Sub-contractor's obligations*

### General obligations

Clause 2.1 of SBCSub/C and SBCSub/D/C sets out the general obligations of the sub-contractor.

These obligations are:

- To carry out and complete the sub-contract works.

    In any building sub-contract, the sub-contractor's primary obligation is to carry out and complete the sub-contract works in accordance with the sub-contract documents. This usually takes the form of an express term in the sub-contract to that effect (as is the case under clause 2.1 of SBCSub/C or SBCSub/D/C); however, even without such an express term, this obligation would be implied.

    It is generally accepted, following the judgment in the *London Borough of Merton* v. *Stanley Hugh Leach*[1] case, that there is an implied term that a contractor will not hinder or prevent the sub-contractor from carrying out its obligations in accordance with the terms of the sub-contract.

    In that case, Mr Justice Vinelott said:

    'Where in a written contract it appears that both parties have agreed that something should be done which cannot effectively be done unless both concur in doing it, the construction of the contract is that each agrees to do all that is necessary to be done on his part for the carrying out of that thing though there may be no express words to that effect.'

- To carry out and complete the sub-contract works in a proper and workmanlike manner.

    This means that the sub-contractor must do the work with all proper skill and care.[2] Normally, when deciding what level of skill and care is required the court will consider all the circumstances of the contract including the degree of skill professed (expressly or impliedly) by the sub-contractor.[3]

---

[1] *London Borough of Merton* v. *Stanley Hugh Leach* (1985) 32 BLR 51.
[2] *Young & Marten Ltd* v. *McManus Childs Ltd* (1969) 9 BLR 77.
[3] *Young & Marten Ltd* v. *McManus Childs Ltd* (1969) 9 BLR 77.

Breach of this duty includes the use of materials containing patent defects, even where the sub-contractor had not chosen the source of those materials. It may also include relying uncritically on an incorrect plan supplied by the contractor where an ordinarily competent sub-contractor should have had serious doubts about the accuracy of the plan.[4] This also links in to a sub-contractor's duty to warn.

Two cases in 1984[5] placed a duty on sub-contractors to warn their employer of design problems that they knew about, and, in certain circumstances this was irrespective of whether or not the sub-contractor had any design liability.

In respect of this matter, it has been found by the Court of Appeal[6] that in a case where there was a major roof collapse the sub-contractor had not done enough to discharge its duty of care even though it had worked to a design instructed by the client, had discussed the matter with the contractor's engineer and had suggested an alternative solution which was unacceptable to the client. In that case there was a risk of personal injury to the sub-contractor's employees, and the Court of Appeal considered that the sub-contractor should have protested more vigorously and pressed its objections on the grounds of safety, perhaps even to the degree that it should have refused to continue to work until the safety of its workmen was addressed.

It is open to question whether the Court of Appeal would have reached the same decision if the sub-contractor's employees were not at risk of personal injury.

However, in a similar case,[7] where it was alleged that the sub-contractor was under a duty to warn in respect of works to be carried out by others *after it had satisfactorily completed its work*, the court found that the sub-contractor did *not* have a duty to warn.

The entire question of duty to warn is far from clear, in particular, in the situations where there is a design defect that does not amount to something dangerous; and where a sub-contractor should have known of the problem of design, but did not; the law is clearly not yet fully developed.

- To carry out and complete the sub-contract works in compliance with the sub-contract documents.

 The sub-contract documents are defined under clause 1.1 of SBCSub/C and SBCSub/D/C as being 'the documents referred to in Article 1', and the documents referred to in article 1 are:

 — the agreement (SBCSub/A or SBCSub/D/A, as appropriate), the sub-contract particulars and the schedule of information;
 — the documents referred to in and annexed to the schedule of information;
 — the Standard Building Sub-Contract conditions (SBCSub/C or SBCSub/D/C) incorporating the JCT amendments stated in the sub-contract particulars,

[4] *Lindenberg* v. *Canning* (1992) 62 BLR 147. In that case the plan incorrectly showed obvious load-bearing walls as non-load-bearing walls.
[5] *Equitable Debenture Assets Corporation Limited* v. *William Moss* (1984) 2 Con. LR 1 and *Victoria University of Manchester* v. *Wilson* (1984–1985) 1 Const LJ 162.
[6] *Plant Construction plc* v. *(1) Clive Adams Associates (2) J.M.H. Construction Services Ltd* [2000] BLR 137, CA.
[7] *Aurum Investments Ltd* v. *Avonforce Ltd* (in liquidation) *and Knapp Hicks & Partners and Advanced Underpinning Ltd (Part 20 defendants)* (2001) 17 Const LJ 145.

and subject to any schedule of modifications included in the numbered documents; and
— the numbered documents annexed to the SBCSub/A or SBCSub/D/A.

- To carry out and complete the sub-contract works in compliance with the health and safety plan.

    The health and safety plan is defined under clause 1.1 of SBCSub/C and SBCSub/D/C as being:

    'where all the CDM Regulations apply, those parts of Health and Safety Plan for the Main Contract applicable to the Sub-Contract Works and annexed to the Schedule of Information, together with any developments of it by the Principal Contractor notified to the Sub-Contractor before or during the progress of the Sub-Contract Works.'

    The health and safety plan is developed in two parts:

    — The pre-construction health and safety plan, which is usually put together by the planning supervisor, and which brings together the health and safety information from all parties involved at the pre-construction stage. The pre-construction health and safety plan must include information from the employer about inherent risks which reasonable enquiry would reveal, and this plan forms the basis of the development of a construction stage health and safety plan by the principal contractor.

        — The construction stage health and safety plan, which is developed by the principal contractor. This plan is developed to include:

            — risk and other assessments prepared by the principal contractor and other contractors/sub-contractors;
            — the health and safety policy of the principal contractor.
            — safe method of work statements, etc.

            The construction stage health and safety plan forms the basis for the health and safety management of the project and continues to evolve through construction. There is usually a requirement that the construction stage health and safety plan is updated at regular intervals. For further commentary on health and safety, the planning supervisor, the principal contractor, and the CDM Regulations, refer to Chapter 7.

- To carry out and complete the sub-contract works in compliance with the statutory requirements, and to give all notices required by the statutory requirements in relation to the sub-contract works.

    The statutory requirements are defined under clause 1.1 of SBCSub/C and SBCSub/D/C as being:

    'Any statute, statutory instrument, regulation, rule or order made under any statute or directive having the force of law which affects the Sub-Contract Works or performance of any obligations under this Sub-Contract and any regulation or bye-law of any local authority or statutory undertaker which has any jurisdiction with regard to the Sub-Contract Works or with whose systems they are, or are to be, connected.'

    These statutory requirements would include:

    — Defective Premises Act 1972
    — Health and Safety at Work, etc. Act 1974

— Sale of Goods Act 1979
— Supply of Goods and Services Act 1982
— Building Act 1984
— Latent Damage Act 1986
— Insolvency Act 1986
— Consumer Protection Act 1987
— Town and Country Planning Act 1990
— Building Regulations Act 1991
— Housing Grants, Construction and Regeneration Act 1996
— Party Wall Act 1996
— Scheme for Construction Contracts (England and Wales) Regulations 1998 (SI 1998/649)
— Human Rights Act 1998
— Late Payment of Commercial Debts (Interest) Act 1998
— Freedom of Information Act 2000
— Enterprise Act 2002

It should be noted that nearly all building work has to comply with the Building Regulations which are updated, amended or changed from time to time, and which stipulate the standards that must be met when carrying out building works. The Public Health Acts 1875 and 1936 enabled local authorities to make bye-laws regulating the construction of buildings. The Public Health Act 1961 provided for the replacement of local building bye-laws by the Building Regulations which, when they came into force in 1966, applied throughout England and Wales with the exception of Inner London for which the London Building Acts remained in force. The current consolidating statute is the Building Act 1984. The Public Health Act 1936 remains in force in relation to drains and sewers.

Certain buildings are, or may be, exempt from the Building Regulations, and a requirement of the Building Regulations may be relaxed or dispensed with upon application to the Secretary of State.

The Building Regulations are currently split into 14 different approved documents that cover different areas of construction. These approved documents are regularly updated and a list of the current approved documents together with the latest update date is provided below:

— Structure, loadings, ground movement and disproportionate collapse, etc. (2004 edition)
— Fire safety, etc. (2000 edition, amended in 2000 and in 2004)
— Dealing with site preparation and resistance to moisture, etc. (2004 edition)
— Toxic substances, cavity insulation, etc. (1992 edition, amended in 2004)
— Resistance to passage of sound, etc. (2003 edition, amended in 2004)
— Ventilation, etc. (2006 edition)
— Hygiene, sanitary conveniences and washing facilities, etc. (1992 edition, amended in 1992 and in 2000)
— Drainage and waste disposal, etc. (2002 edition)
— Heat producing appliances and combustion appliances, etc. (2002 edition)
— Protection from falling, collision and impact, etc. (1998 edition, amended in 2000)
— Conservation of fuel and power, etc. (2006 edition)

— Access and facilities for disabled people, etc. (1998 edition, amended in 2000)
— Glazing, etc. (1998 edition, amended in 2000)
— Electrical safety – dwellings (2006 edition)

It is an express requirement (under clause 2.1 of SBCSub/C and SBCSub/D/C) that the sub-contractor complies with the requirements of the Building Regulations, and therefore a sub-contractor who builds in contravention of the Building Regulations may be in breach of the sub-contract. It is therefore sensible for the sub-contractor to request written confirmation at a pre-contract meeting that all of the required sub-contract works have been approved by building control (building control either being the relevant department of the appropriate local authority, or an equivalent department within an approved external agency (e.g. the NHBC)).

However, where a sub-contractor builds in accordance with the contractual design but in contravention of the Building Regulations, it is considered that his liability may turn on whether or not he was aware of the contravention.[8]

● To carry out and complete the sub-contract works in conformity with directions given in accordance with clause 3.4 (of SBCSub/C and SBCSub/D/C) and all other reasonable requirements of the contractor (so far as they apply) regulating, for the time being, the carrying out of the main contract works. The contractor's power to issue instructions under clause 3.4 of SBCSub/C and SBCSub/D/C is covered in Chapter 7.

It should be noted that the second part of clause 3.2 of SBCSub/C and SBCSub/D/C make it clear that, if the contractor gives his written consent to the sub-contractor to sub-let the whole or any part of the sub-contract works, the sub-contractor is to remain wholly responsible for carrying out and completing the sub-contract works in all respects in accordance with the sub-contract; whilst the second part of clause 3.2 of SBCSub/D/C (*only*) notes, in addition to the foregoing, that the contractor's consent to any sub-letting of design must not in any way affect the obligations of the sub-contractor under clause 2.13.1 (which relates to the sub-contractor's design liabilities) of SBCSub/D/C or any other provision of SBCSub/D/C.

## Sub-contractor's designed portion

The sub-contractor's obligations in respect of the sub-contractor's designed portion (SCDP) only relates to the SBCSub/D/C form and *not* to the SBCSub/C form.

The sub-contractor's obligations are covered under clause 2.2 of SBCSub/D/C, and may be summarized as shown below. The sub-contractor is to:

● Complete the design for the SCDP works (in accordance with the numbered documents to the extent that they are relevant).

---

[8] *Equitable Debenture Assets Corporation Limited* v. *William Moss* (1984) 2 Con LR 1.

- Select any specifications for the kinds and standards of the materials, goods and workmanship to be used in the SCDP works, so far as not stated in the contractor's requirements or the sub-contractor's proposals (clause 2.2.1 of SBCSub/D/C).
- Comply with the directions of the contractor for the integration of the design of the SCDP with the design of the main contract works as a whole, subject to the provisions of clause 3.5.3 of SBCSub/D/C (clause 2.2.2 of SBCSub/D/C). Clause 3.5.3 of SBCSub/D/C relates to the sub-contractor's objections to an instruction. This matter is covered in Chapter 10.
- Comply with regulation 13 of the CDM Regulations including cooperation with the planning supervisor and others as required by CDM Regulation 13(2)(c) (clause 2.2.3 of SBCSub/D/C).

### What are the sub-contractor's designed portion works?

The SCDP works are described in the third recital of SBCSub/D/A. This is done either by stating the nature of work in the SCDP works or by referring to the document(s) (i.e. the design document(s)) that more fully describe it.

The process normally followed (in respect of the design documents) is:

- The contractor provides documents (included as numbered documents under sub-contract particulars item 16 of SBCSub/D/A) showing and describing or otherwise stating the requirements of the contractor for the design and construction of the SCDP works. These documents together comprise the contractor's requirements.
- In response to the above, the sub-contractor provides documents (which should also be included as numbered documents under sub-contract particulars item 16 of SBCSub/D/A) showing and describing the proposals of the sub-contractor for the design and construction of the SCDP works. These documents together comprise the sub-contractor's proposals.

## Materials, goods and workmanship

The normal understanding of the word 'workmanship' in the construction industry relates to the skill and/or care exercised by a sub-contractor (or a contractor) in the physical execution of the works.

Irrespective of whether the sub-contractor is working under SBCSub/C or SBCSub/D/C conditions, to the extent that the choice of materials is left to the sub-contractor, 'workmanship' may also mean design (or at least suitability of the materials for the purpose for which they have been used).

In the absence of express terms to the contrary, the law has, for many years, implied a term that building work will be carried out in a proper and workmanlike manner.[9]

In respect of materials and goods, the position generally is that a sub-contractor is to ensure compliance with the specification where materials or goods are specified in the contract. If the specification states a brand name or a particular supplier

---

[9] *Test Valley Borough Council v. Greater London Council* (1979) 13 BLR 63.

of the material, the sub-contractor would still be under a warranty that those materials or goods are of good quality when used.[10]

A warranty of fitness for purpose (i.e. a warranty that the materials or goods will be fit for the purpose for which they are intended to be used) may also apply if the circumstances indicate that there was a reliance on the sub-contractor's skills regarding the suitability of the materials or goods.[11]

However, where the circumstances indicate that there is no reliance whatsoever on the skill and care on the part of the sub-contractor on the issue of quality (for example, the terms of the sub-contract compel the sub-contractor to accept particular materials or goods from a particular supplier that, for example, had exclusion clauses in respect of liability), the sub-contractor will not be liable for defects in them (i.e. will not be liable if those materials or goods are not of good quality).[12]

This latter position is, however, very much the exception, and the general position remains that the sub-contractor retains liability for materials and goods that are not of good quality when used, and, where there was a reliance on the sub-contractor's skills regarding the suitability of the materials or goods, the sub-contractor retains liability that the materials or goods are reasonably fit for the purpose for which they will be used.

The reason for this is because the law considers that there is a need to maintain a chain of liability from the employer down to the manufacturer. Without such a chain of liability, a sub-contractor would only be able to recover nominal damages from the next party down the chain (the supplier or manufacturer). Generally, the law adopts the view that society is not well served by allowing those causing loss or damage to escape liability whilst those who suffer the loss are denied any remedy. Wherever possible, therefore, the courts interpret contracts in such a way that the chain of liability is maintained.

Against this background we can now consider the contractual position under SBCSub/C and SBCSub/D/C conditions.

Clause 2.4.1 of SBCSub/C states:

'All materials and goods for the Sub-contract Works shall, so far as procurable, be of the kinds and standards described in the Sub-contract Documents. The Sub-contractor shall not substitute any materials or goods so described without the written consent of the Contractor, which shall not be unreasonably delayed or withheld but shall not relieve the Sub-contractor of his other obligations.'

Clause 2.4.1 of SBCSub/D/C states:

'All materials and goods for the Sub-Contract Works excluding the SCDP Works shall, so far as procurable, be of the kinds and standards described in the Sub-Contract Documents. Materials and goods for the SCDP Works shall, so far as procurable, be of the kinds and standards described in the Contractor's Requirements or, if not there specifically described, as described in the Sub-Contractor's Proposals or documents referred to in clause 2.6.1. The Sub-Contractor shall not substitute any materials or goods so described without the written consent of the Contractor, which shall not be unreasonably delayed or withheld but shall not relieve the Sub-Contractor of his other obligations.'

---

[10] Refer generally to the Sale of Goods Act 1979, and the Supply of Goods and Services Act 1982.
[11] *Young & Marten Ltd v. McManus Childs Ltd* (1969) 9 BLR 77.
[12] *Gloucestershire CC v. Richardson* (1968) 1 AC 480.

## Materials and goods

The requirement to work to the sub-contract documents implies that, in respect of materials and goods, the fact that the contractor consents to substituted materials or goods does not give the sub-contractor a defence to its liability that those materials or goods will be of good quality when used.

From the general principles already explained, the sub-contractor is not under a warranty of fitness for purpose in respect of materials or goods specified in the sub-contract documents (but is normally liable that these materials or goods are of good merchantable quality).

However, some specifications contain lists of different types of materials or goods for the same purpose, with the choice left to the sub-contractor as to which type to use. In such a case, a sub-contractor would not take on the fitness for purpose liability simply by making a choice of one of the listed materials or goods, since, in such a case, it could not be considered that any reliance had been placed on the sub-contractor's skills regarding the suitability of the materials or goods in question.[13]

In clause 2.4.1 of SBCSub/C and SBCSub/D/C, the phrase 'so far as procurable' is used. Therefore, because the sub-contract does not specify any geographical limits of procurability, an argument could be raised that materials and goods that are only procurable abroad are still 'procurable' in line with clause 2.4.1. If the materials or goods are not procurable (at all), then the substitution of the materials or goods should be specified by the contractor, and such substitution would be treated as a variation to the sub-contract works.

Under SBCSub/D/C the position is somewhat different to the extent that materials and goods form part of the SCDP works. In that case, clause 2.4.1 of SBCSub/D/C makes it clear that materials and goods for the SCDP works must, (again) so far as procurable, be of the kinds and standards described in the contractor's requirements or, if not specifically described there, as described in the sub-contractor's proposals or the other sub-contract design documents referred to in clause 2.6.1 of SBCSub/D/C.

In the case where materials and goods form part of the SCDP works, the sub-contractor would almost certainly be liable for both the good quality and the fitness for purpose of those materials and goods (but not for design generally – see clause 2.13.1 of SBCSub/D/C, dealt with later in this chapter).

## Workmanship

Clause 2.4.2 of SBCSub/C states that workmanship for the sub-contract works must be of the standards described in the sub-contract documents, whilst clause 2.4.2 of SBCSub/D/C states that workmanship for the sub-contract works must be of the standards described in the contractor's requirements or, if not specifically described there, as described in the sub-contractor's proposals.

The sub-contractor's obligation on standards of workmanship is not qualified by the phrase 'so far as procurable', which, in theory at least, means that it is no

---

[13] *Rotherham Metropolitan Borough Council v. Frank Haslam Milan & Co. Ltd and M.J. Gleeson (Northern) Ltd* (1996) 78 BLR 1.

defence to the sub-contractor's obligation that the human skills or equipment to achieve that standard of workmanship are not procurable.

Clause 2.4.3 of SBCSub/C and SBCSub/D/C states that where and to the extent that approval of the quality or materials or the standards of workmanship is a matter for the opinion of the architect/contract administrator, such quality and standards must be to his reasonable satisfaction.

Obviously, in such a situation, the sub-contractor should obtain confirmation from the contractor of the architect/contract administrator's reasonable satisfaction where it applies to any materials, goods or workmanship in the sub-contract works, as the sub-contract does not provide for any direct communication between the sub-contractor and the architect/contract administrator. It is often appropriate for a sub-contractor to suggest that a sample of its work is approved by the architect/contract administrator so that a 'yardstick' for the standard of future works is provided.

In respect of the architect/contract administrator's reasonable satisfaction, the architect/contract administrator must use an objective standard, and it is therefore submitted that he has no power to demand quality or workmanship of the highest standard, unless that is reasonable in the circumstances of the case.

Interestingly, a decision of the Court of Appeal[14] suggests that the architect/contract administrator may be entitled (or may be required) under certain circumstances to take into account the competitiveness of the contractor's prices for the relevant materials or goods in deciding whether they are to his reasonable satisfaction. Unfortunately, given the competitive nature of the construction industry, this rather tenuous argument is unlikely to have any impact on a sub-contractor that has given a very competitive price in the circumstances where the contractor's price is not so competitive.

Another point to note is that the sub-contract refers to materials or goods being to the reasonable satisfaction of the architect/contract administrator and not to the reasonable satisfaction of the contractor.

In addition to the above, clause 2.4.3 of SBCSub/C and SBCSub/D/C also states that, to the extent that the quality of materials or the standards of workmanship are neither described in the manner referred to in clause 2.4.1 or 2.4.2 of SBCSub/C and SBCSub/D/C (see above) nor stated to be a matter for the opinion or satisfaction of the architect/contract administrator, they must be of a standard appropriate to the sub-contract works.

Clause 2.4.4 of SBCSub/C and SBCSub/D/C requires the sub-contractor (upon the request of the contractor) to provide to the contractor reasonable proof that the materials and goods used by him comply with clause 2.4 of SBCSub/C or SBCSub/D/C as appropriate.

Finally, clause 2.4.5 of SBCSub/C and SBCSub/D/C requires the sub-contractor to take all reasonable steps to encourage the sub-contractor's persons to be registered cardholders under the Construction Skills Certification Scheme (CSCS) or to be qualified under an equivalent recognized qualification scheme.

The Construction Skills Certification Scheme (CSCS) is managed by CSCS Ltd and is administered by the CITB. The board members of CSCS Ltd are from the Construction Confederation; the Federation of Master Builders; the GMB Union;

---

[14] *Cotton v. Wallis* [1955] 1 WLR 1168, CA.

the National Specialist Contractors' Council; the Transport & General Workers' Union (Building Crafts Section); the Union of Construction, Allied Trades and Technicians; the CITB-Construction Skills; the Construction Industry Council; and the Construction Clients' Group. CSCS is supported by organizations such as the Major Contractors' Group (MCG); the National Contractors' Federation (NCF); the Major Home Builders' Group (MHBG); and the Civil Engineering Contractors' Association (CECA).

The purpose of the CSCS is to ensure that those involved in the construction industry are competent in their occupation and have health and safety awareness. The CSCS operates by issuing CSCS cards to those that are qualified to receive the cards in different colours to suit different occupations.

## Compliance with the main contract and indemnity

Under clause 2.5.1 of SBCSub/C and SBCSub/D/C, the sub-contractor is required to observe, perform and comply with the contractor's obligations under the main contract, as identified in or by the schedule of information (included within SBCSub/A or SBCSub/D/A), insofar as those obligations relate and apply to the sub-contract works (or any part of them).

Those obligations include:

- those under clause 2.10 of the main contract conditions (relating to levels and setting out);
- those under clause 2.21 of the main contract conditions (relating to fees or charges legally demandable);
- those under clause 2.22 and 2.23 of the main contract conditions (relating to royalties and patent rights);
- those under clause 3.22 and 3.23 of the main contract conditions (relating to antiquities).

In addition, the sub-contractor is to indemnify and hold harmless the contractor in respect of his obligations under the above main contract liabilities, against and from:

- any breach, non-observance or non-performance by the sub-contractor or his employees or agents of any of the provisions of the main contract (refer to clause 2.5.1.1 of SBCSub/C and SBCSub/D/C); and
- any act or omission of the sub-contractor or his employees or agents which involves the contractor in any liability to the employer under the provisions of the main contract (refer to clause 2.5.1.2 of SBCSub/C and SBCSub/D/C).

Also, under clause 2.5.2 of SBCSub/C and SBCSub/D/C (subject only to the exceptions contained in insurance clauses 6.4 and 6.7.1 of SBCSub/C and SBCSub/D/C which are dealt with in Chapter 11) the sub-contractor is to indemnify and hold harmless the contractor against and from any claim, damage, loss or expense due to or resulting from any negligence or breach of duty on the part of the sub-contractor, his employees or agents (including any misuse by him or them of scaffolding or other property belonging to or provided by the contractor).

An indemnity clause is a clause where one party agrees to make good a loss suffered by the other party in respect of damage or claims arising out of various matters.

It should be noted that because clause 2.5.1 and 2.5.2 of SBCSub/C and SBCSub/D/C contains a general indemnity to the contractor, the sub-contractor's liability could include the contractor's costs, liquidated damages due to the employer, and payments to third parties to whom the contractor is liable. Further, being an indemnity, this may increase the liability of the sub-contractor to the contractor for costs, losses and expenses beyond that which might otherwise be recoverable for breach of the sub-contract and is, potentially, unlimited.

The contractor and the sub-contractor may expressly agree that the sub-contractor's liability is limited, and this may be done by way of an overall cap on liability of the contractor's costs and damages. However, there is no provision for such a cap in SBCSub/A (or SBCSub/D/A) or SBCSub/C (or SBCSub/D/C) and therefore if any overall cap on liability is agreed this would need to be recorded on a numbered document listed and attached to sub-contract particulars item 15 of SBCSub/A or item 16 of SBCSub/D/A (as appropriate).

In addition, the indemnity as set out above has the effect of potentially extending the sub-contractor's period of liability because the cause of action (that sets the liability period running) under an indemnity does not generally arise until the loss in question is actually suffered.

## Supply of documents, setting out, etc.

### Sub-contractor's designed portion information

Clause 2.6 of SBCSub/D/C details the information that the sub-contractor is required to provide to the contractor in respect of the sub-contractor's designed portion (SCDP), and also deals with the timing of its supply.

Clause 2.6.1 of SBCSub/D/C notes first that the sub-contractor is required to comply with Regulation 13 of the CDM Regulations (where the CDM Regulations apply). The CDM Regulations are dealt with in more detail in Chapter 7.

CDM Regulation 13 (requirements on designers) applies to all construction projects (where the CDM Regulations apply), and in the context of SBCSub/D/C, where a sub-contractor is responsible for design works, the sub-contractor is a designer.

The requirements on designers under CDM Regulation 13 are that they should:

- give adequate regard to risk control of health and safety hazards when designing;
- eliminate hazards where feasible (e.g. if specifying roof lights specify non-fragile materials), or reduce risks from those hazards that cannot be eliminated;
- give adequate regard to how the structure can be maintained and repaired safely once built (e.g. specify maintenance-free coatings for materials at height to reduce the need/frequency for repainting, etc.);
- ensure that the design includes adequate information about health and safety (e.g. provide details if a set sequence of assembly or demolition is required to maintain structural stability);

- cooperate with the planning supervisor and any other designer involved in the project;
- provide design information for inclusion in the health and safety plan and the health and safety file.

Clause 2.6.1 of SBCSub/D/C notes that the sub-contractor is to provide the contractor (free of charge) with three copies of:

- such sub-contractor's design documents and (if requested) calculations that are reasonably necessary to explain or amplify the sub-contractor's proposals; and
- all levels and setting out dimensions which the sub-contractor prepares or uses for the purposes of carrying out and completing the SCDP.

Clause 2.6.2 of SBCSub/D/C deals with the timing of the issue of the sub-contractor's design documents.

This sub-clause states that the sub-contractor is to provide the documents 'as and when necessary from time to time' to enable the contractor to observe and perform his obligations:

- under clause 2.9.3 of the main contract conditions and schedule 5 of SBCSub/D/C (i.e. the contractor's design submission procedure); or
- as otherwise stated in the sub-contract documents (if any other such procedure is stated in the sub-contract documents, these would need to be listed and attached as a numbered document under sub-contract particulars item 16 of SBCSub/D/A).

### The contractor's design submission procedure

The contractor's design submission procedure as detailed under schedule 5 of the SBCSub/D/C is as follows:

- The contractor prepares and submits two copies (in such format as is stated in the employer's requirements or the contractor's proposals) of the relevant design documents (including, where appropriate, the sub-contractor's design documents) to the architect/contract administrator in sufficient time to allow any comments of the architect/contract administrator to be incorporated prior to the relevant design document being used for procurement and/or for the carrying out of the works.

No actual time limit is set for the submission of the design documents, but the contractor (and the sub-contractor) need to be aware of the submission procedure to be followed and need to ensure that the various design documents are issued at an early enough date to allow the submission procedure to be completed without causing a delay to the main contract works and/or the sub-contract works.

This is particularly relevant because clause 2.6.2 of SBCSub/D/C makes it clear that the sub-contractor must not commence any work until the relevant design document has satisfactorily completed the submission procedure; and clause 2.14 of SBCSub/D/C states that no extension of time will be given to the sub-contractor where it has failed to provide in time any necessary design documents in line with clause 2.6.2 of SBCSub/D/C.

- Within 14 days from the date of receipt of the design documents referred to above (or, if later, 14 days from either the date of or the expiry of the period for submission of the design documents as stated in the contract documents), the architect/contract administrator must return one copy of the contractor's design documents to the contractor marked either 'A', 'B' or 'C'.

**Documents marked with an 'A'**
If documents are marked with an 'A' that means that the sub-contractor can carry out the work in accordance with those documents.

**Documents marked with a 'B'**
If documents are marked with a 'B' that means that the sub-contractor can carry out the work in accordance with the submitted documents, provided that the architect/contract administrator's comments are incorporated and provided that an amended copy of the document in question is promptly submitted to the architect/contract administrator.

**Documents marked with a 'C'**
If documents are marked with a 'C' that means that the sub-contractor cannot carry out the work in accordance with the submitted documents without following a further procedure, as outlined below.

Where design documents are marked with a 'B' or a 'C', the architect/contract administrator is to identify why he considers that the document is not in accordance with the main contract.

If the architect/contract administrator does not respond to the design documents noted above within the required time period (also as noted above), then the documents will be regarded as though they had been marked with an 'A'.

Clearly, because of this positive default position, it is vitally important that sub-contractors maintain a strict control over the management and register of documents issued and received.

It is particularly significant to note that nowhere in the above submission procedure does the architect/contract administrator or the contractor 'approve' the sub-contractor's design documents.

In any event, it is fairly well established in law that, even if some form of 'approval' was given, this would be unlikely to detract from the sub-contractor's general liability as the designer of the SCDP works.

The further procedure in respect of design documents marked with a 'C', as referred to above, is outlined below.

First, the sub-contractor is not to carry out any work in accordance with a design document marked with a 'C', and the employer (and the contractor, by inference) will not be liable to make payment for any such work that is executed by the sub-contractor.

Then, if the sub-contractor agrees with the architect/contract administrator's comments, he simply amends the design document and resubmits it to the contractor to allow it to go through the contractor's design submission procedure (as outlined above) again.

Alternatively, if the sub-contractor does not agree with the architect/contract administrator's comments, then he is to notify the architect/contract administrator with reasons (presumably through the contractor) within seven days of receipt of the architect/contract administrator's comments that he considers that the compliance with the comments would give rise to a variation.

Upon receipt of such a notification, the architect/contract administrator must, within seven days, either confirm or withdraw the comment.

Where the comment is withdrawn by the architect/contract administrator, the design document (previously marked with a 'C') will assume the status of a design document marked with an 'A'.

Alternatively, where the comment is confirmed by the architect/contract administrator, the sub-contractor is to amend the design document and is then to resubmit it to the contractor to allow it to go through the contractor's design submission procedure (as outlined above) again.

Of course, if the sub-contractor still does not agree with the architect/contract administrator's comment, the option of adjudication to resolve the dispute is available to the sub-contractor.

In line with the sub-contractor's strict liability for design, the confirmation or withdrawal by the architect/contract administrator of a comment under this procedure does not signify acceptance by the employer or the architect/contract administrator that the design document (or the amended design document, as appropriate) is in accordance with the main contract (or the sub-contract) requirements or that it gives rise to a variation (refer to SBCSub/D/C, Schedule 5, paragraph 8.1).

The sub-contractor's strict liability for design is reinforced by clause 8.3 of schedule 5 (of SBCSub/D/C) which states:

> 'Neither compliance with the design submission procedure in this Schedule nor with the Architect/Contract Administrator's comments shall diminish the Contractor's obligations to ensure that the Contractor's Design Documents and CDP Works are in accordance with the Main Contract.'

Because of the way that the contractor's design submission procedure has been incorporated into the SBCSub/D/C, it is submitted that in the above extract the word 'contractor' is interchangeable with the word 'sub-contractor', the 'CDP works' (contractor's design portion works) must include the 'SCDP works' (sub-contractor's design portion works), and the 'main contract' must incorporate the 'sub-contract'.

In a similar vein, clause 2.13.4 of SBCSub/D/C makes it clear that, although the contractor is to give notice to the sub-contractor specifying anything which appears to him to be an inadequacy in the sub-contractor's design documents, no such notice (nor any failure to give such a notice) shall relieve the sub-contractor of his obligations in connection with the design.

However, if the sub-contractor (via the contractor) does not state that he considers that the architect/contract administrator's comments constitute a variation (on a design document marked with a 'C') within seven days of receipt of the architect/contract administrator's comments, then the comments in question must *not* be treated as giving rise to a variation (refer to SBCSub/D/C, schedule 5, paragraph 8.2).

## Further drawings, details and directions (clause 2.7 of SBCSub/C and SBCSub/D/C)

Although, in theory, the numbered documents are intended to show and describe the sub-contract works in full, in practice there is often the need for further drawings and details to be issued to the sub-contractor, or to be issued by the sub-contractor.

Clause 2.7.1 and 2.7.2 of SBCSub/C and SBCSub/D/C deals with this situation, and these clauses require:

- The contractor to provide without charge to the sub-contractor:

  — two copies of such further drawings or details as are reasonably necessary to explain and amplify the numbered documents; and/or
  — such directions (including those for or in regard to the expenditure of provisional sums);

  as are necessary to enable the sub-contractor to carry out and complete the sub-contract works in accordance with the sub-contract.

  The contractor's obligations are, of course, conditional upon the receipt by the contractor of any relevant drawings and/or information from the architect/contract administrator.

- The sub-contractor to provide to the contractor working/setting out drawings and other information necessary for the contractor to make appropriate preparations to enable the sub-contractor to carry out and complete the sub-contract works in accordance with the sub-contract.

  Although clause 2.7.2 of SBCSub/C and SBCSub/D/C does not state that the information to be provided by the sub-contractor is to be free of charge to the contractor, this would probably be the case by implication. Also, as no number of copies is expressly stated, it could be argued that only one copy needs to be provided, although it is more likely that two copies would be implied in keeping with the more general contractor's design submission procedure.

Clause 2.7.3 and 2.7.4 of SBCSub/C and SBCSub/D/C (which are referred to later in this book in relation to clause 2.19 and clause 4.20 of SBCSub/C and SBCSub/D/C in Chapters 5 and 9 respectively) require that:

- such further drawings, details, information and directions referred to in clauses 2.7.1 and 2.7.2 must be provided or given at the time it is reasonably necessary for the recipient party to receive them, having regard to the progress of the sub-contract works and the main contract works; and
- where the recipient party has reason to believe that the other party is not aware of the time by which the recipient needs to receive such further drawings, details, information or directions, he shall, so far as is reasonably practicable, advise the other party sufficiently in advance to enable him to comply with the requirements of clause 2.7 of SBCSub/C or SBCSub/D/C.

A worked example may assist to explain the above clauses.

(1) Polyester powder coated handrails are due to be installed on site in week 50 of the programme.

(2) The sub-contractor that is to supply and install the polyester powder coated handrails asks for the colour of the polyester powder coating in week 10 of the programme.

(3) The contractor advises the sub-contractor that the required information is not available at that time, and, in all likelihood, will not be available until week 35 of the programme. The contractor adds that, allowing for an expected 12 week manufacturing period, the colour of the polyester powder coating will not need to be advised to the sub-contractor until week 38 of the programme.

(4) The sub-contractor advises the contractor that because of the particular configuration of the handrails, coupled with a factory shutdown period, the actual anticipated manufacturing period was 26 weeks, and the colour of the polyester powder coating would therefore need to be advised to the sub-contractor by week 24.

(5) In the early stages, the main contract works are delayed by some 10 weeks, and it is now anticipated that the polyester powder coated handrails will not be installed on site until week 60 of the programme.

(6) In week 25 of the programme, the sub-contractor has not received the colour of the polyester powder coating, and complains to the contractor and states that as the information was required in week 24 (see point 4 above) this may cause a delay to the works.

(7) The contractor retorts that, in line with clause 2.7.3 of SBCSub/C or SBCSub/D/C, the information only needs to be provided having regard to the *progress* of the sub-contract works and main contract works, and not having regard to the *original programme* for the sub-contract works and main contract works; and therefore, because the works are 10 weeks in delay, the information does not now need to be provided until week 34.

## Errors, discrepancies and divergencies

### Preparation of bills of quantities and contractor's requirements (clause 2.8 of SBCSub/C and SBCSub/D/C)

#### Bills of quantities

Clause 2.8 of SBCSub/C (clause 2.8.1 of SBCSub/D/C) states that unless specifically stated otherwise, any bills of quantities (including any bills of quantities prepared for the purpose of obtaining a schedule 2 quotation) must be prepared in accordance with the 'Standard Method of Measurement'.

Clause 1.1 of SBCSub/C and SBCSub/D/C defines Standard Method of Measurement as:

'The Standard Method of Measurement of Building Works, 7th edition, produced by the Royal Institution of Chartered Surveyors and the Construction Confederation, current, unless otherwise stated in the Sub-contract Documents, at the Sub-contract Base Date (references in that publication to "the Appendix" being read as references to the Sub-Contract Particulars).'

The Standard Method of Measurement 7th edition (normally simply referred to as 'SMM7') provides a uniform basis for measuring building works.

SMM7 provides 22 different sections, each offering detailed information, classification tables and supplementary rules. It contains sections for:

- preliminaries/general conditions;
- demolition/alteration/renovation;
- groundwork;
- in-situ concrete/large precast concrete;
- masonry;
- structural/carcassing metal/timber;
- cladding/covering;
- waterproofing;
- linings/sheathing/dry partitioning;
- windows/doors/stairs;
- surface finishes;
- furniture/equipment;
- building fabric sundries;
- paving/planting/fencing/site furniture;
- disposal systems;
- piped supply systems;
- mechanical heating/cooling/refrigeration systems;
- ventilation/air conditioning systems;
- electrical supply/power/lighting systems;
- communications/security/control systems;
- transport systems;
- mechanical and electrical services.

In addition to descriptions and quantities, SMM7 requires drawn information to be provided to give:

- an overall picture of the project to allow assessment of the cost significance of the design and to allow decisions to be made about methods of construction;
- detailed information about parts of the work to be communicated graphically where this is more effective, rather than by a lengthy description in the bill of quantities.

The following types of drawings are referred to in SMM7:

- location drawings, i.e.:

  — block plans;
  — site plans; and
  — plans, sections and elevations.

- component drawings (which are required by general rule 5.2 of SMM7) to show the information necessary for the manufacture and assembly of components;
- dimensioned diagrams (which are required by general rule 5.3 of SMM7) to show the shape and dimensions of the work covered by an item.

Strangely, clause 1.1 in SMM7 is often overlooked:

'The Standard Method of Measurement provides a uniform basis for measuring building works and embodies the essentials of good practice. Bills of quantities shall fully describe and accurately represent the quantity and quality of the works to be carried out. *More detailed information than is required by these rules shall be given where necessary in order to define the precise nature and extent of the required work.*'

The italics in the above quotation have been added to highlight that pure compliance with the specific trade measurement rules of SMM7 will not always be sufficient in the situation where those rules do not adequately cover the description of the work required.

### Contractor's requirements

Clause 2.8.2 of SBCSub/D/C makes it clear that, subject only to clause 2.12 of SBCSub/D/C (dealt with below) the sub-contractor is not responsible for the contractor's requirements or for verifying the adequacy of the design contained within them.

This clause would appear largely to protect the sub-contractor from the more normal position as noted in the *Co-operative Insurance Ltd* v. *Henry Boot* case,[15] where it was found that in a standard design and build contract the contractor was not only responsible for any design specifically carried out by it but, more onerously, was also liable for the completed scheme as a whole, even where the design work for a specific element was carried out earlier by others.

## Bills of quantities and SCDP documents – errors and inadequacy (clause 2.9 of SBCSub/C and SBCSub/D/C)

Clause 2.9.1 of SBCSub/C and SBCSub/D/C states that if, in respect of any bills of quantities (as referred to above), any of the following list apply, then the departure, error or omission must not vitiate (i.e. will not make invalid or ineffectual) the sub-contract, but must be corrected and, where the description of a provisional sum for defined work does not provide the information required by SMM7, the description must be corrected so that it does provide that information:

- SMM7 has not been used in the preparation of the bills of quantities, but this departure has not been stated; or
- there is any error in description or quantity; or
- there is any omission of items that should have been measured; or
- there is any error in, or omission of information in, any item which is the subject of a provisional sum for defined work.

The question of whether there may be some limit to the nature and extent of variations which can be ordered, without the sub-contract being vitiated, is considered in Chapter 10.

---

[15] *Co-operative Insurance Society Ltd* v. *Henry Boot Scotland Limited* [2002] EWHC 1270 (TCC).

Clause 2.9.2 of SBCSub/D/C states that if an inadequacy is found in any design in the contractor's requirements then if, or to the extent that, the inadequacy is not dealt with in the sub-contractor's proposals, the contractor's requirements shall be altered or modified accordingly.

From clause 2.9.2, it is therefore clear that in the event of a discrepancy between the contractor's requirements and the sub-contractor's proposals, the sub-contractor's proposals will (in general terms) take precedence.

Clause 2.9.2 of SBCSub/C (and clause 2.9.3 of SBCSub/D/C) notes that, subject only to the provisions of clause 2.12 of SBCSub/C (which relates to notices to be issued by the sub-contractor and directions to be given by the contractor in respect of divergencies from statutory requirements), any correction, alteration or modification under clause 2.9.1 of SBCSub/C or SBCSub/D/C must be treated as a variation.

Clause 2.9.4 of SBCSub/D/C notes that the final sub-contract sum will not be adjusted to take into account any variations that are issued arising from errors in description or in quantity in the sub-contractor's proposals or in the sub-contractor's design portion analysis.

## Notification of discrepancies, etc.

Under clause 2.10 of SBCSub/C and SBCSub/D/C, if the sub-contractor finds any error or discrepancy in or between any of the following documents then the sub-contractor is required to immediately give a written notice to the contractor, and the contractor must issue a direction as to how the error or discrepancy is to be dealt with:

- the sub-contract documents (i.e. the documents referred to in article 1 of SBCSub/A or SBCSub/D/A, as appropriate);
- the main contract;
- any direction issued by the contractor under the SBCSub/C or SBCSub/D/C (except for a direction requiring a variation); and
- (in respect of SBCSub/D/C, only), the sub-contractor's designed portion documents as detailed under clause 2.6.1 of SBCSub/D/C.

## Discrepancies in sub-contractor's design portion documents (SCDP)

Discrepancies in SCDP documents are dealt with under clause 2.11 of SBCSub/D/C.

Where there are discrepancies or divergences (i.e. differences or inconsistencies) within or between the SCDP documents (excluding the contractor's requirements), the procedure is:

- the sub-contractor is to give a notice to the contractor (under clause 2.10 as noted above);
- the sub-contractor is to send a statement setting out his proposed amendments to remove this discrepancy or divergence;
- the contractor is then to issue his directions accordingly;

- the sub-contractor must comply with those directions;
- to the extent that those directions relate to the removal of a discrepancy or divergence, they will not be taken into account in the calculation of the final sub-contract sum.

Where the discrepancy or divergence is within the contractor's requirement, and the sub-contractor's proposals deal with that discrepancy or divergence, then the sub-contractor's proposals are to prevail (on the assumption that they comply with statutory requirements, which is dealt with further below). In that case no adjustment will be taken into account in the calculation of the final sub-contract sum. In other words, if the sub-contractor's proposals already deal with the discrepancy or divergence, then the sub-contract sum (or the sub-contract tender sum) is deemed to include for the resolution of the discrepancy or divergence within the contractor's requirements.

If, however, the sub-contractor's proposals do not deal with the discrepancy or divergence within the contractor's requirements, then the sub-contractor is to inform the contractor in writing of his proposed amendment to deal with it. The contractor will either agree to the proposed amendment or will decide how it wishes the discrepancy or divergence to be dealt with. In either event, the agreement or the decision of the contractor will be treated as a variation.

Of course, the above comments need to be considered against the background of the sixth recital of SBCSub/D/A. That recital confirms that the contractor has examined the sub-contractor's proposals and, subject to the sub-contract conditions, is satisfied that they appear to meet the contractor's requirements. It is important to note that the sixth recital does not require the contractor to check that the sub-contractor's proposals satisfy the contractor's requirements, but merely that 'they appear to meet' them.

## Divergencies from statutory requirements

Under clause 2.12 of SBCSub/C and SBCSub/D/C, if the sub-contractor *or the contractor* finds any divergence between the statutory requirements and any item in the list below then the sub-contractor or the contractor, as appropriate, is required to immediately give a written notice to the other specifying the divergence:

- the sub-contract documents (i.e. the documents referred to in article 1 of SBCSub/A or SBCSub/D/A, as appropriate);
- the main contract;
- any direction issued by the contractor under the SBCSub/C or SBCSub/D/C (except for a direction requiring a variation);
- (in respect of SBCSub/D/C, only), the sub-contractor's designed portion documents as detailed under clause 2.6.1 of SBCSub/D/C; and
- any direction requiring a variation issued under clause 3.4 of SBCSub/C or SBCSubD/C issued by the contractor under the SBCSub/C or SBCSub/D/C conditions.

In addition, where the divergence is between the statutory requirements and any of the sub-contractor's designed portion documents, the sub-contractor, at the time of issuing the notice of divergence, is to inform the contractor in writing of his proposed amendment for removing the divergence.

Except in the cases where the sub-contractor has informed the contractor in writing of his proposed amendment for removing the divergence (dealt with below), then, within ten days of finding a divergence, or of receiving a notice from the sub-contractor in respect of a divergence, the contractor is to issue directions in respect of that divergence, and insofar as those directions require the sub-contract works to be varied, those directions must be treated as directions requiring a variation to be issued.

In the cases where the sub-contractor has informed the contractor in writing of his proposed amendment for removing the divergence, the contractor is to issue directions in respect of that divergence within 17 days of receipt of the sub-contractor's proposed amendment, and, again, insofar as those directions require the sub-contract works to be varied, those directions must be treated as directions requiring a variation.

In the situation where there is a divergence between the statutory requirements and any of the sub-contractor's designed portion documents, the normal position is that the sub-contractor must comply with the contractor's directions at no additional cost to the contractor.

However, if there is a change to the statutory requirements after the sub-contract base date (i.e. the date stated as such in the sub-contract particulars (item 3)) which necessitates an alteration or modification to the sub-contractor's designed portion, then that alteration or modification is to be treated as being a direction requiring a variation to the contractor's requirements.

# Sub-contractor's design portion (SCDP) design work

## Design liabilities and limitation

Most standard form contracts produced by the professional bodies of the construction industry contain an express term which stipulates that the professional will carry out his duties using 'reasonable skill and care'.

For example, the RIBA Standard Form of Agreement for the Appointment of an Architect (SFA/99) sets out at condition 2.1:

'The Architect shall in performing the Services and discharging all the obligations under this Part 2 of these Conditions, exercise reasonable skill and care in conformity with the normal standards of the Architect's profession.'

One of the best definitions for skill and care has been provided by the Court of Appeal in the *Eckersley* v. *Binnie*[16] case, where Lord Justice Bingham stated:

'The law requires of a professional man that he live up in practice to the standards of the ordinary skilled man exercising and professing to have his specialist professional skills. He need not possess the highest expert skill; it is enough if he exercises the ordinary skill of an

---

[16] *Eckersley* v. *Binnie & Partners* (1988) 18 Con LR 1, CA.

ordinary competent man exercising his particular art. In deciding whether a professional man has fallen short of the standards observed by ordinary skilled competent members of his profession, it is the standard prevailing at the time of the acts or omissions which prove the relevant yardstick. He is not to be judged by the wisdom of hindsight.'

With regard to the professional man the standard of reasonable skill and care is also implied by Section 13 of the Supply of Goods and Services Act 1982.

It needs to be remembered that a designer (including the sub-contractor, where the sub-contractor carries out design work) may also have tortuous duties.

Negligence as a tort may be defined as the breach of a duty of care owed by the defendant to the claimant which results in damage or loss to the claimant.

The loss may arise through acts or omissions whether with intent or not and may consist of personal injury, damage to property or what may be categorized as 'pure economic loss'.[17]

Tort needs *no agreement or contractual relationship* and is largely judge-made law. The injured party may sue a designer when the following three conditions are satisfied:

(1) The designer owes a duty of care to that person.
(2) The designer has acted in such a way as to breach that duty of care (usually by failing to exercise reasonable skill and care in the way that he carried out his work).
(3) The injured party has suffered relevant damage as a result of that breach of duty.

The duty of care owed by a person who claims to have a special skill is judged, not by the standard of the reasonable man in the street, but according to the standard of the reasonable person enjoying the skill which he claims to possess.

The ground rules were laid down by Mr Justice McNair in the case of *Bolam* v. *Friern Hospital*,[18] where he said:

'Where you get a situation which involves the use of some special skill or competence, then the test as to whether there has been negligence or not is not the test of the man on the top of the Clapham omnibus, because he has not got that special skill. The test is the standard of the ordinary skilled man exercising and professing to have that special skill . . . he is not guilty of negligence if he has acted in accordance with a practice accepted as proper by a responsible body of . . . men skilled in that particular art.'

---

[17] The entire question of 'pure economic loss' is a very complicated area, and it is not something that needs to be further investigated within this book. Generally speaking, damage to 'the thing itself' (i.e. the building) is pure economic loss, whereas damage to 'other property' (i.e. other than 'the thing itself') is physical damage; the position is further complicated by the 'complex structure theory'. The issues surrounding these matters have been developed in a number of court cases, for example: *Anns v. Merton London Borough Council* [1977] 2 WLR 1029; *D & F Estates Limited v. Church Commissioners* [1989] AC 177, HL; *Murphy v. Brentwood District Council* [1991] 1 AC 398, HL; *Jacobs v. Morton and Partners* [1994] 72 BLR 92; *Tunnel Refineries Limited v. [1] Bryan Tonkin Co. Limited [2] Alsthom SA [3] GEC Alsthom Electromecanique SA* [1998] CILL 1392; *Bellefield Computer Services Limited and Others v. E. Turner & Sons Ltd* [2000] BLR 97, CA.

[18] *Bolam v. Friern Hospital Management Committee* [1957] 1 WLR 582.

As well as contract and tort, the law imposes a statutory obligation on house builders (and designers). This statute is the Defective Premises Act 1972.

The Defective Premises Act sets down obligations in respect of new homes for occupation. The important aspect of the Act is that anyone designing a new dwelling owes a duty to anyone who later purchases the dwelling to ensure that his work is done in a professional manner. However, it should be noted that in order to be in breach of the duty the failure must cause the dwelling to be unfit for habitation when completed. This situation is covered by clause 2.13.2 of SBCSub/D/C.

Despite the above comments regarding a designer's normal obligation, where there is not an express obligation to the contrary in the sub-contract, the sub-contractor's liability would normally be far more onerous. This liability would be to meet the standard of 'fitness for purpose', which means that the sub-contractor would be liable to ensure that the finished product was reasonably fit for its intended purpose, which is a far higher obligation than that normally imposed on professional designers of using 'reasonable skill and care'.

*However, despite the above 'usual' position, clause 2.13.1 of SBCSub/D/C makes it clear that when carrying out design work under SBCSub/D/C, the level of care expected of the sub-contractor is one of 'reasonable skill and care' rather than one of 'fitness for purpose'.*

Obviously, it is far easier to pursue an action in contract than it is to pursue an action in tort, and it is for this reason that the employer will normally require a collateral warranty to be in place which establishes a direct contractual link between the designing sub-contractor and the employer (and others). Collateral warranties are dealt with under clause 2.26 of SBCSub/C and SBCSub/D/C, and this matter is dealt with in Chapter 6.

Although actions under tort may be more difficult to pursue (principally because there is a need to establish that a duty of care existed in the first place), there are benefits in pursuing an action in tort in terms of limitation periods.

The principle behind limitation periods is to provide a cut-off date after which no claim can be brought (i.e. the claim becomes time barred). The reason behind this is that the law regards it as undesirable that liability could exist without time limit. Limitation periods are set down for both contract and tort in the Limitation Act 1980.

In contract, the limitation period for a breach is six years for a simple contract (i.e. a contract executed under hand), and twelve years for one executed as a deed. This is the reason why most employers will require the contractor to fill in the 'executed as a deed' option in a construction contract and, consequently, why most contractors will require the sub-contractor to execute a sub-contract (including the Standard Building Sub-Contract) as a deed.

The time for limitation runs from the date of breach and not from the date of discovery of the defect. In an entire contract (such as the Standard Building Sub-Contract) it is usual that the breach of contract will be deemed to have occurred once practical completion has been given even if the defective act occurred earlier.

In contrast to the above, in tort (i.e. negligence) the limitation period is six years from when the claimant suffered damage (i.e. from when the cause of action accrued).

The Latent Damage Act 1986 amended the Limitation Act 1980 in the light of dissatisfaction resulting from the *Pirelli* v. *Oscar Faber*[19] case. It was held in that case

---

[19] *Pirelli General Cable Works* v. *Oscar Faber & Partners* [1983] 2 AC 1, HL.

that the date of accrual of a cause of action in tort caused by the negligent design or construction of a building was the date *when the damage came into existence*, and not the date when the damage was discovered or could with reasonable diligence have been discovered.

The amendments effected by the Latent Damage Act 1986 extended the limitation period for claims in tort to add in a new provision that the limitation period would also be three years starting from the date when the claimant had both the knowledge required for bringing an action for damages in respect of the relevant damage and a right to bring such an action. Thus, a party could be pursued outside the primary limitation period.

However, there is an overriding time limit for actions for negligence (other than for personal injury) of 15 years from the act of negligence to which the damage is alleged to be attributed. Refer to the Limitation Act 1980, sections 14A and 14B, as amended. The said sections do not apply if there is deliberate concealment within section 32(1)(b) of the Limitation Act 1980 – see section 32(5).

Difficulties were also appreciated with successive owners, and the Latent Damage Act, in certain circumstances, gives a fresh cause of action in such a situation.[20]

As outlined earlier within this chapter, the effectiveness of the above is questionable in respect of latent damages to buildings themselves which reduce the value of the building only. This is what is known as 'economic loss' or 'pure economic loss' and the courts have held in the case of *Murphy* v. *Brentwood District Council*[21] that this is not recoverable as part of a claim for negligence unless there is 'negligent misstatement'.

A negligent misstatement may give rise to an action for damages for pure economic loss since the law will imply a duty of care when a party seeking information from a party possessed of special skill trusts him to exercise due care, and that party knew or ought to have known that reliance was being placed on his skill and judgment.

Clause 2.13.3 of SBCSub/D/C deals with the financial limitation of liability, and states that if a financial limit of liability has been included in the main contract (under clause 2.19.3) then that same limit will also apply to the sub-contractor. In the situation where the sub-contractor is not solely liable in respect of this matter, then the sub-contractor would only be expected to make a proportional contribution towards any financial limit set under the main contract.

## Design development

One of the major areas of dispute in respect of changes to design works relates to what is called 'design development'.

Generally, a sub-contractor may be called upon to design an entire component, such as a cladding system where it will be required to consider all relevant matters (e.g. location, wind forces, anticipated snow loadings, etc.), or it may be called upon to design to a performance specification (where another designer has already made the basic design considerations and calculations), for example a central heating system that complies with a performance specification provided.

---

[20] See Latent Damage Act 1986 section 3; also *Perry* v. *Tendring District Council* (1984) 30 BLR 118.
[21] *Murphy* v. *Brentwood District Council* [1991] 1 AC 398, HL.

In the latter case, in particular, if the sub-contractor does not meet the required performance specification, it cannot rely on the defence that the product supplied is 'fit for purpose'. In other words, the central heating system designed by the sub-contractor may achieve an acceptable level for the building in question (i.e. it is fit for purpose), but if it does not reach the specification level required by the contractor, then it will be unacceptable.

However, what frequently happens is that a design put forward by a sub-contractor is rejected by a contractor (normally because it has been rejected by the employer) because it is not to the contractor's liking or does not meet the contractor's perceived requirements.

In such a situation, the contractor will often ask for a change to the sub-contractor's design on the basis that this is simply design development (i.e. that the sub-contractor will not get paid any extra for the development to his design to meet the requirements of the contractor). The sub-contractor will naturally resist on the basis that the changes required by the contractor are a variation to the scope of the sub-contract works.

In such a situation, it will be for the sub-contractor to demonstrate that its submitted design satisfied the requirements of the design brief, or for the contractor to demonstrate that the sub-contractor's submitted design would not satisfy the design brief. Of course, the contractor's objection to the sub-contractor's design would be unsuccessful if it was based upon considerations of price (i.e. savings effected by the sub-contractor that were not being passed onto the contractor) rather than on any technical failings of the sub-contractor's design.

There is not a great deal of case law on this particular subject, but the *Skanska* v. *Egger*[22] Court of Appeal case considered the matter of design development. In that particular case the parties entered into a contract whereby Skanska agreed, for a so-called guaranteed maximum price, to develop the design, manage, procure and construct a factory building.

Disputes arose as to various claims by Skanska (in particular in respect of additional steelwork). Skanska argued that the claims arose from changes in the employer's requirements, whilst Egger contended that they were merely instances of design development comprehended within the employer's requirements.

Skanska was responsible for the completion of the design of the project as outlined in the contract documents and the contract drawings.

The Court of Appeal found that although Skanska was obliged to install more steel than was indicated at tender stage, the tender drawings *did show a requirement for steel* and there was sufficient evidence for the first instance judge to have decided that the contract provided for more detailed information to be provided post-contract. No further payment was therefore due for the additional work.

The Skanska case illustrates the dangers to a contractor of entering into a design and build contract on a guaranteed maximum price basis where design issues have not been fully formulated at the time of contracting.

Of course, much turns upon the individual contract terms and the facts of a given case, but the principle that a contractor (or a sub-contractor) who takes on design responsibilities may find himself responsible for a good deal of development costs is of universal application.

---

[22] *Skanska Construction UK Ltd* v. *Egger (Barony) Ltd* [2002] BLR 236.

Finally, sometimes a sub-contractor is required to carry out a design that is 'to the entire satisfaction' of the contractor (or the employer). This would appear to be an entirely open-ended obligation on the sub-contractor. However, it is submitted that in respect of any such requirement the contractor's (or the employer's) scope for satisfaction is limited by the design specification provided to the sub-contractor, and that, unless there are any express provisions in the sub-contract to the contrary, the sub-contractor will be entitled to an additional payment, and, if appropriate, additional time, if the design level that satisfies the contractor (or the employer) exceeds the design specification level that was initially provided to the sub-contractor. This principle was established in the case of *Dodd v. Churton*.[23]

## Errors and failures – other consequences

Clause 2.14 of SBCSub/D/C states that no extension of time will be given, and no loss and expense will be allowed, and no (contractual) termination for non-payment will be permitted, where the cause of the progress of the sub-contract works having been delayed, affected or suspended is:

- any error, divergence, omission or discrepancy in the sub-contractor's proposals;
- any error, divergence, omission or discrepancy in any of the SCDP documents;
- failure of the sub-contractor in completing the SCDP documents to comply with regulation 13 of the CDM Regulations;
- failure of the sub-contractor to provide in due time any SCDP documents or related calculations or information either:

  — as required by clause 2.6.2 of SBCSub/D/C (see commentary above); or
  — in response to a written application from the contractor specifying the date (having due regard to the progress of the sub-contract works) when certain relevant documents or other information is required.

## *Unfixed materials and goods*

This issue is dealt with in Chapter 8.

---

[23] *Dodd v. Churton* [1897] 1QB 562.

# Chapter 5
# Time

## *Sub-contractor's obligations*

### Time of commencement and completion

Clause 2.3 of SBCSub/C and SBCSub/D/C sets out the sub-contractor's obligations in respect of carrying out and completing the sub-contract works.

Most programming issues are considered under the commentary for sub-contract particulars item 5 (programme) of SBCSub/A and SBCSub/D/A which is dealt with in Chapter 2.

Clause 2.3 of SBCSub/C and SBCSub/D/C makes it clear that the sub-contractor's obligations to completion may be affected by clauses 2.16 to 2.19 of SBCSub/C and SBCSub/D/C. These clauses are considered further within this chapter under the heading of 'Adjustment of period for completion'.

The sub-contractor's obligation is to carry out the sub-contract works reasonably in accordance with the progress of the main contract works, and it is therefore submitted that, after the commencement of the sub-contract works on site, the onus is on the sub-contractor to remain aware of the progress of the main contract works to ensure that he carries out the sub-contract works reasonably in accordance with that progress. This requirement would be consistent with the sub-contractor's obligation (under clause 2.17.1 of SBCSub/C and SBCSub/D/C) to give notice that the commencement, progress or completion of the sub-contract works is being or is likely to be delayed.

Despite the above, it is, of course, good practice for the contractor to regularly inform and update sub-contractors on the progress of the main contract works, particularly prior to the commencement on site of the sub-contract works.

#### *Time for completion*

It should be noted that although there is nothing to prevent a sub-contractor from completing before the completion date, this does not amount to an obligation on the part of the contractor to provide information to enable the sub-contractor to complete early.

# Adjustment of period for completion

## Generally: the purpose of clauses allowing for the adjustment of the period for completion

Clearly, the principal purpose of clauses allowing for the adjustment of the period for completion is to allow the contractor to extend the date for completion.

In the absence of such clauses the contractor could not extend the date for completion. This may seem self-evident, but as explained below, this point is of some significance.

If clauses allowing for the adjustment (extension) of the period for completion did not exist and a situation arose where the delay was caused by the contractor or the employer, or for which the contractor or employer was responsible in law, then the contractor would be unable to extend the period for completion. This inability by the contractor may cause the time for completion to become 'at large'.

When the time for completion becomes at large it means that, in effect, there is no end completion date and the sub-contractor must simply use his best endeavours to complete in a reasonable time. This position was stated in the case of *Peak v. McKinney*[1] (a case between a contractor and an employer) where it was said that an extension of time clause and the granting of it must take account of the breaches of contract by the employer – otherwise time becomes at large.

The main reason for the above is that extension of time clauses are to be construed strictly *contra proferentum*. This means that if there is an ambiguity in the sub-contract which cannot be resolved by other methods of interpretation, then the clauses are to be construed against the party putting forward the sub-contract that seeks to rely on the clauses in question.

A contractor cannot get around this principle by adjusting the period for completion, notwithstanding the fact that no express clause allows for that. This action is not permitted by law, and the effect of the law as it stands explains why there are very comprehensive extensions of time clauses in most standard contracts and sub-contracts, which attempt to cover most matters which normally affect the completion date and which can either be said to be the employer's (or, as applicable, the contractor's) fault or at least certainly not the contractor's (or, as applicable, the sub-contractor's) fault.

Most standard (sub-contract) extension of time clauses cover two sets of situations:

- delays for which the employer or his agents (or the contractor or his agents, as applicable) are responsible;
- delays, otherwise arising, over which the sub-contractor has no control.

## Notice by sub-contractor of delay to progress

Clause 2.17 of SBCSub/C and SBCSub/D/C deals with the matter of notices by the sub-contractor of delay to progress.

---

[1] *Peak Construction (Liverpool) Ltd* v. *McKinney Foundations Ltd* (1970) 1 BLR 111.

Clause 2.16 of SBCSub/C and SBCSub/D/C makes it clear that in clauses 2.17 to 2.19 of SBCSub/C and SBCSub/D/C and, so far as is relevant, in the other clauses of SBCSub/C or SBCSub/D/C conditions, any reference to delay or extension of time includes any further delay or further extension of time.

Under clause 2.17.1 of SBCSub/C and SBCSub/D/C a notice is required to be given: 'if and whenever it becomes reasonably apparent that the commencement, progress or completion of the Sub-Contract Works or of such works in any Section is being or is likely to be delayed'.

Obviously, there could be disagreements about when the delay 'becomes reasonably apparent' and this matter would need to be viewed, effectively on a common-sense basis, against the circumstances of any particular case.

Also, the delay or likely delay is to be considered against the commencement (actual) progress or completion of the sub-contract works, which may be different to the programmed dates for the execution of the sub-contract works. This position was established from the *Glenlion* v. *The Guinness Trust*[2] case.

In line with clause 2.17.1 of SBCSub/C and SBCSub/D/C, the notice is required to be given 'forthwith', which, although not defined in the conditions, would normally be taken as being 'immediately' or 'without delay'.

Clause 2.17.1 of SBCSub/C and SBCSub/D/C also requires the notice to contain:

- the material circumstances (i.e. the relevant background matters);
- the cause or causes of delay (insofar as the sub-contractor is able);
- identification of any event which in the sub-contractor's opinion is a 'relevant sub-contract event'.

A relevant sub-contract event, which is dealt with in further detail below, is an event that entitles the sub-contractor to an adjustment of the period for completion; assuming that the event in question causes a delay to the sub-contract works, or of such works in any section, beyond the relevant period or periods for completion stated in the sub-contract particulars (item 5) or any previously revised period or periods.

Because the sub-contractor is to identify in the notice which events are relevant sub-contract events, the intention of the sub-contract clearly appears to be that the sub-contractor should give a notice of delay in respect of both those delay events that are of its own making (e.g. shortage of resources) and those which are not its liability (e.g. instructions for additional works). Experience shows that sub-contractors often give a notice of delay for events that are not their liability, but rarely give a notice of delay for events that are their liability.

In addition to the above, under clause 2.17.2 of SBCSub/C and SBCSub/D/C, the sub-contractor is to give particulars of the expected effects of the notified delay event, including an estimate of any expected delay in the completion of the sub-contract works, or of such works in any section, beyond the relevant period or periods for completion stated in the sub-contract particulars (item 5) or any previously revised period or periods. (SBCSub/C and SBCSub/D/C are silent on the question of how (or if) a sub-contractor should take into account the question of concurrent delays when considering any periods of delay.)

---

[2] *Glenlion Construction Ltd* v. *The Guinness Trust* (1987) 39 BLR 89.

The above particulars are to be provided in writing either:

- with the notice (if this is practicable); or, alternatively
- as soon as possible thereafter.

Clause 2.17.3 of SBCSub/C and SBCSub/D/C adds that the sub-contractor must forthwith (i.e. immediately):

- notify the contractor in writing of any material change in the estimated delay or in any other particulars;
- supply such further information as the contractor may at any time reasonably require.

Two particular questions may arise from the above, namely:

- Is the sub-contractor bound in any way by the estimate of delay that he provides?
- Is the issuing of a notice a condition precedent to the contractor giving an extension of time?

### Is the sub-contractor bound in any way by the estimate of delay that he provides?

All clause 2.17.2 of SBCSub/C and SBCSub/D/C requires is for the sub-contractor to give 'an estimate of any expected delay'.

The sub-contractor is not bound in any way by this estimate, and nor should he be.

If, for example, a sub-contractor is instructed to supply and install a lift as variation to the sub-contract works, and considers that the anticipated delivery period of eight weeks for the lift will cause a delay to the completion of the sub-contract works of six weeks, then his estimate of delay at that time would be six weeks.

However, if when making further enquiries to his supplier it becomes apparent to the sub-contractor that the actual delivery period will be 16 weeks (perhaps because of the purpose-built nature of the lift), then it would be quite in order for the sub-contractor to revise his estimate of delay to 14 weeks.

If, following this, the sub-contractor considered that it could reprogramme its work and thereby mitigate the delay to some degree, it could again revise its estimate of delay to, for example, ten weeks.

The sub-contractor should not be concerned about revising its estimate of delay if circumstances arise that affect its earlier estimate (providing, of course, that both its initial estimate and any subsequent estimates are 'reasonable'), and, in fact, clause 2.17.3 of SBCSub/C and SBCSub/D/C requires the sub-contractor to carry out this review constantly.

Of course, if an estimate of delay is revised by the sub-contractor without circumstances arising that affect the earlier estimate, doubts may be expressed about the sub-contractor's judgment and/or ability to properly programme the works.

In respect of the above, it should be particularly noted that it is sometimes very difficult to forecast the effects of a future possible delay, and in making any such

estimate the sub-contractor needs to be aware of the principles of delay analysis
that are dealt with later in this chapter.

### Is the issuing of a notice a condition precedent to the contractor giving an extension of time?

A 'condition precedent' means an event that *must* take place or a condition that
*must* be fulfilled before a party to a contract is required to perform or do their part.

The answer to the above question is yes and no.

Put simply, if the sub-contractor does not give a notice of delay, then the contractor does not need (and, in fact is not empowered) to give an extension of time
before the expiry of the period or periods for the completion of the sub-contract
works or of such works in a section.

However, *after* the expiry of the period or periods for the completion of the sub-
contract works or of such works in a section, then, subject only to the timescales
set out under clause 2.18.5 of SBCSub/C and SBCSub/D/C (which is further considered below), the contractor is to give an extension of time (if such an extension
of time is due) irrespective of whether or not a notice has been given by the
sub-contractor.

It is considered that the reason for this 'exception' is that if, because of a lack of
a notice, a contractor is not empowered to give an extension of time in respect of
a delay event that was clearly the contractor's liability, then this may cause the time
for completion to become at large (following the *Peak* v. *McKinney* principle outlined above). Notwithstanding the fact that a notice has not been issued by the sub-
contractor, by permitting the contractor to give an extension of time in respect of
a delay event that was clearly the contractor's liability (for example), this potential problem can been circumvented.

## Fixing period for completion

### Fixing period for completion before the expiry of the period or periods (in item 5 of the sub-contract particulars) for the completion of the sub-contract works or of such works in a section

The 'trigger' for consideration of an extension of time at this stage is the receipt by
the contractor of a notice and particulars from the sub-contractor under clause 2.17
of SBCSub/C and SBCSub/D/C.

When the contractor receives such a notice, he is to take the following
action:

- in line with clause 2.18.1.1 of SBCSub/C and SBCSub/D/C, he is to consider
  whether the event which is stated to be a cause of delay is a relevant sub-
  contract event as clause 2.19 of SBCSub/C and SBCSub/D/C (further commentary on relevant sub-contract event is provided later in this chapter);
- then (as clause 2.18.1.2 of SBCSub/C and SBCSub/D/C) he is to consider
  whether the event is likely to delay the completion of the sub-contract works
  beyond the period or periods for the completion of the sub-contract works or of
  such works in a section.

Following this (except where SBCSub/C or SBCSub/D/C expressly provides otherwise) the contractor is to:

- give an extension of time to the sub-contractor by fixing such revised period or periods for completion as he then estimates to be fair and reasonable (clause 2.18.1 of SBCSub/C and SBCSub/D/C); or
- notify the sub-contractor that he does not consider that the sub-contractor is entitled to an extension of time (clause 2.18.2 of SBCSub/C and SBCSub/D/C).

Whichever response the contractor makes, clause 2.18.2 of SBCSub/C and SBCSub/D/C requires him to:

- issue his decision as soon as is reasonably practicable;
- issue his decision within 16 weeks of receipt of the notice or the required particulars;
- in the case where the date from the receipt of the notice or the required particulars, to the date of expiry of the period or periods (in item 5 of the sub-contract particulars) for the completion of the sub-contract works or of such works in any relevant section, is less than 16 weeks, then the contractor should endeavour to issue his decision prior to the expiry date given.

When issuing his first (and subsequent) decisions in respect of an extension of time, in line with clause 2.18.3.1 of SBCSub/C or SBCSub/D/C, the contractor is to state the extension of time that he has attributed to each relevant sub-contract event.

It should be noted that the first extension of time may be by way of a 'Pre-agreed adjustment'.

A pre-agreed adjustment is the fixing of a revised period or periods for completion in respect of a variation or other work referred to in clause 5.2 of SBCSub/C and SBCSub/D/C by the acceptance of a schedule 2 quotation. (Schedule 2 quotations are considered in detail in Chapter 10.)

In either event, whether the first extension of time is granted under clause 2.18.1 of SBCSub/C or SBCSub/D/C, or as a result of a pre-agreed adjustment, when issuing subsequent extensions of time, the contractor is to state:

- the extension of time that he has attributed to each relevant sub-contract event (in line with clause 2.18.3.1 of SBCSub/C and SBCSub/D/C); and
- any *reduction in time* that he has attributed to a 'Relevant sub-contract omission' (in line with clause 2.18.3.2 of SBCSub/C and SBCSub/D/C). A relevant sub-contract omission is defined under clause 2.16.3 of SBCSub/C and SBCSub/D/C as being the omission of any work or obligation:

  — through a direction for a variation under clause 3.4 of SBCSub/C, or SBCSub/D/C; or
  — in regard to a provisional sum for defined work in any bills of quantities.

Note that a relevant sub-contract omission only relates to an omission in respect of a defined provisional sum; it does not relate to an omission in respect of an undefined provisional sum. The reason for this is that a sub-contractor is deemed to have made allowance in its programme for the effect of defined provisional sums but not for undefined provisional sums. (For further commentary regarding defined provisional sums and undefined provisional sums refer to Chapter 10.)

In respect of the above possible reduction in time, the contractor can only take into account any relevant sub-contract omission that has occurred since the last occasion when a new period for completion was fixed by the contractor (refer to clause 2.18.4 of SBCSub/C and SBCSub/D/C).

Also, although the contractor can reduce the extension of time previously given, clause 2.18.6.3 of SBCSub/C and SBCSub/D/C makes clear that under no circumstance can the period or periods for completion of the sub-contract works or of such works in any section, be reduced from that stated in the sub-contract particulars (item 5).

Further, the contractor cannot reduce any extension of time previously given by way of a pre-agreed adjustment, unless the relevant variation that related directly to the pre-agreed adjustment was the subject of the relevant sub-contract omission in question (as clause 2.18.6.4 of SBCSub/C and SBCSub/D/C).

Finally, there are two basic conditions that the sub-contractor must comply with:

- clause 2.18.6.1 of SBCSub/C and SBCSub/D/C requires the sub-contractor to: 'constantly use his best endeavours to prevent delay in the progress of the Sub-Contract Works or of such works in any Section, however caused, and to prevent their completion being delayed or further delayed beyond the relevant period for completion';
- clause 2.18.6.2 of SBCSub/C and SBCSub/D/C requires the sub-contractor to 'do all that may reasonably be required to the satisfaction of the Contractor to proceed with the Sub-contract Works'.

The wording of clause 2.18.6.1 and clause 2.18.6.2 of SBCSub/C and SBCSub/D/C is fairly typical JCT wording, and there has been much debate over the years as to what the wording 'constantly use his best endeavours to prevent delay' and 'do all that may reasonably be required' means in effect.

Does it mean accelerate the works?

Does it mean bring in extra resources and work non-productively?

Or, does it mean re-sequence the works to suit the delays encountered?

Unfortunately, there appears to be no direct case law available to shed any light on this matter. The wording is generally considered to mean that the sub-contractor would be expected to reprogramme the works, reschedule material deliveries and information request schedules, and keep all parties advised of the situation; and it may even mean that the sub-contractor may need to expend some monies to meet this obligation; but it is not normally considered that the sub-contractor would be expected to expend substantial sums of money in respect of this obligation.

Some commentators consider that the wording does not require a sub-contractor to expend any monies at all, but this is probably taking too extreme a view.

However, a sub-contractor almost certainly would not be expected to take acceleration measures or incur disruption costs (without clear instructions being issued by the contractor) to satisfy its obligation to 'constantly use his best endeavours' and to do 'all that may reasonably be required'.

The sub-contractor's obligation, as set out above, must be balanced with the sub-contractor's obligation to 'proceed regularly and diligently with the Sub-Contract Works', which, if it fails to do, can be considered as being a default by the sub-contractor that may lead to the termination of the sub-contractor's employment under the sub-contract. (Refer to clause 7.4.1.2 of SBCSub/C and SBCSub/D/C which is dealt with in Chapter 12.)

*Fixing period for completion after the expiry of the period or periods (in item 5 of the sub-contract particulars) for the completion of the sub-contract works or of such works in a section*

The 'trigger' for consideration of an extension of time at this stage is simply the expiry of the period or periods for completion of the sub-contract works or of such works in a section.

At that time (as noted above) the contractor does not require a notice from the sub-contractor at all, and, irrespective of any notice being issued by the sub-contractor, clause 2.18.5 of SBCSub/C and SBCSub/D/C requires the contractor to review the sub-contractor's extension of time entitlement and confirm his decision regarding that entitlement in writing to the sub-contractor. The contractor *may* do this after the expiry of the period or periods for completion of the sub-contract works or of such works in a section, if this occurs before practical completion of the sub-contract works. The use of the word *may* makes this *optional* for the contractor.

However, the sub-contract states that the contractor must do this no later than 16 weeks after the date of practical completion of the sub-contract works. This obligation is not optional but is, in effect, mandatory for the contractor.

In carrying out the above review, the contractor may either:

- extend the relevant period for completion previously fixed if he considers that is fair and reasonable having regard to any relevant sub-contract event, irrespective of whether or not this is reviewing an earlier decision, and irrespective of whether or not the particular relevant sub-contract event has been notified by the sub-contractor (refer to clause 2.18.1 of SBCSub/C or SBCSub/D/C); or
- he may reduce the extension of time previously given, but he can only take into account any relevant sub-contract omission that occurred since the last occasion on which he set a new period for completion; or
- he may confirm the period or periods for completion previously fixed.

In respect of this matter the same general provisos as outlined above continue to apply; that is, in no circumstance can the period or periods for completion of the sub-contract works or of such works in any section, be reduced from that stated in the sub-contract particulars (item 5) (refer to clause 2.18.6.3 of SBCSub/C or SBCSub/D/C), and the contractor cannot reduce the extension of time previously given by way of a pre-agreed adjustment, unless the relevant variation that related directly to the pre-agreed adjustment was the subject of the relevant sub-contract omission in question (refer to clause 2.18.6.4 of SBCSub/C and SBCSub/D/C).

## Relevant sub-contract events

The relevant sub-contract events are listed under clause 2.19 of SBCSub/C and SBCSub/D/C, and each of these events is considered below.

*Clause 2.19.1*

'Variations and other matters or directions which under these Conditions are to be treated as, or as requiring, a Variation.' This clause is self-explanatory; variations are dealt with in Chapter 10.

*Clause 2.19.2*

'Directions of the Contractor, including those which pass on instructions of the Architect/Contract Administrator:

2.19.2.1  given in order to comply with such instructions under any of clauses 2.15, 3.15, 3.16 (excluding an instruction for expenditure of a 'Provisional Sum' for defined work), 3.23 or 5.3.2 of the Main Contract Conditions;

2.19.2.2  for the opening up for inspection or testing of any work, materials or goods under clause 3.17 or 3.18.4 of the Main Contract Conditions (including making good), unless the inspection or test shows that the work, materials or goods comprised in the Sub-Contract Works are not in accordance with this Sub-Contract; or

2.19.2.3  for the opening up for inspection or testing of any work, materials or goods under clause 3.10 (including making good), unless the inspection or test shows that the work, materials or goods were not in accordance with this Sub-Contract.'

Again, this clause is fairly self-explanatory; but the following points should particularly be noted.

Under clause 2.19.2.1 of SBCSub/C and SBCSub/D/C, the exclusion of an instruction for the expenditure of a provisional sum for defined work is because the sub-contractor is deemed to have allowed for works covered by defined provisional sums within its on-site period.

An instruction for the opening up for testing of any work, materials or goods, is not a relevant sub-contract event in the situation where such opening up/testing, etc. reveals that the works, materials or goods were not in accordance with the sub-contract (refer to clauses 2.19.2.2 and 2.19.2.3 of SBCSub/C and SBCSub/D/C).

*Clause 2.19.3*

'Deferment of the giving of possession of the site or any Section under clause 2.5 of the Main Contract Conditions.' The deferment of the giving of possession of the site or any section under clause 2.5 of the main contract conditions may, obviously, have a direct effect upon the commencement and/or progress of the sub-contract works.

It should be noted that (unless expressly stated otherwise within the main contract particulars under clause 2.5) the period of any such deferment is not to exceed six weeks.

*Clause 2.19.4*

2.19.4.1  The execution of work for which an Approximate Quantity is included in the Contract Bills which is not a reasonably accurate forecast of the quantity of work required;

2.19.4.2  where there are Bills of Quantities, the execution of work for which an Approximate Quantity included in those bills is not reasonably accurate forecast of the quantity of work required.'

Unfortunately, there is no definition of 'reasonably accurate forecast' within the sub-contract, and the required accuracy of that forecast would need to be considered on a case-by-case basis.

Therefore, if an approximate quantity of $2000\,m^2$ was allowed for brickwork in the contract bills or in the bills of quantities, but the final total of brickwork was $2010\,m^2$, then the original forecast would most likely be considered to be a reasonably accurate forecast, even though it was $10\,m^2$ less than the actual total.

However, using the same example of an addition of $10\,m^2$ of brickwork, if the approximate quantity only allowed for $1\,m^2$ or even $10\,m^2$ of brickwork then the original forecast would most likely not be considered to be a reasonably accurate forecast.

The reasonableness therefore seems to relate (in part) to the increased quantity as a proportion of the original approximate quantity allowance. Therefore, in the first case cited above, the increase equates to a 0.5% increase (which would seem reasonable); whereas in the second and third cases the increase equates to a 1000% increase and a 100% increase, respectively (which would seem to be unreasonable).

There are no hard and fast rules governing this, although anecdotal evidence would suggest that a limit of up to plus (or minus) 10% of the approximate quantity would generally be considered to be a reasonably accurate forecast.

However, this 'guide' percentage may not always apply.

If, for example (in an unusual case), an approximate quantity allowance of 20 escalators was made, but 21 escalators needed to be installed, would the one additional escalator (i.e. 5% increase on the approximate quantity allowance) be considered to be a reasonably accurate forecast?

It is submitted that this is unlikely to be the case.

Of course, the latter example is an unlikely scenario; however, the example has been provided to illustrate that each case needs to be considered on its merits.

The other point to consider is that both an increase *and a decrease* in the approximate quantity above a reasonable level are relevant sub-contract events.

An increase may result in an entitlement to an increase in the sub-contractor's extension of time entitlement, whereas a decrease may result in a reduction in the sub-contractor's extension of time entitlement (subject, of course, to the provision noted previously that in no circumstance can the period or periods for completion of the sub-contract works or of such works in any section be reduced from that stated in the sub-contract particulars (item 5); and the contractor cannot reduce the extension of time previously given by way of a pre-agreed adjustment, unless the relevant variation that related directly to the pre-agreed adjustment was the subject of the relevant sub-contract omission in question (refer generally to clauses 2.18.6.3 and 2.18.6.4 of SBCSub/C and SBCSub/D/C).

### Clause 2.19.5

'Suspension by the Sub-contractor under clause 4.11 of the performance of his obligations under this Sub-Contract.' The sub-contractor's right of suspension is dealt with in more detail in Chapter 8.

Section 112 of Part II of the Housing Grants, Construction and Regeneration Act 1996 provides the right of suspension because of non-payment to parties on a qualifying construction contract (for details of a qualifying construction contract see sections 104, 105 and 106 of Part II of the Act), and the SBCSub/C and SBCSub/D/C conditions deal with the matter of suspension under clause 4.11.

If the sub-contractor properly suspends the performance of his obligations pursuant to clause 4.11, then the period of such suspension is a relevant sub-contract event. However, it should be noted that the extension of time period is *only* for the period from when the suspension starts (which can only start after the seven-day notice has been given) to the time when payment in full occurs; and it is submitted that no lead in time would be permitted for the sub-contractor to return to site after the period of suspension.

Although there is no guidance regarding the words 'the time when payment in full occurs' it is submitted that this must be when the sub-contractor has received cleared funds through its account.

However, given that it is a well-settled principle of law that a cheque is to be treated as cash, and thus is not susceptible to set-off,[3] it is considered that the courts are likely to decide that payment in full is made when the sub-contractor receives a cheque rather than (the later date) of when the cheque has been cleared through the sub-contractor's account. It should be noted, however, that in rare cases the courts have accepted that an employer could cancel an issued cheque to a contractor.[4]

### Clause 2.19.6

'Suspension by the Contractor under clause 4.14 of the Main Contract Conditions of the performance of his obligations under the Main Contract.' Clause 3.21 of SBCSub/C and SBCSub/D/C states that if the contractor, under clause 4.14 of the main contract conditions, gives the employer a written notice of his intention to suspend the performance of his obligations under the main contract because of non-payment, he is immediately to copy that notice to the sub-contractor, and if he then suspends his obligations under the main contract, he is immediately to notify the sub-contractor.

However, it is most important for the sub-contractor to note that he is not to suspend his performance in respect of the sub-contract works simply because he has been notified by the contractor that the contractor has suspended its performance in respect of the main contract works. In fact, following the notice from the contractor advising that the contractor has suspended his performance in respect of the main contract works, the sub-contractor should (where possible) simply proceed regularly and diligently with its sub-contract works, until and/or unless he receives an express direction from the contractor (under clause 3.22.1 of SBCSub/C or SBCSub/D/C) to cease carrying out the sub-contract works. This entire matter is dealt with in more detail in Chapter 7.

The relevant sub-contract event dealt with under clause 2.19.6 of SBCSub/C and SBCSub/D/C relates specifically to the suspension by the contractor and not necessarily the suspension by the sub-contractor; it is therefore submitted that clause 2.19.6 would be applicable in the case where a sub-contractor itself suspended its performance (as a result of an express direction from the contractor following the contractor's suspension of its performance), *and* in the case where the sub-

---

[3] *Nova (Jersey) Knit Ltd* v. *Kammgarn Spinnerei* [1977] 1 WLR 713, HL; *Isovel Contracts Ltd* (in administration) v. *ABB Building Technologies Ltd* (formerly ABB Steward Ltd) [2002] 1 BCLC 390.
[4] *Willment Brothers Ltd* v. *North West Thames Regional Health Authority* (1984) 26 BLR 51.

contractor did not suspend performance but was, in any event, delayed in its progress as a result of the contractor suspending performance.

A case in point could be, for example, where a sub-contractor is to supply and install studwork partitions, but the contractor is to supply and install electrical services within the studwork partitions.

Because of non-payment, the contractor suspends the performance of its obligation under the main contract, but the contractor does not require the sub-contractor to suspend the performance of its own obligations under the sub-contract.

In this situation, the sub-contractor would be expected to proceed regularly and diligently with the sub-contract works, and in fact may not be delayed in the com-mencement and progress of the sub-contract works relating to the erection of the studwork and the fixing of one face of the partition.

However, because of the suspension by the contractor this would prevent the installation of the electrical services within the studwork partitions, and this would consequently have the effect of stopping the sub-contractor from fixing the second face, thereby closing in the studwork partitions. This would therefore delay the progress of the sub-contract works, even though the sub-contractor had not suspended its own performance.

### Clause 2.19.7

'Any impediment, prevention or default, whether by act or by omission, by the Employer, the Architect/Contract Administrator, the Quantity Surveyor or any of the Employer's Persons except to the extent caused or contributed to by any default, whether by act or omission, of the Sub-Contractor or of any of the Sub-Contractor's Persons.'

This clause is self-explanatory, and simply has the caveat that the relevant sub-contract event does not include for the position where any impediment, preven-tion or default is caused by or is contributed to by any default, whether by act or omission, of the sub-contractor or of any of the sub-contractor's persons.

### Clause 2.19.8

'Any impediment, prevention or default, whether by act or by omission, by the Contrac-tor or any of the Contractor's Persons (including, where the Contractor is the Principal Contractor, any default, whether by act or omission, in that capacity) except to the extent caused or contributed to by any default, whether by act or omission, of the Sub-contractor or of any of the Sub-contractor's Persons.'

This clause is, again, self-explanatory, and has the caveat that the relevant sub-contract event does not include for the position where any impediment, preven-tion or default is caused by or is contributed to by any default, whether by act or omission, of the sub-contractor or of any of the sub-contractor's persons.

As noted earlier within this chapter, the purpose of this clause is to prevent the *Peak v. McKinney*[5] doctrine from having effect.

Although doubts exist regarding whether the wording of a clause along the lines of 'beyond the Sub-Contractor's control' will be considered to be legally

---

[5] *Peak Construction (Liverpool) Ltd* v. *McKinney Foundations Ltd* (1970) 1 BLR 111.

acceptable to circumvent the *Peak* v. *McKinney* doctrine, there seems to be no reason why a more explicit, but still wide clause, such as 'any impediment, prevention or default, whether by act or by omission, by the Contractor', should not be valid.

In respect of this relevant sub-contract event, clauses 2.7.3 and 2.7.4 of SBCSub/C and SBCSub/D/C require that:

- the contractor is required to provide or give such further drawings, details, information and directions referred to in clause 2.7.1 and 2.7.2 of SBCSub/C or SBCSub/D/C at the time it is reasonably necessary for the sub-contractor to receive them, having regard to the progress of the sub-contract works and the main contract works; and
- where the sub-contractor has reason to believe that the contractor is not aware of the time by which the sub-contractor needs to receive such further drawings, details, information or directions, he must, so far as is reasonably practicable, advise the contractor sufficiently in advance to enable him to comply with the requirements of clause 2.7 of SBCSub/C or SBCSub/D/C.

Therefore, if the sub-contractor does not take a proactive approach in respect of this process, it could be argued that he has defaulted on his obligations. If it could be shown that this default has contributed to the delay suffered, the sub-contractor may not be entitled to an extension of time for the period of this default.

### Clause 2.19.9

'The carrying out by a Statutory Undertaker of work in pursuance of its statutory obligations in relation to the Main Contract Works, or the failure to carry out such work.' A 'Statutory undertaker' is defined under clause 1.1 of SBCSub/C and SBCSub/D/C as being 'any local authority or statutory undertaker when executing work solely in pursuance of its statutory obligations, including any persons employed, engaged or authorized by it upon or in connection with that work'.

A statutory undertaker (sometimes referred to as a public utility company) is a body authorized by statute to carry out the laying of services (e.g. water, electricity, gas, telephones).

A point that could easily be overlooked is that this relevant sub-contract event only applies when the statutory undertaker is undertaking work in pursuance of its statutory obligations.

If a statutory undertaker is carrying out a statutory obligation then this is covered by this event. If they are not carrying out statutory obligations then they will normally be considered to be sub-contractors.

Therefore, although clause 1.1 of SBCSub/C and SBCSub/D/C defines 'contractor's persons' as: 'excluding the Architect/Contract Administrator, the Quantity Surveyor, the Employer, Employer's Persons, the Sub-Contractor, Sub-Contractor's Persons and any Statutory Undertaker', it is submitted that when a statutory undertaker is not carrying out statutory obligations, it will not be considered to be a statutory undertaker pursuant to the definition under clause 1.1 of SBCSub/C and SBCSub/D/C, but will actually be considered to be simply a contractor's person.

If, as a contractor's person, the statutory undertaker causes any impediment, prevention or default, then this would be covered by the relevant sub-contract event covered under clause 2.19.8 of SBCSub/C and SBCSub/D/C.

## Clause 2.19.10

'Exceptionally adverse weather conditions.' This relevant sub-contract event relates to *exceptionally* adverse weather conditions. Exceptional means 'unusual' or 'not typical', whilst 'adverse' means 'unfavourable' or 'harmful'.

Therefore, this relevant sub-contract event is concerned only with:

(1) weather conditions that are both 'unfavourable' or 'harmful' to the sub-contractor's progress on site; *and*
(2) weather conditions that are 'unusual' or are 'not typical'.

It is fairly easy to determine weather conditions that may be 'unfavourable' or 'harmful' to the sub-contractor's progress on site. For example:

• heavy rain may prevent external rendering from proceeding;
• snowfall may prevent road surfacing from being carried out;
• high winds may prevent cranes from operating;
• low temperatures may prevent bricklayers from using 'frozen' mortar; and
• high temperatures may prevent external painting from being executed.

The above points should simply be matters of fact.

The problem comes in trying to decide if the adverse weather conditions are 'unusual' or are 'not typical'.

What yardstick should be used when deciding this matter?

For example, if a day of heavy rain and strong winds was experienced in November, would that be unusual?

Or, if snow fell, or if temperatures were at or below freezing point on one day in January, would that be untypical?

Of course not, and a sub-contractor would be expected to allow within its sub-contract period for 'normal' adverse weather conditions.

SBCSub/C and SBCSub/D/C does not provide any guidance as to how to judge whether the adverse weather conditions encountered are exceptional.

The method normally adopted is to compare the actual adverse weather conditions encountered on site against the records of the previous five, ten or fifteen years' weather conditions maintained by a meteorological centre that is in the same locality as the site.

In such a situation, the site records may show that nine days of rain, that prevented work on site from proceeding, were actually encountered in the month of June; whilst the local meteorological centre's records may show that (as an average of the previous ten years, for example) only six days of rain are normally encountered in the month of June.

In such a situation, the argument that would be presented would be that there were three days (i.e. nine days less six days) of *exceptionally* adverse weather conditions in the month of June that prevented work on site from proceeding, and a claim for three days' extension of time would therefore be made.

The above method (or a variation on it) is frequently used, but, at best, it is only a very crude guide to weather conditions and trends. For example, it takes no account of the fact that the nearest meteorological centre may be many miles from the site; it takes no account of any local weather abnormalities (e.g. working near the coast, or the vortex effect of working near tall buildings); and it takes little account (historically) of the timing of the adverse weather conditions (i.e. day or night weather conditions).

### Clause 2.19.11

'Loss or damage occasioned by any of the Specified Perils.' This relevant sub-contract event relates to clause 6.7.4 of SBCSub/C and SBCSub/D/C., and relates to the delay effect of any loss or damage by any of the specified perils.[6]

It should be noted that although this matter is a relevant sub-contract event in respect of extensions of time, it is not a relevant sub-contract matter in respect of loss and expense.

### Clause 2.19.12

'Civil commotion or the use or threat of terrorism and/or the activity of the relevant authorities in dealing with such event or threat.' A civil commotion is normally considered to be a riot or a similar disturbance, and where such an event did occur, and where that event affected the progress of the sub-contract works on site, then this would be covered by this relevant sub-contract event.

In respect of terrorism, it is not only the actual use of terrorism that is covered, it is also:

- the threat of use of terrorism (for example where an area, including the site area, is evacuated because a warning has been given that a bomb has been planted nearby); and
- the activity of the relevant authorities in dealing with an actual terrorist incident or a threatened terrorist incident (for example where entry to an area, including the site area, is prohibited by the police because of actual or threatened terrorist activity).

### Clause 2.19.13

'Strike, lock-out or local combination of workmen affecting any of the trades employed upon the Main Contract Works or any of the trades engaged in the preparation, manufacture or transportation of any of the goods or materials required for the Main Contract Works.'

This matter relates to clause 3.23 of SBCSub/C and SBCSub/D/C dealt with in Chapter 7.

A strike has variously been interpreted in statute and in case law as being:

- 'an agreement between persons who are working for a particular employer not to continue working for him';[7]
- 'any concerted stoppage of work'.[8]

---

[6] 'Specified perils' are defined under clause 6.1 of SBCSub/C and SBCSub/D/C as 'fire, lightning, explosion, storm, tempest, flood, escape of water from any water tank, apparatus or pipe, earthquake, aircraft or other aerial devices, or articles dropped therefrom, riot and civil commotion, but excluding excepted risks.'

[7] *Lyons* v. *Wilkins* [1896] 1 Ch 811.

[8] Trade Unions and Labour Relations (Consolidation) Act 1992 Part V (sections 219–246).

A lock-out is where employees are excluded by their employer from their place of work during an industrial dispute, as a means of imposing certain conditions.

A local combination of workmen would normally be where workmen take unofficial industrial action in respect of site works.

Based upon similar wording in earlier JCT forms, it is considered that the trades employed upon the main contract works or the trades engaged in the preparation, manufacture or transportation of any of the goods or materials required for the main contract works, must themselves be *directly* involved with a strike, a lock-out, or a local combination of workmen, before relevant sub-contract event clause 2.19.13 of SBCSub/C or SBCSub/D/C has any effect.

Because of this, it is submitted that a regional or national fuel blockade, for example, would not be covered by this relevant sub-contract event.

### Clause 2.19.14

'The exercise after the Main Contract Base Date by the United Kingdom Government of any statutory power which directly affects the execution of the Main Contract Works.' A statutory power is simply a power that is required, permitted or enacted by a Statute of the United Kingdom Government. This could, for example, relate to a statute that in some way restricts or limits working hours because of some fuel conservation requirement.

Such an event would only be a relevant sub-contract event in the situation where:

(1) the statutory power is only exercised after the main contract base date. In other words, this would apply not only to statutes that are enacted after the main contract base date but also to statutes that were enacted before the main contract base date, but did not come into effect until after the main contract base date; and
(2) the statutory power that is exercised directly affects the execution of the main contract works.

### Clause 2.19.15

'Force majeure.' This is a French law term and is generally wider than the common law term act of God.

In terms of SBCSub/C and SBCSub/D/C the term has a narrower meaning because matters such as strikes, fire, weather, etc. are dealt with under specific relevant sub-contract events.

There is no direct English authority on the JCT definition, but the usual English authority of the term[9] states broadly that it covers all 'matters independent of the will of man and which it is not in his power to control'.

Because the *force majeure* clause is not also a relevant sub-contract matter for loss and expense purposes, it will possibly sometimes be used (purely on a commercial basis, and not on a contractual basis) as a clause to absorb any unresolved issues that are not easily incorporated into another relevant sub-contract event.

[9] *Lebeaupin v. Crispin* [1920] 2 KB 714.

## Delay analysis

Under clause 2.17.2 of SBCSub/C and SBCSub/D/C the sub-contractor is expected to provide an estimate of any expected delay in the completion of the sub-contract works or of such works in any section beyond the relevant period or periods for completion stated in the sub-contract particulars (item 5).

Under clause 2.18.1 and 2.18.5 of SBCSub/C and SBCSub/D/C, the contractor may give an extension of time if it considers that such an extension of time is justified.

To carry out either of the above exercises, a certain level of knowledge of delay analysis is required.

The most important and indeed difficult task in terms of determining delay is to establish the nexus of cause and effect; the mere happening of an event which happens to be a relevant sub-contract event confers no entitlement to an extension to the contract period.

The test is whether the event in question is 'such as may fairly entitle the Sub-Contractor to an extension of time' for completion of the sub-contract works, or for any section of it.

It is the sub-contractor's actual progress which must be delayed. The sub-contractor's originally planned/programmed progress is not generally relevant.

What needs to be considered are the actual facts on site to determine whether a particular event affected operations critical to the sub-contractor's completion date.

Such matters are best proved with contemporaneous records. Indeed without records, and evidence going some way to establish such matters, a case will not be made out.

What has to be established in determining an extension of time entitlement can briefly be described as follows:

- that an event occurred at all;
- that the event occurred in the manner asserted by the sub-contractor;
- that the event comes within a category providing an express entitlement to an extension of time within the sub-contract (i.e. that it is a relevant sub-contract event);
- that the sub-contractor has complied with the notice requirements within the sub-contract;
- that the event outlined has had a particular delaying effect on the ability of the sub-contractor to complete within the periods stipulated.

In order to establish cause and effect, good record keeping and evidence is essential.

Delay analysis is sometimes referred to colloquially as a 'black art', mostly because of the numerous techniques used in analysing delay and the various tests used to determine causation.

For example, in 2002, the Society of Construction Law published a *Delay and Disruption Protocol* (that has no contractual effect) which set out four basic generic methods of carrying out a delay analysis:

- As-planned programme compared to the as-built programme – in this analysis the as-planned programme periods are compared to the as-built programme periods and the differences are then analysed to find the reasons for the delay.

- Impacted as-planned programme – in this analysis the impact of delay events is added on to the as-planned programme to show their projected delay effect.
- Collapsed as-built programme – in this analysis the impact of delay events is deducted from the as-built programme to show that 'but for' those delay events the works would have finished in the original time period.
- Time impact analysis (or time slice analysis) – in this analysis the 'slippage' against the programme is considered at a set period of time (often monthly) throughout the project. That slippage is then analysed to determine the cause, or causes, of the delay.

Obviously, within this book it is not possible to provide a full exposé of delay analysis techniques; however it is possible to show the basis of delay analysis principles in a simple format.

The point that should be noted is that the principles of delay analysis in its simplest form are no different to the principles of delay analysis in a more complicated form, and once those principles are understood, then the application of those principles can be made to any programme.

The following simplified programmes explain the steps and the problem areas of delay analysis. The examples below deliberately ignore matters relating to resourcing of the works since the principles are easier to understand if the resources are assumed to remain constant.

### What is the critical path?

Refer to programme 1, which shows three activities only:

- studwork for partitions;
- plasterboard for partitions; and
- taping and jointing of partitions.

In the circumstances of this case, the plasterboard for the partitions cannot commence until the studwork for the partitions is entirely complete; and the taping and jointing of the partitions cannot commence until the plasterboard for the partitions is entirely complete.

Therefore, there is what is known as a 'finish to start' link between:

- the finish of the studwork for the partitions, and the start of the plasterboard for the partitions; and between

| PROGRAMME 1 | DAYS | | | | | | | | | | | | | |
|---|---|---|---|---|---|---|---|---|---|---|---|---|---|---|
| | 1 | 2 | 3 | 4 | 5 | 6 | 7 | 8 | 9 | 10 | 11 | 12 | 13 | 14 |
| Studwork for partitions | | | | | | | | | | | | | | |
| Plasterboard for partitions | | | | | | | | | | | | | | |
| Taping and jointing of partitions | | | | | | | | | | | | | | |

- the finish of the plasterboard for the partitions, and the start of the taping and jointing of the partitions.

In this case, these 'finish to start' links form the basis of the 'critical path' through the programme.

(For the ease of understanding finish to start links only have been used; however, in addition to finish to start links there can be 'finish to finish' links, 'start to start' links and 'start to finish' links. All of these various links can incorporate 'leads' and 'lags'. A lead in respect of a finish to start link is where the following activity can start before the preceding activity is finished. Therefore, there could be a finish to start link with a five-day lead which would mean, in effect, that the following activity could commence five days before the preceding activity was complete. A lag in respect of a finish to start link, on the other hand, is where the following activity cannot start until a certain period after the preceding activity is finished. Therefore, there could be a finish to start link with a five-day lag which would mean, in effect, that the following activity could not commence until five days after the preceding activity was complete.)

The 'critical path' through a programme simply shows the sequence of activities which result in the minimum time necessary for the execution of the works.

The critical path for programme 1, as described above, is shown in programme 2.

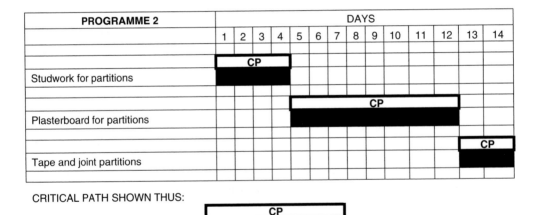

## Can there be two critical paths?

There is a lot of nonsense spoken about the fact that you can only have one critical path through a programme. It is quite clear, that in certain circumstances, two critical paths can run through a programme.

Programme 3 illustrates this position; and shows four activities, namely:

- studwork for partitions;
- plasterboard for partitions;
- timber sheathing for partitions; and
- decorative finishing to the face of the partitions.

| PROGRAMME 3 | DAYS | | | | | | | | | | | | | |
|---|---|---|---|---|---|---|---|---|---|---|---|---|---|---|
| | 1 | 2 | 3 | 4 | 5 | 6 | 7 | 8 | 9 | 10 | 11 | 12 | 13 | 14 |
| Studwork for partitions | ████ | | | | | | | | | | | | | |
| Plasterboard for partitions | | | | | ████████ | | | | | | | | | |
| Timber sheathing for partitions | | | | | ████████ | | | | | | | | | |
| Decorative finish to partitions | | | | | | | | | | ████ | | | | |

| PROGRAMME 4 | DAYS | | | | | | | | | | | | | |
|---|---|---|---|---|---|---|---|---|---|---|---|---|---|---|
| | 1 | 2 | 3 | 4 | 5 | 6 | 7 | 8 | 9 | 10 | 11 | 12 | 13 | 14 |
| Studwork for partitions | CP ████ | | | | | | | | | | | | | |
| Plasterboard for partitions | | | | | CP ████████ | | | | | | | | | |
| Timber sheathing for partitions | | | | | CP ████████ | | | | | | | | | |
| Decorative finish to partitions | | | | | | | | | | CP ████ | | | | |

CRITICAL PATH SHOWN THUS:

    ┌─────────────────────┐
    │          CP         │
    └─────────────────────┘

In the circumstances of this case, neither the plasterboard for the partitions nor the timber sheathing for the partitions can commence until the studwork for the partitions is entirely complete; and the decorative finishing to the face of the partitions cannot commence until both the plasterboard and the timber sheathing for the partitions is entirely complete.

In this case finish to start links are formed between:

- the finish of the studwork for the partitions, and the start of the plasterboard for the partitions;
- the finish of the studwork for the partitions, and the start of the timber sheathing for the partitions;
- the finish of the plasterboard for the partitions, and the start of the decorative finishing to the face of the partitions; and
- the finish of the timber sheathing for the partitions, and the start of the decorative finishing to the face of the partitions.

These finish to start links form the basis of the two critical paths running through the programme. The two critical paths described above are shown in programme 4.

### What happens when a delay event impacts upon a critical path?

If we return to the example of a single critical path, as shown in programme 2, it is clear that if another activity is inserted onto the critical path this will, all things being equal, cause a delay to the completion date.

Therefore, if, for example, the specification of the plasterboard changed, and the result of this change was that it took longer to fix the plasterboard, this would have the effect of increasing the length of the programme bar for the plasterboard.

Because the programme bar for the plasterboard was on the critical path, the extra time required due to the change of the specification (i.e. four days, being days eleven to fourteen inclusive) would have a direct impact upon the completion date (which is now delayed from day 12 to day 16).

This scenario is illustrated in programme 5.

As noted at the commencement of this section on delay analysis, the question of resources has been ignored for the ease of understanding.

However, the parties need to be aware that a counter argument to delay is often linked to resources.

Take, for example, a brick wall that was on the critical path and which was programmed to be completed in one week using a 2/1 gang of bricklayers/labourers. If a further brick wall was added (of equal size) it would normally be argued that this impacted directly onto the critical path and would therefore add one week onto the programme period.

Of course, the alternative argument is simply to say that if the resources were increased to a 4/2 gang of bricklayers/labourers, this would prevent delay from

| PROGRAMME 5 | | | | | | | | | | DAYS | | | | | | |
|---|---|---|---|---|---|---|---|---|---|---|---|---|---|---|---|---|
| | 1 | 2 | 3 | 4 | 5 | 6 | 7 | 8 | 9 | 10 | 11 | 12 | 13 | 14 | 15 | 16 |
| Studwork for partitions | | CP | | | | | | | | | | | | | | |
| Plasterboard for partitions | | | | | CP | | | | | | | | | | | |
| Revised specification for plasterboard | | | | | | | | | | | | CP | | | | |
| Tape and joint partitions | | | | | | | | | | | | | | | CP | |
| PERIOD OF DELAY | | | | | | | | | | | | DELAY | | | | |

CRITICAL PATH SHOWN THUS:

CP

PERIOD OF DELAY SHOWN THUS:

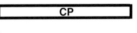

DELAY

occurring because twice the amount of work could be executed in the same programme period.

This increase in resources would normally be considered to be in line with the sub-contractor's obligation, under clause 2.18.6.1 of SBCSub/C and SBCSub/D/C, to 'constantly use his best endeavours to prevent delay'.

### What is 'float'?

Programme 6 shows four activities, namely:

- studwork for partitions;
- plasterboard for partitions;
- timber sheathing for partitions; and
- decorative finishing to the face of the partitions.

| PROGRAMME 6 | DAYS | | | | | | | | | | | | | |
|---|---|---|---|---|---|---|---|---|---|---|---|---|---|---|
| | 1 | 2 | 3 | 4 | 5 | 6 | 7 | 8 | 9 | 10 | 11 | 12 | 13 | 14 |
| Studwork for partitions | | | | | | | | | | | | | | |
| Plasterboard for partitions | | | | | | | | | | | | | | |
| Timber sheathing for partitions | | | | | | | | | | | | | | |
| Decorative finish to partitions | | | | | | | | | | | | | | |

In the circumstances of this case, neither the plasterboard for the partitions nor the timber sheathing for the partitions can commence until the studwork for the partitions is entirely complete; and the decorative finishing to the face of the partitions cannot commence until both the plasterboard and the timber sheathing for the partitions is entirely complete.

However, it is programmed that the plasterboard will take six days to complete, whereas the timber sheathing will only take four days to complete.

When the critical path is superimposed (see programme 7) this shows that the studwork for the partitions, the plasterboard for the partitions and the decorative finishing to the face of the partitions are all on the critical path.

However, the timber sheathing to the partitions is not on the critical path. In fact the timber sheathing to the partitions can take two days longer than programmed without impacting upon the critical path. The period of time before an activity impacts upon the critical path is known as a period of 'float'.

In this case finish to start links are formed between:

- the finish of the studwork for the partitions, and the start of the plasterboard for the partitions;
- the finish of the studwork for the partitions, and the start of the timber sheathing for the partitions;
- the finish of the plasterboard for the partitions, and the start of the decorative finishing to the face of the partitions; and

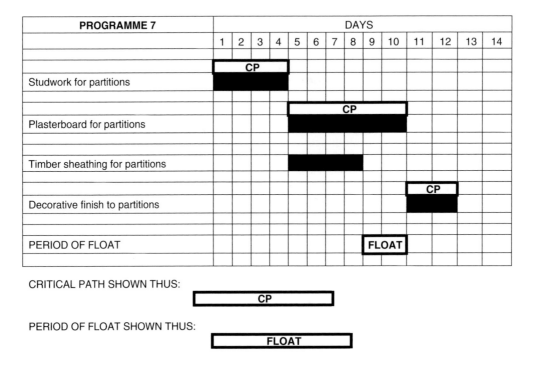

CRITICAL PATH SHOWN THUS:

PERIOD OF FLOAT SHOWN THUS:

- the finish of the timber sheathing for the partitions, and the start of the decorative finishing to the face of the partitions, with a two-day float period.

The illustration of this period of float is shown in programme 7.

There is often some debate about who 'owns' the float in respect of a construction programme. Is it the sub-contractor or is it the contractor?

The normal approach (in respect of JCT contracts) is that whoever uses the float first, effectively owns the float. Therefore, if a delay event occurs, whether that be a delay event for which the sub-contractor is responsible, or a delay event for which the contractor is responsible, then that delay event will use any available period of float and will effectively own that part of the float.

### Delay to an activity not on the critical path

In certain instances, a delay may occur to an activity that is not on the critical path. Depending on the circumstances, this may or may not delay the completion date.

If we use programme 7 as a base, then the activity with the float may be delayed without affecting the completion date.

Thus, for example, if the timber sheathing for the partitions was delayed by one day (i.e. day 9), this would not have an effect upon the critical path and would not affect the completion date (as illustrated in programme 8).

However, if the timber sheathing for the partitions was delayed by three days (i.e. days 9 to 11 inclusive), this would have an effect upon the critical path and would affect the completion date – but by one day only (i.e. day 13) (as shown in programme 9).

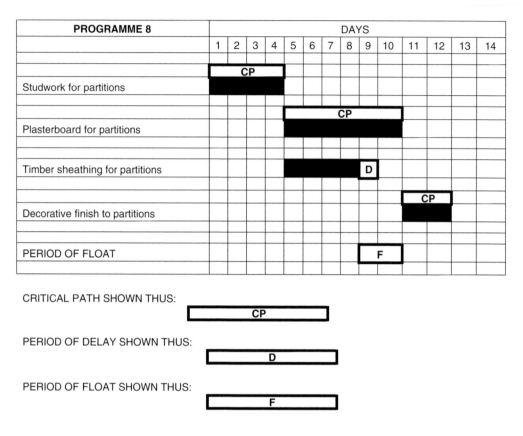

CRITICAL PATH SHOWN THUS:

PERIOD OF DELAY SHOWN THUS:

PERIOD OF FLOAT SHOWN THUS:

*Could loss and expense be recovered even if no delay to completion occurs?*

This matter can most easily be explained by returning to programme 8.

On that programme, the timber sheathing was delayed by one day, but because there was a float of two days on that activity, the delay of one day did not impact upon the critical path and did not delay the completion date.

However, if we assume the following, then it may be the case that loss and expense for an extra one day's hire for the scaffolding for the timber sheathing may be recoverable even though the sub-contractor does not have the right to an extension to the completion date:

● the delay to the timber sheathing was due to a variation issued by the contractor, and
● the scaffolding needed to be in place to enable the timber sheathing to be fixed.

Therefore, in such a situation, loss and expense could be recovered even though no delay to completion occurs.

*What are concurrent delays?*

When programmes become extremely involved and complicated there can be a great deal of debate about what concurrent delays are. (For example, if two delay

| PROGRAMME 9 | DAYS | | | | | | | | | | | | | |
|---|---|---|---|---|---|---|---|---|---|---|---|---|---|---|
| | 1 | 2 | 3 | 4 | 5 | 6 | 7 | 8 | 9 | 10 | 11 | 12 | 13 | 14 |
| Studwork for partitions | | CP | | | | | | | | | | | | |
| Plasterboard for partitions | | | | | | CP | | | | | | | | |
| Timber sheathing for partitions | | | | | | | | CP D | | | | | | |
| Decorative finish to partitions | | | | | | | | | | | | CP | | |
| PERIOD OF FLOAT | | | | | | | | | F | | | | | |
| PERIOD OF DELAY | | | | | | | | | | | D | | | |

CRITICAL PATH SHOWN THUS: CP

PERIOD OF DELAY SHOWN THUS: D

PERIOD OF FLOAT SHOWN THUS: F

events do not start at exactly at the same time and do not run for the same period of time, can they be truly referred to as concurrent? Also, if the delay events are of marked different causative potency then it may be obvious that one delay event on its own would not have caused a delay to the completion date.)

However, in the scenario outlined in the above programmes, the position can be kept relatively simple.

Programme 4 showed the position where two parallel critical paths ran through the programme.

The critical path ran through the activities of plasterboard for partitions; and of timber sheathing to partitions.

If both of those activities were delayed by delay events of an equal length, then the completion date would be delayed by what is known as concurrent events (as shown in programme 10).

If both of the delay events described above are as a result of variations, for example, then the fact that the delay events are concurrent does not cause a major concern.

Similarly, if both of the delay events are delays for which the sub-contractor is liable (for example lack of resources) then the fact that the delay events are concurrent, again, does not cause a major concern.

| PROGRAMME 10 | DAYS | | | | | | | | | | | | | | | |
|---|---|---|---|---|---|---|---|---|---|---|---|---|---|---|---|---|
| | 1 | 2 | 3 | 4 | 5 | 6 | 7 | 8 | 9 | 10 | 11 | 12 | 13 | 14 | 15 | 16 |
| Studwork for partitions | | CP | | | | | | | | | | | | | | |
| Plasterboard for partitions | | | | | CP | | | | | | | D | | | | |
| Timber sheathing to partitions | | | | | CP | | | | | | | D | | | | |
| Tape and joint partitions | | | | | | | | | | | | | | | CP | |
| PERIOD OF CONCURRENT DELAY | | | | | | | | | | | CD | | | | | |

CRITICAL PATH SHOWN THUS: CP

PERIOD OF DELAY SHOWN THUS: D

PERIOD OF CONCURRENT DELAY SHOWN THUS: CD

The problem arises where one of the delay events is due to an event that the subcontractor is liable for (e.g. lack of resources, etc.) and the other one for which he is not liable for (e.g. variations, etc.). The question then arises as to how the allocation of the delay to completion is allocated between the two concurrent delay events.

*How are concurrent delays dealt with?*

There are at least six approaches (or variants on those approaches) that are used. Those approaches may be grouped under the following headings:

- the 'apportionment' approach;
- the 'American' approach;
- the 'chronology of events' approach;
- the 'dominant cause' approach;
- the application of the 'but for' test; and
- the 'Malmaison'[10] approach

[10] Named after the *Henry Boot Construction (UK) Limited* v. *Malmaison Hotel (Manchester) Limited* [1999] 70 Con LR 32.

### The apportionment approach

This approach relies upon the apportionment of any overrun (and its consequences) between the contractor and the sub-contractor. However, this approach has attracted very limited support in the courts.

To some degree it is thought that this is because the courts tend to apply the principles of causation in an all or nothing way.[11]

Historically, the courts have sought to attribute any one event to a single cause with the result that a plaintiff succeeds completely, or not at all.

In respect of SBCSub/C and SBCSub/D/C an apportionment approach is not expressly provided as an option under the terms of the sub-contract conditions.

It should be noted, however, that clause 2.18.1 of SBCSub/C and SBCSub/D/C requires the contractor to grant an extension of time that is 'fair and reasonable', and it may be that by pursuing an apportionment route this may be perceived as being 'fair and reasonable'.

### The American approach

Under the American approach, in cases of concurrent critical delay, neither party will recover financial recompense unless and to the extent that they can separate delay associated with each competing cause and prove the delay upon which it relies.

Therefore, where excusable delay occurs concurrently with inexcusable delay, neither the contractor not the employer recovers compensation.

This result is achieved by allowing the contractor an extension of time but denying him financial recompense. The effect of allowing the contractor an extension of time is to deprive the employer of financial recompense (because liquidated damages cannot be recovered).

However, it should again be noted that clause 2.18.1 of SBCSub/C and SBCSub/D/C requires the contractor to grant an extension of time that is 'fair and reasonable', and it therefore may well be the case that the courts may decline to apply the American approach on the basis that it does not comply with the intentions of the parties, as evidenced by the express terms of SBCSub/C and SBCSub/D/C.

### The chronology of events approach

Under the chronology of events approach, the delay event that occurs first (whether that is a delay event that the sub-contractor is liable for, or not) remains as the dominant (and, in practice, the only) cause of delay until it runs its course, at which time any concurrent delay event that started at a later date becomes the applicable delay event.

The reasoning behind the chronology of events approach is that when the second (concurrent) delay event came into effect, it added nothing to the existing situation because the sub-contractor was already going to be delayed by the first delay event, in any case.

The chronology of events is inconsistent with the dominant cause approach and the Malmaison approach outlined below.

---

[11] Refer to *Quinn v. Burch Brothers (Builders) Limited* [1966] 2 QB 370, CA.

### The dominant cause approach

For many years, it has commonly been suggested that the correct approach to the matter of causation in determining concurrent delays is to apply what is called the dominant cause approach.

Using this approach, the contractor would be required to choose between competing causes of concurrent delay according to which is dominant or predominant.

Therefore, under the dominant cause approach, if there are two causes, one the contractual responsibility of the contractor and the other the contractual responsibility of the sub-contractor, the sub-contractor succeeds if he establishes that the cause for which the contractor is responsible is the effective, dominant cause.

Although the establishment of which cause is dominant is to be determined as a question of fact, in a case where the competing causes are of approximately equal causative potency, the task of making such a choice is extremely difficult.

In terms of judicial authority, the dominant cause approach has almost entirely escaped attention. It is mentioned only in *H. Fairweather v. London Borough of Wandsworth*,[12] where Mr Justice Fox-Andrews appeared to conclude that the dominant cause approach was inappropriate.

In addition, it would appear that the dominant cause approach would not be entirely compatible with the discretion which SBCSub/C and SBCSub/D/C confer on the contractor when considering extensions of time.

For these reasons, it is thought that the courts are unlikely to adopt the dominant cause approach in respect of disputes regarding extensions of time under SBCSub/C and SBCSub/D/C.

### The application of the but-for test

The so-called but-for test is likely to be preferred by sub-contractors as it will usually give them the most favourable result.

The but-for test uses the principle that (irrespective of any delays caused by the sub-contractor), but for a delay event for which the contractor was liable, the sub-contract works would have been completed on time.

There do not appear to be any reported English cases where it has been suggested that the but-for test could be applied to the determination of contractual claims, and it is thought that the English courts are most unlikely to apply the but-for test of causation to a contractual claim unless the wording of the contract clearly demands it (which is not the case in respect of SBCSub/C and SBCSub/D/C).

### The Malmaison approach

In the *Henry Boot v. Malmaison*[13] case, Mr Justice Dyson determined an appeal relating to a dispute on the pleadings in an arbitration, as to the extent of the inquiry which the arbitrator was entitled to undertake to resolve one of the contractor's extension of time claims.

In the introductory paragraphs of his judgment, Mr Justice Dyson recorded certain matters of common ground which had been agreed between the parties before him. They included the following:

---

[12] *H. Fairweather and Company Limited v. London Borough of Wandsworth* (1987) 39 BLR 106.
[13] *Henry Boot Construction (UK) Limited v. Malmaison Hotel (Manchester) Limited* [1999] 70 Con LR 32.

'It is agreed that if there are two concurrent causes of delay, one of which is a relevant event, and the other is not, then the contractor is entitled to an extension of time for the period of delay caused by the relevant event notwithstanding the concurrent effect of the other event. Thus, to take a simple example, if no work is possible on a site for a week not only because of exceptionally inclement weather (a relevant event), but also because the contractor has a shortage of labour (not a relevant event), and if the failure to work during that week is likely to delay the works beyond the completion date by one week, then if he considers it fair and reasonable to do so, the architect is required to grant an extension of time of one week. He cannot refuse to do so on the grounds that the delay would have occurred in any event by reason of the shortage of labour.'

The above approach recognizes that any one period of delay may properly, as a matter of causation, be attributed to more than one delaying event; and this factor will be sufficient to allow the contractor to use its 'fair and reasonable' discretion to grant an extension of time if one of the delaying events affords grounds for extension of time.

The rationale that may be advanced for the above approach is that it does no more than reflect the allocation of risk agreed upon by the parties when they enter into the sub-contract.

In advancing the above argument, reliance will no doubt be placed upon the observations of Mr Justice Edgar Fay concerning the allocation of risk under the equivalent of the JCT standard form applicable at that time, in the *Henry Boot* v. *Central Lancashire*[14] case:

'The suggestion is, that in allocating risks as between themselves, the parties may be taken, first, to have recognized that any one delay or period of delay might well be attributable to more than one cause but, secondly, to have agreed nevertheless that provided one of those causes affords grounds for relief under the contract, then the contractor should have his relief.'

More recently, in *Royal Brampton Hospital* v. *Hammond*[15] further support for the Malmaison approach was afforded by Mr Justice Seymour when he said:

'However, if Taylor Woodrow was delayed in completing the works both by matters for which it bore the contractual risk and by relevant events, within the meaning of that term in the Standard Form, in the light of the authorities to which I have referred, it would be entitled to extensions of time by reason of the occurrence of the relevant events notwithstanding its own defaults.'

In view of the foregoing comments, it is considered likely that the Malmaison approach would be the one that is most appropriate for use with SBCSub/C and SBCSub/D/C.

### What is the 'dot on' principle?

The 'dot on' principle emanates from the case of *Balfour Beatty* v. *Chestermount Properties*.[16]

[14] *Henry Boot Construction Limited* v. *Central Lancashire New Town Development Corporation* (1980) 15 BLR 1.
[15] *Royal Brompton Hospital NHS Trust* v. *Hammond (No. 7)* [2001] 76 Con LR 148, QBD (TCC).
[16] *Balfour Beatty* v. *Chestermount Properties* (1993) 62 BLR 1.

This principle relates to relevant sub-contract events, which occur after any previously fixed completion date and/or those that occur in a period of culpable delay.

In such a situation, the sub-contractor will be entitled to an extension of time on the net method of extension (commonly referred to as the dot on principle), which allows the incremental time lost for the new event to be added back to the previously extended completion date, rather than allowing an extension of time to the date on which the late instructed work is completed.

Therefore, if we take, for example, that a carpentry sub-contractor was due to complete its work by no later than 31 January, but, because of its own default, was still working on site on 20 February, then the period after 31 January would be a period of the sub-contractor's culpable delay.

If the contractor issued an instruction on 20 February for the carpentry sub-contractor to install an additional door; and if that additional door could be procured and installed in four days, then the sub-contractor would be entitled to an extension of four days onto the date of 31 January (on the dot on principle) rather than an extension of four days onto the date of 20 February (the date when the instruction was issued).

In general terms, the dot on principle mirrors the Malmaison approach outlined above.

# Practical completion and lateness

## Date of practical completion

### What is practical completion?

SBCSub/C and SBCSub/D/C do not attempt to define practical completion, which is perhaps unfortunate given the lack of a clear definition arising from the court cases that have dealt with this matter.

In general terms, the following may be considered as being the correct analysis of practical completion:

(1) The works can be practically complete notwithstanding that there are latent defects.
(2) A certificate of practical completion may not be issued if there are patent defects. The defects liability period (rectification period) is provided in order to enable defects not apparent at the date of practical completion to be remedied.[17]
(3) Practical completion means the completion of all the construction work that has to be done.[18]
(4) However, the architect (in this case the contractor) has a discretion to certify practical completion where there are very minor items of work left incomplete, on *de minimis* principles.[19]

---

[17] *Jarvis & Sons v. Westminster Corp.* [1970] 1 WLR 637; *H.W. Nevill (Sunblest) v. William Press* (1981) 20 BLR 78.
[18] *Jarvis & Sons v. Westminster Corp.* [1970] 1 WLR 637.
[19] *H.W. Nevill (Sunblest) v. William Press* (1981) 20 BLR 78.

Further explanation of the above analysis is given below.

(1) Latent defects are defects that are effectively dormant at the date of practical completion. For example, a pipe connection may not have been installed properly. At the date of practical completion, this incorrectly installed pipe connection does not cause any problems, and effectively remains dormant. However, one week after the practical completion date the pipe starts leaking because of the incorrectly installed pipe connection. It is quite clear that the leak started because of the incorrectly installed pipe connection, and it is equally clear that the incorrectly installed pipe connection was a defect. However, at the date of practical completion the defect was not apparent and was effectively dormant, or, in other words was a latent defect. It is well known that defects may appear during the rectification period (previously known as the defects liability period, which is dealt with in more detail in Chapter 6), and the above analysis simply states the fairly obvious point that the works can be practically complete notwithstanding that there may be latent defects.

(2) Patent defects are defects that are apparent at the date of practical completion. Taking the above example of an incorrectly installed pipe connection, if this incorrectly installed pipe connection was apparent at the date of practical completion (perhaps because the pipe was leaking at that time) this would be a patent defect and, on the above analysis, practical completion cannot be achieved.

(3) Despite the above, the contractor has discretion to accept practical completion where there are very minor items of work left incomplete, on *de minimis* principles. Therefore, based upon the above analysis, the contractor could, in theory, accept that practical completion had been achieved if a pipe run and tap were missing entirely (particularly if this pipe run and tap was *de minimis* in the context of the overall project) but could not accept that practical completion had been achieved if the pipe run and tap had been installed but the joint between the pipe and the tap was leaking.

In reality, the contractor's discretion is wider than the above analysis, and he may agree that practical completion has been achieved when he is of the opinion that the sub-contract works are practically complete. Additionally, a 'qualified' practical completion is often agreed by a contractor, whereby a list of those items of work where practical completion has not been achieved is issued at the same time as practical completion is confirmed for the balance of the works.

### Date of practical completion

Clause 2.20.1 of SBCSub/C and SBCSub/D/C requires the sub-contractor to notify the contractor in writing when, in his opinion, the sub-contract works as a whole or such works in a section are practically complete, and he has complied sufficiently with:

- clause 2.24 (i.e. the provision of as-built drawings, etc.) in respect of SBCSub/D/C only; and
- clause 3.20.4 (i.e. information for the health and safety file) in respect of both SBCSub/C and SBCSub/D/C.

The important point to note from the above is that this notice relates to the sub-contractor's *opinion* that the works are complete, and his *opinion* that he has complied *sufficiently* with the provision of as-built drawings, etc. and information for the health and safety file.

Because practical completion gives the sub-contractor certain benefits (e.g. release of retention (where applicable); commencement of rectification period; cessation of sub-contract period on site in terms of any liability for culpable delays and liability for contractor's loss and/or expense; commencement of liability period, etc.) then it may well be the case that the sub-contractor has a vested interest in making the sub-contract practical completion date as early as possible, and it may therefore be that the sub-contractor's opinion may be influenced by these factors.

With this in mind, it is important that the contractor notes that clause 2.20.1 of SBCSub/C and SBCSub/D/C states that 'if the Contractor does not dissent in writing to the sub-contractor's notice as given above within 14 days of receipt of same, practical completion of such work shall be deemed for all the purposes of this sub-contract to have taken place on the date so notified.'

If the contractor does dissent, and he must provide reasons for this (refer to clause 2.20.1 of SBCSub/C and SBCSub/D/C), then clause 2.20.2 of SBCSub/C and SBCSub/D/C states that as soon as the contractor is satisfied that the works are practically complete and he is satisfied that the requirements of the above noted clauses 2.24 and 3.20.4 (as appropriate) have been sufficiently complied with, then he is to notify the sub-contractor in writing of this fact as soon as practicable, and for all of the purposes of the sub-contract practical completion will be deemed to have taken place on the date notified by the contractor.

If the sub-contractor does not agree with the contractor's opinion in respect of the date of practical completion, and if the contractor and the sub-contractor are unable to reach an agreement regarding the date of practical completion, then clause 2.20.2 of SBCSub/C and SBCSub/D/C makes clear that either party is free to take the dispute regarding the date of practical completion to be determined by the dispute resolution procedures (e.g. adjudication, etc.) in the sub-contract.

Clause 2.20.2 of SBCSub/C and SBCSub/D/C also makes clear that the date of practical completion of the sub-contract works must not, under any circumstances, be any later than the certified date of practical completion of the main contract works or any relevant section.

In line with clause 2.20.2 of SBCSub/C and SBCSub/D/C, the contractor is to notify the sub-contractor in writing of the certified date of practical completion of the main contract works or any relevant section.

## Failure of the sub-contractor to complete on time (refer to clause 2.21 of SBCSub/C and SBCSub/D/C)

If the sub-contractor fails to complete the sub-contract works within the relevant period for completion, and the contractor gives the requisite notice within a reasonable time after the expiry of that period, the sub-contractor is required to pay the contractor his loss and expense resulting from the failure.[20]

This loss and expense may include:

- the contractor's loss and expense;
- the contractor's other sub-contractors' loss and expense;
- (where applicable) any liquidated damages suffered by the contractor under the main contract. (For commentary on damages and liquidated damages generally, refer to Chapter 9.)

An alternative to the sub-contractor paying for the above is for the sub-contractor to allow the amount of the contractor's loss and expense to be deducted from payments otherwise due to it.

If the contractor follows this latter option then it is important that the withholding notices required by clauses 4.10.3 and 4.12.3 of SBCSub/C and SBCSub/D/C are issued timeously (refer to Chapter 8).

With regard to the above matter, reference should also be made to clause 4.21 of SBCSub/C and SBCSub/D/C (dealt with in Chapter 9) which deals with the recovery by the contractor of any direct loss and/or expense caused to the contractor where the regular progress of the main contract works is materially affected by an act, omission or default of the sub-contractor, or any of the sub-contractor's persons.

---

[20] Refer to clause 4.21 of SBCSub/C and SBCSub/D/C (dealt with in Chapter 9) which deals with the recovery by the contractor of any direct loss and/or expense caused to the contractor where the regular progress of the main contract works is materially affected by an act, omission or default of the sub-contractor, or any of the sub-contractor's persons.

# Chapter 6
# Defects, Design Documents and Warranties

## Defects

A defect is an imperfection, a shortcoming or a failing.

In construction industry terms this may, for example, be paint flaking from walls, a room thermostat not operating correctly, or a water pump not operating at all.

## Sub-contractor's liability

Clause 2.22 of SBCSub/C and SBCSub/D/C requires the sub-contractor to make good at his own cost and in accordance with any direction of the contractor all defects, shrinkages, and other faults in the sub-contract works or in any part of them due to materials or workmanship not in accordance with the sub-contract.

Clause 2.22 of SBCSub/D/C adds that the sub-contractor's liability extends to any failure of his to comply with his obligations in respect of the sub-contractor's designed portion.

Although the above clauses impose obligations upon the sub-contractor, it is generally considered that most defects liability clauses (including those clauses referred to above) will be construed so as to give the sub-contractor the *right* (as well as the obligation) to remedy defects which come within the remit of the clause. If a contractor did not give notice to the sub-contractor to clear defects, but simply cleared the defects with his own (or other) resources, he may not be able to recover more than the amount that it would have cost the sub-contractor to clear the defects.[1]

Clause 2.22 of SBCSub/C and SBCSub/D/C does not state the period during which the sub-contractor's liability for the rectification of defects applies.

However, it is submitted that (in line with clause 4.15.2 of SBCSub/C and SBCSub/D/C) this would be the period noted as the rectification period in the main contract, which, unless specifically stated otherwise in a numbered document, would, by default, be six months (although, in reality, it is quite common for services installations, in particular, to have a twelve-month rectification period).

---

[1] *Pearce & High Ltd* v. *Baxter* [1999] BLR 101; *William Tomkinson* v. *Parochial Church Council* (1990) 6 Const LJ 319.

The rectification period commences from the date of practical completion of the main contract works (or a section of it) and not from the date of practical completion of the sub-contract works (or a section of it).

Therefore, if a sub-contractor achieves practical completion some considerable time before the contractor achieves practical completion, the sub-contractor's liability for defects could be extended for a period considerably longer than the rectification period (in terms of months) in respect of the main contract works.

This latter point could be particularly significant in the situation where a sub-contractor is relying on manufacturer's warranties to cover for the rectification period of, say, 12 months, when (because of the above situation) the period of liability exceeds 12 months.

Under clause 2.38.1 of the main contract conditions it is noted that a list of defects to be rectified by the contractor is to be provided by the employer to the contractor not later than 14 days after the expiry of the rectification period, and adds that no instructions for the rectification of defects must be issued in respect of any defect lists issued after the expiry of that 14-day period.

Even though no such express provision is provided for under SBCSub/C or SBCSub/D/C it is considered that similar implied conditions would equally apply in respect of the sub-contractor.

SBCSub/C and SBCSub/D/C do not stipulate how often defects lists may be provided during the rectification period, nor do they provide for any timetable or protocol for the clearance of the defects, etc. that are listed. It is for this reason that contractors may wish to address (in a numbered document) how defects should be prioritized and the timetable to be complied with, in respect of such prioritized defects.

Another interesting point is that in earlier editions of the JCT main contract form, the contractor's liability for defects, shrinkages and other faults in the main contract works was specifically stated to include for defects, shrinkages and other faults resulting from frost occurring before the date of practical completion of the main contract works. The JCT Standard Building Contract no longer allows for such a provision, and therefore no such provision can be implied into SBCSub/C or SBCSub/D/C.

Whether this factor will have any impact on the contractor's liability for defects (and the sub-contractor's liability for defects) in real terms, is open to question, in any event.

It is considered that if a contractor (or a sub-contractor) argued that it was not liable for shrinkages, for example, because those shrinkages had been caused by frost occurring before practical completion, the counter-argument would be that the contractor (and/or the sub-contractor) had not complied with their workmanship obligations to adequately protect their works before practical completion was achieved.

## Deductions under the main contract conditions

Clause 2.10 of the main contract conditions allows the architect/contract administrator to instruct the contractor not to carry out remedial works arising from the contractor's inaccurate setting out, and adds that if such an instruction is issued an appropriate deduction must be made in the calculation of the ascertained final sum in respect of those errors not amended.

In addition, clause 2.38 of the main contract conditions allows the architect/ contract administrator to instruct the contractor not to make good defects, shrinkages or other faults that have been notified to the contractor on a defects list. If such an instruction is issued then the said clause 2.38 adds that an appropriate deduction must be made in the calculation of the ascertained final sum in respect of the defects, shrinkages or other faults not made good.

Although not stated by either clause 2.10 or clause 2.38 of the main contract conditions, it is considered that, in the normal course of events, the 'appropriate deduction' would be based upon an abatement to the contractor's value for the works rather than on the cost of completing the defects, etc. by others.

In respect of the sub-contractor, clause 2.23 of SBCSub/C and SBCSub/D/C notes that where an 'appropriate deduction' has been made under clause 2.10 or 2.38 of the main contract conditions, then, to the extent that such deduction is attributable to inaccurate setting out by the sub-contractor, or to defects or other faults in the sub-contract works, the deduction (or an appropriate proportion of the deduction) must be borne by the sub-contractor and shall be taken into account in the calculation of the final sub-contract sum or shall be recoverable by the contractor from the sub-contractor as a debt.

## Sub-contractor's design documents

### As-built drawings

Clause 2.24 of SBCSub/D/C requires the sub-contractor to provide to the contractor (at no charge) the following documents before practical completion of the sub-contract works or such works in any relevant section, which show or describe the sub-contractor's design portion (SCDP) as-built information, and which relate to the maintenance and operation of that portion, including any installations forming part of it.

The documents to be provided are:

- such SCDP documents as may be specified in the sub-contract documents; and/or
- such SCDP documents as the contractor may reasonably require.

The above requirement does not detract from the sub-contractor's obligations under clause 3.20.4 of SBCSub/C and SBCSub/D/C in respect of the health and safety file.

### Copyright and use

Clause 2.25 of SBCSub/D/C sets out the rights/licences given to the contractor and the employer in respect of the copyright in SCDP documents.

*It should be particularly noted that the listed rights are conditional upon the sub-contractor being paid all monies due under the sub-contract in full.*

The copyright in all SCDP documents remains vested in the sub-contractor.

However, the contractor must have an irrevocable, royalty-free, non-exclusive licence with the full right to sub-license the employer, to copy and reproduce the SCDP documents for any purpose relating to the main contract works, including, without limitation, those listed below:

- the construction of the main contract works;
- the completion of the main contract works;
- the maintenance of the main contract works;
- the letting of the main contract works;
- the sale of the main contract works;
- the promotion of the main contract works;
- the advertisement of the main contract works;
- the reinstatement of the main contract works;
- the refurbishment of the main contract works;
- the repair of the main contract works.

The contractor also has an irrevocable, royalty-free, non-exclusive licence, with the full right to sub-license the employer, to copy *but not to reproduce* (which must mean using the SCDP documents as the basis of a similar design issue) the SCDP documents for any extension of the main contract works.

## Collateral warranties

As noted in Chapter 4, it is normally far easier to pursue an action in contract than it is to pursue an action in tort, and it is for this reason that the employer will normally require a collateral warranty to be in place which establishes a direct contractual link between a sub-contractor (particularly those that have a design liability) and the employer (and/or other interested parties).

Clause 2.26 of SBCSub/C and SBCSub/D/C notes that part 2 of the main contract particulars (as annexed to the schedule of information) and/or other sub-contract documents allows for the provision by the sub-contractor to give collateral warranties to:

- a purchaser; and/or to
- a tenant or funder; and/or to
- the employer.

In the case of SBCSub/C and SBCSub/D/C the sub-contractor is to execute and deliver a collateral warranty within 14 days from receipt of a notice from the contractor that identifies the beneficiary and requires execution of a collateral warranty.

Clause 2.26.1 of SBCSub/C and SBCSub/D/C states that if the sub-contractor requires any (reasonable) amendment to the proposed collateral warranty he is to notify the contractor within seven days of receipt of the proposed collateral warranty, of the proposed amendments. The contractor would then obviously need to seek approval for the proposed changes from the employer.

Pursuant to clause 2.26.1 of SBCSub/C and SBCSub/D/C, the sub-contractor is then to execute and deliver the collateral warranty (with any approved amendments, as applicable) within seven days of being notified of the decision of the employer in respect of the request made for any amendments (irrespective of whether the requested amendments were agreed to or not).

Clause 2.26.2 of SBCSub/C and SBCSub/D/C makes clear that where the main contract is executed as a deed (i.e. where a 12-year limitation of liability period applies), the collateral warranties are to be executed as a deed; and where the main contract is executed under hand (i.e. where a six-year limitation of liability period applies), the collateral warranties are to be executed under hand.

SBCSub/C and SBCSub/D/C do not stipulate any particular form of collateral warranty to use; however, SBCSub/G notes that if the collateral warranties are not the standard collateral warranties that the JCT produces (see below), then copies of the required forms of collateral warranties should be annexed to the schedule of information forming part of SBCSub/A or SBCSub/D/A.

Obviously, if the sub-contractor has any queries regarding, or requires any changes to, the terms of the proposed collateral warranty (whether this be the standard warranty or an ad hoc warranty) these should be raised by the sub-contractor before the sub-contract is entered into, since, if not, it will be more difficult for the sub-contractor to insist on the required changes at a later date.

Collateral warranties normally consist of the following parts:

(1) The parties:

    (a) the warrantor, that is the sub-contractor in this case;
    (b) the warrantor's employer, that is the contractor in this case;
    (c) the beneficiary, that is the purchaser and/or a tenant; and/or the funder; and/or the employer.

(2) Recitals
These set the scene so as to enable anyone reading the document to understand its background. They should briefly describe the project, the roles of the parties and the reason for the warranty being entered into.

(3) The warranty
This is the core and most important part of the warranty. It usually begins with words such as 'The [warrantor] warrants as at and with effect from' and then goes on, by means of a number of subclauses, to set out the different obligations of the warrantor.

The most common warranty is that the warrantor has and will continue to perform his obligations under whichever contract the warranty is in respect of.

The normal effect of such warranties is to repeat for the beneficiary (e.g. the employer) the obligations that the warrantor (i.e. the sub-contractor) has to his client (i.e. the contractor).

(4) Supplementary warranties
Matters covered by supplementary conditions include:

    (a) that the warrantor has exercised reasonable skill and care in the selection of materials (a list of deleterious materials is usually included);
    (b) that the works designed by the warrantor will satisfy any performance specification contained in the contract with the warrantor's client;

(c) that the works will be of sound manufacture and workmanship and (for design and build contracts) will satisfy the requirements of the employer.

(5) Qualification

Collateral warranties are often qualified by a number of provisions. It is common to see a provision that the warrantor will have no greater liability to the beneficiary, nor a liability of longer duration, than the warrantor would have had if the beneficiary had been named as the warrantor's client in the principal contract to which the warranty refers. The purpose of such a qualification is to ensure that the warranty does not extend the underlying obligations. This is important to the warrantor, as any extension of his liabilities may fall outside the limits of his insurances.

(6) Professional indemnity insurance

Collateral warranties usually contain provisions that require the warrantor to maintain professional indemnity insurance. The level of cover will be specified and it will also be specified that it must be maintained for six or twelve years from practical completion (normally of the main contract works), provided that the insurance is available at commercially reasonable rates and on reasonable terms. This proviso is sufficient protection to the warrantor and it is not necessary to qualify the requirement by asking it to use best or reasonable endeavours to maintain the insurance.

The warrantor will often be required to produce evidence that the insurance is being maintained and renewed, but care should be taken here, because many policies only allow the disclosure of a certificate from the warrantor's broker confirming the amount of the insurance, the basis of cover, the amount of any excess and the period covered.

One issue that can arise in relation to professional indemnity insurance is whether the insurance covers each and every claim or if it is on an aggregate basis for all claims within any year. Most beneficiaries will be advised not to accept aggregate insurance, as there is a danger that there will be no cover for some claims made in the same year as others because of an aggregate limit in value being reached.

It should also be borne in mind that many professional indemnity insurance policies are provided on a 'claims made' basis which means that the applicable insurance is that which covers the period when the claim was made, not that which covers the period when the alleged defect occurred.

In respect of professional indemnity insurance it is in the interest of both the warrantor and the beneficiary that the terms of the warranty fall within the terms of the warrantor's policy. In this regard it should be noted that the purpose of the professional indemnity insurance is to provide a fund of monies upon which the beneficiary can call in the event that a claim needs to be made. If the terms of the warranty do not fall within the terms of the warrantor's policy, that fund of money will not exist.

A reality of the construction industry is that many smaller contractors and professional consultants have very limited assets and, in the absence of satisfactory insurance, would be unable to meet any significant claim under a warranty.

It is, therefore, in the interests of both the warrantor and the beneficiary that the terms of the warranty fall within the risks usually covered by a professional indemnity insurance policy.

The warrantor obviously does not wish to be faced with claims that it cannot meet, and the beneficiary seeks the comfort of knowing that a source of accessible funds is available. Common sense dictates that it is necessary to strike a balance between these two demands.

Beneficiaries should be aware that there are two common exclusions to warrantor's professional indemnity insurance policies:

(a) obligations in warranties that exceed those taken on in the building contract or appointment; and
(b) obligations that amount to an 'express guarantee'.

(7) 'Step-in rights'
Such provisions are popular where the beneficiary is a funder or freehold purchaser. The purpose of 'step-in-rights' is to allow the beneficiary to take over the contract with the warrantor in the event that either the warrantor has a right to terminate that contract or the beneficiary's own contract has been breached.

Technically speaking, 'step-in' has to take place by means of a novation of the contract to the beneficiary. Many collateral warranties that contain such clauses express such rights in this way, which otherwise may be difficult to enforce.

The main features of a step-in provision are:

(a) a requirement that the warrantor gives notice to the beneficiary if the warrantor has the right to terminate its contract;
(b) a right for the beneficiary to give counter-notice to the warrantor requiring the warrantor to carry on working for the beneficiary;
(c) a right for the beneficiary to give notice to the warrantor requiring the warrantor to work for the beneficiary in place of the warrantor's client (to cover the warrantor's client's breach of its own contract with the beneficiary);
(d) an obligation on the beneficiary, if notice is given to the warrantor, to become responsible for sums payable to the warrantor; and
(e) a right for the beneficiary to require that the contract be novated to a third party instead of itself, but the warranty will usually require the beneficiary to guarantee any such third party.

Any of the above provisions that require a notice to be given should state the period of that notice, particularly as the warrantor has to continue working during the notice period thereby incurring costs that will not be reimbursed, if the beneficiary decides not to step in.

On the other hand the beneficiary (in the same period) will require sufficient time to consider whether to make other arrangements such as selling the development, and/or requiring that the contract be novated to the purchaser.

(8) Copyright licence
Collateral warranties, as a general rule, contain an irrevocable royalty-free copyright provision in favour of the beneficiary. Such a licence enables the beneficiary to reproduce and use the warrantor's design information. The rights of the beneficiary are normally limited to the use of the information for purposes related to the development.

(9)  Contracts (Rights of Third Parties) Act 1999
     The Contracts (Rights of Third Parties) Act is almost invariably excluded in
     most standard and bespoke collateral warranty forms.

   The important factor with collateral warranties, from the warrantor's viewpoint,
is that he should have no greater liability to the beneficiary than he has to his client
and he has the same defences against a claim from the beneficiary as he would
have against his client.
   It is important to note that the terms of a collateral warranty that a party to a
contract is required to give to a third party must be known and/or available at the
time the principal contract is formed. In the event that this is not the case then there
is unlikely to be sufficient certainty over the particular terms requiring the giving
of a collateral warranty, making such a requirement difficult to enforce.
   In respect of the JCT Standard Building Sub-Contract, as noted above, the normal
situation would be that the standard collateral warranties that the JCT produces
would be used. In this case, many of the above issues (where applicable) would be
automatically covered.
   The two standard collateral warranties produced by the JCT are:

• the sub-contractor collateral warranty for a funder (SCWa/F); and
• the sub-contractor collateral warranty for a purchaser or tenant (SCWa/P&T).

   These collateral warranties are each provided with a three-page guidance notes
section.
   In light of the general comments regarding collateral warranties above, the
collateral warranties produced by the JCT are reasonably self-explanatory, and
are not replicated in this book.
   It should be noted, however, that in respect of the JCT standard collateral warranties, the section of the collateral warranty that relates to professional insurance
should be deleted if the sub-contractor has no design liability.

# Chapter 7
# Control of the Sub-Contract Works

## *Assignment and sub-letting*

### Non-assignment

Clause 3.1 of SBCSub/C and SBCSub/D/C states that the sub-contractor must not assign the sub-contract or any rights without the written consent of the contractor.

In law, the general principle is that benefits under a contract may be assigned (that is, legally transferred from the original contracting party to a third party) without the consent of the other party, but burdens under a contract may not. Therefore, without the consent of the contractor, a sub-contractor could not normally assign his liability to complete, as this is a burden, not a benefit.[1]

Strictly speaking, if both the benefit and the burden of a contract are legally transferred to a third party, then that is a novation, and not an assignment. A novation is a tripartite agreement by which an existing contract between party X and party Y is legally discharged and a new contract, usually on the same terms as the first contract, is made between party X and (third) party Z.

Therefore, the reference in clause 3.1 of SBCSub/C and SBCSub/D/C that the sub-contractor must not assign the sub-contract does not, in truth, make a great deal of sense.

However, in the two consolidated appeals of *Linden Gardens* v. *Lenesta Sludge* and *St Martins* v. *Sir Robert McAlpine*[2] Lord Browne-Wilkinson said:

'Although it is true that the phrase "assign this contract" is not strictly accurate, lawyers frequently use those words inaccurately to describe an assignment of the benefit of a contract since every lawyer knows that the burden of a contract cannot be assigned.'

In the light of the above, what clause 3.1 of SBCSub/C and SBCSub/D/C does is to make clear by express terms that, irrespective of the general principle of law, no benefit (or right) (in addition to any burden) under the sub-contract may be assigned to any third party without the written consent of the contractor.

---

[1] For this principle, see *Nokes* v. *Doncaster Amalgamated Collieries Ltd* [1940] AC 1014.
[2] *Linden Gardens* v. *Lenesta Sludge Disposals* [1993] 3 WLR 408; *St Martins Property Corporation Ltd* v. *Sir Robert McAlpine Ltd* [1993] 3 All ER 417.

## Consent to sub-letting

Following on from the non-assignment clause noted above, clause 3.2 of SBCSub/C notes that the sub-contractor must not sub-let the whole or any part of the sub-contract works, without the written consent of the contractor; whilst clause 3.2 of SBCSub/D/C adds, in addition to the above, that the sub-contractor must not sub-let the design for the sub-contractor's designed portion, without the written consent of the contractor.

In neither instance is the contractor's consent to be unreasonably delayed or withheld.

It should be noted that these clauses do not say that the contractor's consent is *not* to be delayed or withheld, but that such consent is not to be *unreasonably* delayed or withheld.

Therefore, if a contractor had serious doubts about the ability of a proposed sub-subcontractor (i.e. a proposed sub-contractor to the sub-contractor) to deal with a particular aspect of the sub-contract works or of the sub-contractor's designed portion, then it may be entirely reasonable for him to delay his consent to the use of the proposed sub-subcontractor until he is provided with documentary evidence (perhaps references, etc.) that he may have requested the sub-contractor to provide to him, before he gives his consent to the use of the proposed sub-subcontractor.

Alternatively, if the contractor had received or was receiving a very poor service from the proposed sub-subcontractor on another project, and therefore had serious doubts that the proposed sub-subcontractor would be capable of carrying out the sub-contract works (or a part of them) to the required standard, then it may be entirely reasonable for him to withhold his consent to the use of the proposed sub-subcontractor entirely.

The second part of clause 3.2 of SBCSub/C makes it quite clear that, even if the contractor does give his written consent to the sub-contractor to sub-let the whole or any part of the sub-contract works, the sub-contractor is to remain wholly responsible for carrying out and completing the sub-contract works in all respects in accordance with the sub-contract; whilst the second part of clause 3.2 of SBCSub/D/C notes, in addition to the foregoing, that the contractor's consent to any sub-letting of design must not in any way affect the obligations of the sub-contractor under clause 2.13.1 of SBCSub/D/C (which relates to the sub-contractor's design liabilities, which is dealt with in detail in Chapter 4) or any other provision of SBCSub/D/C.

## Conditions of sub-letting

There are only two specific conditions of sub-letting to a sub-subcontractor. These are that:

- the sub-subcontractor's employment under the sub-subcontract must terminate immediately upon the termination (for any reason) of the sub-contractor's employment under the sub-contract (as clause 3.3.1 of SBCSub/C and SBCSub/D/C); and

- the sub-contractor must provide for simple interest to be paid to the sub-subcontractor at the level of, and subject to the terms equivalent to those of clauses 4.10.5 and 4.12.4 of SBCSub/C and SBCSub/D/C, in respect of any payment, or any part of it, not properly paid to the sub-subcontractor by the final date for payment. The interest rate stated in SBCSub/C and SBCSub/D/C, being a rate 5% per annum above the official dealing rate of the Bank of England current at the date that a payment due under the sub-contract becomes overdue (as clause 3.3.2 of SBCSub/C and SBCSub/D/C).

Apart from the above, a sub-contractor is at liberty to agree any terms (that are enforceable) with his sub-subcontractor as he wishes. The JCT has actually issued a sub-subcontract form for use in such a situation (denoted as 'SubSub'), but the sub-contractor is not obliged to use that particular form and may use any form that it wishes provided that the two specific conditions of sub-sub-letting as outlined above are met.

## Contractor's directions

Issues relating to the contractor's power to issue directions, the sub-contractor's compliance with directions, the procedure that applies in the event of non-compliance with directions, and matters relating to directions to be in writing, are all dealt with in Chapter 10.

### Person in charge

Clause 3.8 of SBCSub/C and SBCSub/D/C requires the sub-contractor to ensure that at all times during the execution of the sub-contract works he has on site a competent person in charge. Clause 3.8 also adds that any directions given to that competent person in charge will be deemed to have been issued to the sub-contractor.

### Access for contractor and architect/contractor administrator

Clause 3.9 of SBCSub/C and SBCSub/D/C requires the sub-contractor to allow access at all reasonable times to the contractor, to the architect/contract administrator, and to any person authorized by either the contractor or the architect/contract administrator, to any work which is being prepared for or is to be utilized in the sub-contract works.

The only restrictions to this requirement are those that are necessary to protect any proprietary rights. This latter situation may arise where a product involves particular trade secrets and/or the manufacturer wishes to prevent the manufacturing process from becoming known to third parties.

## Opening up and remedial measures

### Directions to open up and test

Clauses 3.10 to 3.15 of SBCSub/C and SBCSub/D/C set out the procedures to be followed in respect of the opening up of works and the remedial measures to works that may be required after the works have been opened up.

Clause 3.10 of SBCSub/C and SBCSub/D/C allows the contractor to issue directions which require the sub-contractor to:

- open up for inspection any work covered up; or
- arrange for or carry out any test of any materials and goods (whether or not already incorporated in the sub-contract works); or
- arrange for, or carry, out any test of any executed work.

To prevent the contractor from using this power indiscriminately, the default position of clause 3.10 of SBCSub/C and SBCSub/D/C is that the cost of any such opening up or testing (including the cost of any resultant making good) shall be taken into account in the final sub-contract sum (i.e. the final account).

However, where the opening up or testing is included in the sub-contract sum in any event, or where the opening up or testing shows that the materials, goods or works are not in accordance with the sub-contract requirements, the cost of such opening up or testing (and the cost of any resultant making good) is to be borne by the sub-contractor.

## Work not in accordance with the sub-contract

In the event that any work, materials or goods are not in accordance with the sub-contract, it is termed 'non-compliant work', and the contractor, in addition to his other powers, may:

- In line with clause 3.11.1 of SBCSub/C and SBCSub/D/C, issue directions requiring the removal from site, or rectification, of all or any of the non-compliant work following the issue of similar instructions under clause 3.18 of the main contract conditions. In any such direction the contractor is to inform the sub-contractor that an instruction under clause 3.18 of the main contract conditions has been issued.

- Also, pursuant to clause 3.11.1 of SBCSub/C and SBCSub/D/C, issue directions requiring the removal from site or rectification of all or any of the non-compliant work (without similar instructions being issued under clause 3.18 of the main contract conditions) but only after the contractor has consulted with the sub-contractor and has had regard to the sub-contract code of practice set out in schedule 1 to SBCSub/C and SBCSub/D/C. The sub-contract code of practice is dealt with later in this chapter.

- After consultation with the sub-contractor, issue directions requiring a variation that is reasonably necessary as a consequence of either of the above directions (as clause 3.11.2 of SBCSub/C and SBCSub/D/C). Clause 3.11.2 makes it clear that, to the extent that such directions are reasonably necessary (as a consequence of either of the directions listed under the first two bullet points above) the sub-contractor will *not* be entitled to be paid for the variation and will not be entitled to an extension of time as a consequence of the issue of the variation.

- Having due regard to the sub-contract code of practice set out in schedule 1 to SBCSub/C and SBCSub/D/C, issue such directions under clause 3.10 of

SBCSub/C or SBCSub/D/C to open up for inspection or to test (as is reasonable in all of the circumstances) to establish to the reasonable satisfaction of the contractor the likelihood or extent (as appropriate to the circumstances) of any further similar non-compliance (refer to clause 3.11.3 of SBCSub/C and SBCSub/D/C).

In this latter case, and irrespective of whether the opening up and/or testing shows the work to be compliant or non-compliant, the sub-contractor will not be paid for the opening up, testing, or subsequent making good; but may be entitled to an extension of time (assuming that the works in question cause a delay to the completion date of the sub-contract works or such works in any section) and any resultant (proven) loss and expense (refer to clause 4.20.2.3 of SBCSub/C or SBCSub/D/C), if the works are found to be compliant.

Despite the above powers of the contractor, the contractor may not issue any of the above directions in respect of materials or goods or workmanship, where the approval of the quality and standards is a matter for the opinion of the architect/contract administrator, unless such direction comprises or reflects an instruction of the architect/contract administrator under the main contract which relates to the sub-contract works (refer to clause 2.4.3 of SBCSub/C and SBCSub/D/C).

## Instructions and deductions under the main contract

Under clause 3.18.2 of the main contract conditions, if any work, materials or goods are not in accordance with the contract then the architect/contract administrator may consult with the contractor and, with the agreement of the employer, may allow all or any of such work, materials or goods to remain; and an appropriate deduction would then be made to the ascertained final (main contract) sum.

Where this consultation relates in whole or in part to the sub-contract works, the contractor is required by clause 3.12.1 of SBCSub/C and SBCSub/D/C to immediately consult with the sub-contractor.

Although the words 'immediately consult with the sub-contractor' are used in the above clause, there is no express requirement that this consultation must be before the architect/contract administrator allows all or any such work, materials or goods to remain. However, by using the word 'consult' (which would normally be taken to mean 'refer to a person for advice', or 'seek permission or approval from'), this would imply that this should be before the architect/contract administrator makes his decision.

When, under clause 3.18.2 of the main contract conditions, the architect/contract administrator allows all or any non-compliant work to remain, the contractor is to notify the sub-contractor in writing and (where applicable) an appropriate deduction would then be made to the final sub-contract sum or, if this was not possible, then the amount of such deduction would be recoverable by the contractor from the sub-contractor as a debt (refer to clause 3.12.2 of SBCSub/C and SBCSub/D/C).

Clause 3.12.3 of SBCSub/C and SBCSub/D/C requires the sub-contractor to comply with any directions of the contractor requiring a variation to the extent that the directions are a reasonably necessary consequence of the architect/contract administrator allowing all or any non-compliant work to remain.

Following the same principle as clause 3.11.2 of SBCSub/C and SBCSub/D/C (as noted earlier within this chapter), if such directions are reasonably necessary then the sub-contractor will not be entitled to be paid for the variation and will not be entitled to an extension of time as a consequence of the issue of the variation.

## Workmanship not in accordance with the sub-contract

Clause 3.13 of SBCSub/C and SBCSub/D/C deals with the situation where 'workmanship' rather than simply 'work' is not in accordance with the sub-contract.

In such a situation, and where there has been a failure to carry out work in a proper and workmanlike manner, and/or in accordance with the health and safety plan, the contractor, in addition to his other powers, may, after consultation with the sub-contractor, issue such directions (whether requiring a variation or otherwise) as are in consequence reasonably necessary. Where the failure to comply is the subject of an instruction issued by the architect/contract administrator under clause 3.19 of the main contract conditions, the contractor is to inform the sub-contractor of this fact when issuing his directions (as clause 3.13.2 of SBCSub/C and SBCSub/D/C).

Again, following the same principle as already noted above, to the extent that such directions are reasonably necessary then the sub-contractor will not be entitled to be paid for the variation but shall not be entitled to an extension of time or to loss and expense as a consequence of the issue of the variation.

## Effect of non-compliant work by others

In the situation where, for any one of the above reasons, the contractor or *another* sub-contractor is required to remove or rectify any of the works, and this necessitates the sub-contractor in question taking down, re-executing, re-fixing and/or re-supplying work that has been properly executed, or materials or goods properly fixed or supplied under the sub-contract, then the sub-contractor is to comply with the contractor's directions requiring the sub-contractor to take down, re-execute, re-fix and/or re-supply such work, provided that such directions are issued before the date of practical completion of the sub-contract works or of such works in any relevant section (as clause 3.14.1 of SBCSub/C and SBCSub/D/C).

Where such directions are issued then:

- In line with clause 3.14.1 of SBCSub/C and SBCSub/D/C, a copy of the direction is to be sent to the sub-contractor whose non-compliant work or failure to comply gave rise to the direction being issued.

- Pursuant to clause 3.14.2 of SBCSub/C and SBCSub/D/C, the sub-contractor must be paid for the work directed on the basis of a fair valuation. This 'fair valuation' may mean that the sub-contract rates are used, or that rates pro rata to the sub-contract rates are used, or that the works would be valued on the basis of 'fair rates and prices', or on the basis of dayworks. All of these alternatives are dealt with in detail in Chapter 10.

  In addition, the sub-contractor would be entitled to an extension of time (assuming that the works in question caused a delay to the completion date of the sub-contract works or such works in any section), and would be entitled to recover any consequential (and proven) loss and expense.

Clause 3.14 of SBCSub/C and SBCSub/D/C is not listed specifically as a relevant sub-contract matter (in respect of loss and expense) under clause 4.20 of SBCSub/C and SBCSub/D/C, and it must therefore be assumed that directions issued under clause 3.14.1 of SBCSub/C and SBCSub/D/C are to be treated as variations, at least for the purposes of clause 4.20.1 of SBCSub/C or SBCSub/D/C.

## Indemnity by sub-contractor

In the situation where there is non-compliant work by the sub-contractor or there is a failure by him to carry out work in accordance with the sub-contract, clause 3.15 of SBCSub/C and SBCSub/D/C requires the sub-contractor to indemnify the contractor in respect of any liability and any costs the contractor incurs as a result of such non-compliance or such failure by the sub-contractor.

## Schedule 1: the sub-contract code of practice

As noted above, the sub-contract code of practice should be referred to when directions in respect of clauses 3.11.1 and 3.11.3 of SBCSub/C and SBCSub/D/C are being considered by the contractor.

This code of practice is contained under schedule 1 of SBCSub/C and SBCSub/D/C.

Although the contractor is not bound to follow the code of practice, if a dispute arose, his case would be greatly weakened unless he could show that the terms of the code had been fully considered before his directions were issued. The purpose of the code is to ensure that directions under clauses 3.11.1 and 3.11.3 of SBCSub/C and SBCSub/D/C are not issued unreasonably.

The code is divided into two parts: A and B.

Part A requires, for the reasonable operation of clauses 3.11.1 and 3.11.3 of SBCSub/C and SBCSub/D/C, that an assessment is made of the non-compliance, the likelihood of it affecting other parts of the sub-contract works (whether already constructed or not) and the reasons for it, including:

- whether the non-compliance is in a primary structural element; or
- whether it indicates an inherent weakness that requires selective testing; or
- whether it is a less significant element, where the non-compliance, etc. is statistically expected and can be simply repaired.

In making that assessment there should be regard of all of the circumstances, including:

Part A .1 The nature of the work.

Part A .2 The significance of the non-compliance (including any likely, similar non-compliance) in terms of the safety of the building, its users, adjoining property, the public and compliance with statutory requirements.

Part A .3  Any applicable codes of practice or similar advice issued by responsible bodies, recognized testing and remedial procedures and technical advice obtained by either party.

Part A .4  Any instructions of the architect/contract administrator (or any decision by him not to issue instructions).

Part A .5  The level of supervision and control of the sub-contract works exercised by the sub-contractor.

Part A .6  Relevant records of the sub-contractor and (where applicable) of any sub-subcontractor. It is fairly common for sub-contractors and/or sub-sub-contractors (particularly those operating under a quality assurance system) to implement a regime of 'Test and inspection plans'. These plans are designed to show and record what tests were taken in respect of installed works, and also what inspections were carried out. Therefore, in the case of paintwork for example, where a 'mist' coat and two full coats of emulsion paint are to be applied to walls, there would be a record of the inspection taken after each coat of paint and also perhaps some confirmation that a slightly different shading was applied in respect of each coat of paint to make the inspection requirements easier to satisfy. If such test and inspection plans were available, these would be of enormous assistance when assessments of non-compliance are made.

Part A .7  Any failure by the sub-contractor to carry out, or secure the carrying out of, any other required opening up, testing or remedial work.

Part B requires that the parties should endeavour to agree the amount and method of any opening up and testing and/or any remedial work but, in any event, in issuing any directions (other than to the extent that they reflect instructions of the architect/contract administrator) the contractor shall have regard to:

Part B .1  the assessment and the matters referred to in items 1 to 7 of part A above;
Part B .2  the obligations of the contractor under the main contract;
Part B .3  in the case of opening up or testing, the practicability of progressive testing to establish the likelihood of similar non-compliance and, where alternative testing methods are available, the efficacy and the likely time and cost of each;
Part B .4  in the case of remedial measures, the practicability of, the time required for and the consequential costs of rectification as opposed to removal;
Part B .5  any proposals of the sub-contractor;
Part B .6  any other relevant matters.

## Attendance and site conduct

### Attendance and temporary buildings

Clause 3.16.1 of SBCSub/C and SBCSub/D/C states that all items of attendance are required to be provided by the sub-contractor, except for either:

- item 6 of the sub-contract particulars (SBCSub/A and SBCSub/D/A); and/or
- a numbered document referred to in sub-contract particulars item 15 of SBCSub/A (item 16 of SBCSub/D/A).

In respect of item 6 of the sub-contract particulars, a basic list of attendance items that will be provided by the contractor free of charge is provided under sub-contract particulars item 6.1 of SBCSub/A and SBCSub/D/A, and this list is to be amended (as appropriate) and added to (if appropriate). If there is insufficient room on SBCSub/A or SBCSub/D/A form for further attendance items to be listed, further sheets should be used and should be annexed to SBCSub/A or SBCSub/D/A as a numbered document.

It is important that the list of attendance items under item 6.1 is carefully considered and correctly amended, because all attendances (other than those finally listed) are to be provided by the sub-contractor.

Clause 3.16.3 of SBCSub/C and SBCSub/D/C provides that the contractor and the sub-contractor will not and will respectively ensure that the contractor's persons and the sub-contractor's persons do not, wrongfully use or interfere with the plant, ways, scaffolding, temporary works, appliances or other property belonging to or provided by others; also that no party should infringe upon any statutory requirements.

However, clause 3.16.3 also makes it clear that the above points do not affect the rights of the contractor or of the sub-contractor to carry out their respective statutory duties and contractual obligations under the sub-contract or the main contract.

Any misuse of scaffolding or other property belonging to or provided by the contractor is covered by the indemnity given by the sub-contractor to the contractor under clause 2.5.2 of SBCSub/C and SBCSub/D/C.

The list of attendances provided under item 6.1 of the sub-contract particulars in SBCSub/A and SBCSub/D/A (i.e. those attendances that the contractor is to provide free of charge) are:

- provision and erection of all necessary scaffolding and scaffold boards for work over 3.30 metres;
- single phase supply of electricity at 240v for temporary site accommodation;
- single phase supply of electricity at 110v for tools and temporary lighting;
- water supplied at the points identified in the sub-contract documents for temporary site accommodation and for carrying out of sub-contract works;
- use of mess rooms;
- use of sanitary accommodation;
- use of welfare facilities;
- all reasonable non-exclusive use of hoisting facilities that the contractor has on site at the time the sub-contract works are being carried out.
- the benefit of all reasonable watching to be provided by the contractor under the main contract.

These points will now be considered in more detail.

- Provision and erection of all necessary scaffolding and scaffold boards for work over 3.30 metres
  The 'free' issue scaffold does not apply for work that is less than 3.30 metres high, and only relates to work over 3.30 metres high, which (depending on the works to be executed) may mean that the working platform is less than 3.30 metres high.
  In addition to this 'free' attendance item, clause 3.16.2 of SBCSub/C and SBCSub/D/C notes that the contractor, the sub-contractor, the contractor's persons and the sub-contractor's persons in common with all others having a

like right must for the purposes of the main contract works (and only for the purposes of the main contract works) be entitled to use any erected scaffolding belonging to or provided by the contractor or the sub-contractor while it remains erected. This clause relates to:

— both scaffolding belonging to or provided by the contractor and scaffolding belonging to or provided by the sub-contractor;
— scaffolding for work both over 3.30 metres high and under 3.30 metres high;
— the use of the scaffolding in question while it remains erected.

This presumably means that scaffold may be used whilst it remains erected for its primary purpose, but that it will not remain erected thereafter for any secondary purpose. Also, the scaffolding would (by implication) need to be used without alteration (e.g. amending lift heights, strengthening for loadings, etc.)

Clause 3.16.2 of SBCSub/C and SBCSub/D/C also adds that the use of this scaffold by either party is on the express condition that no warranty or other liability on the part of the contractor or the contractor's persons or of the sub-contractor or the sub-contractor's persons, as appropriate, is created or implied under this sub-contract in regard to the fitness, condition or suitability of such scaffolding.

This is a very significant point since it is the responsibility of both the contractor and the sub-contractor to ensure that their operatives on site are working from safe scaffolding.

With this point in mind, it may be the case that the sub-contractor (or the contractor, as appropriate) before using the scaffolding referred to under clause 3.16.2 of SBCSub/C and SBCSub/D/C would need to satisfy itself, at least, that the scaffolding:

— was erected, altered, and dismantled (as appropriate) by competent persons;
— had been regularly inspected;
— was secured to the building or structure (if appropriate) in enough places to prevent collapse;
— was strong enough to allow for the work that was proposed to be carried out from it, and for the materials that were proposed to be stored on it.

- Space for temporary site accommodation

It should be noted that this item relates *only* to the *space* for temporary site accommodation, not for the accommodation itself. It is for the sub-contractor (at his own expense) to provide, erect, maintain, move and subsequently remove all temporary site accommodation that it requires (refer to clause 3.17 of SBCSub/C and SBCSub/D/C).

Site accommodation would normally include site offices, workshops, stores and sheds etc. (but not mess rooms, sanitary accommodation, and welfare facilities which are normally provided by the contractor).

Normally, space on site is fairly restricted, and the space for temporary site accommodation is limited. It is therefore usual for the contractor to agree with the sub-contractor upon the number of pieces of individual site accommodation that will be permitted, the size of the various elements of the site accommodation, and any stacking arrangements that will be necessary.

It may also be necessary for the sub-contractor to move its temporary site accommodation to a different location on site during the course of the work.

If the sub-contractor considers that it is being asked to place its site accommodation in an unsuitable location on site, clause 3.17 of SBCSub/C and SBCSub/D/C give it the right to make a reasonable objection.

Clause 3.17 of SBCSub/C and SBCSub/D/C states that for the purposes of the provision, erection, maintenance, moving and/or removal of temporary site accommodation by the sub-contractor, the contractor agrees to give all reasonable facilities.

It is submitted that in the context of the above, all reasonable facilities simply means that the contractor will allow the sub-contractor, or give the opportunity to the sub-contractor to provide, erect, maintain, move, and/or remove temporary site accommodation; it does not mean that the contractor will provide any equipment or resources (e.g. cranage or labour) to the sub-contractor for any of the above listed operations.

Item 6.3 of the sub-contract particulars (SBCSub/A and SBCSub/D/A) may have an impact on the construction of the temporary accommodation where the joint fire code applies, and this matter is dealt with in more detail in Chapter 2.

- Single phase supply of electricity at 240v for temporary site accommodation
  This item relates to a point of single phase 240v electricity supply for the sub-contractor's temporary site accommodation. The actual connection from that point of supply to the temporary site accommodation is to be carried out by (or at the expense of) the sub-contractor.

  This point is emphasized by clause 3.17 of SBCSub/C and SBCSub/D/C, which makes clear that this situation not only relates to the electricity supply, but also includes for all other temporary services (which could include, for example, a water supply and a telephone/fax connection). The only difference in respect of the telephone connection would be that the contractor is not required to provide a point of supply *at all*.

- Single phase supply of electricity at 110v for tools and temporary lighting
  This item again relates to a point of single phase 110 volt electricity supply for the sub-contractor's tools and temporary lighting.

  For safety reasons, electricity for temporary supplies being distributed around construction sites is transformed down from 240 volts to 110 volts. All the contractor is required to do is provide a point of supply of 110 volt electricity for the sub-contractor to connect to; the distribution of the electricity to the area of work, etc. is the responsibility of the sub-contractor.

  On large sites, there can be large distances from the point of electricity supply to the area of work, and there could be many sub-contractors trailing their own leads across the site to the same (or very similar) areas of work.

  For this reason it is common for contractors to provide several points of electricity supply (perhaps on each floor level, at the location of any common entrance areas (e.g. staircase enclosures/lift lobbies, etc.).

  However, this should not be taken for granted by a sub-contractor, and it is preferable for the location of the points of electricity supply to be agreed in advance and for that agreement to be recorded on a numbered document referred to in sub-contract particulars item 15 of SBCSub/A (item 16 of SBCSub/D/A).

  In respect of temporary lighting, it is normal to consider this under two categories, namely:

— safety lighting; and
— task lighting.

Safety lighting means lighting that is needed to allow for safe access around the site (so that steps and obstructions, etc. can clearly be seen). Obviously, if each sub-contractor were to provide its own safety lighting, there would be many runs of safety lights all in the same, or very similar, locations. This would be neither sensible (nor cost effective) and for this reason it is again common for contractors to provide safety lighting in all areas, leaving the sub-contractor to provide task lighting only. However, as noted above in respect of the point of electricity supply, this should not be taken for granted by a sub-contractor, and any agreement of this sort should be recorded on a numbered document referred to in sub-contract particulars item 15 of SBCSub/A (item 16 of SBCSub/D/A).

Task lighting means lighting that is required to carry out the sub-contract works. This is frequently portable lights that can be transported by the sub-contractor to the particular areas where works are being executed. Sub-contractors need to carefully consider the type of task lighting used, particularly in situations where finishing trades (for example) are carried out under task lighting but will be inspected for acceptance under higher lux level permanent lighting. (Lux is the SI (Système International) unit of illumination.)

- Water supplied at the points identified in the sub-contract documents for temporary site accommodation and for the carrying out of sub-contract works
  This item relates to a point (or points) of water supply for the sub-contractor's temporary site accommodation and for the execution of the sub-contract works.

  The actual connection from that point of supply to the temporary site accommodation and to the area of the sub-contract work is to be carried out by (or at the expense of) the sub-contractor.

  As noted above in respect of the electricity supply, on large sites, there can be large distances from the point of supply to the area of work, and many sub-contractors could have their own hoses trailing across the site to the same (or to a very similar) area of work.

  For this reason it is common for contractors to provide several points of water supply (perhaps on each floor level, at the location of any common entrance areas (e.g. staircase enclosures/lift lobbies, etc.). However, this should not be taken for granted by a sub-contractor, and it is preferable for the location of the points of water supply to be agreed in advance and for that agreement to be recorded on a numbered document referred to in sub-contract particulars item 15 of SBCSub/A (item 16 of SBCSub/D/A).

  Another point that should be considered by a sub-contractor is, will the water pressure available be adequate to allow it to carry out the sub-contract works or will additional pumping arrangements (which, unless agreed otherwise, will be at the expense of the sub-contractor) need to be provided?

- Use of mess rooms
  A mess room normally means a site canteen.

- Use of sanitary accommodation
  Sanitary accommodation means the site toilets.

- Use of welfare facilities
  Welfare facilities would include, amongst other items, drying rooms, shower rooms, and/or first aid rooms.

- All reasonable non-exclusive use of hoisting facilities that the contractor has on site at the time the sub-contract works are being carried out
Hoisting facilities could be hoists, tower cranes, mobile cranes, access platforms, mast climbers, forklift trucks, or any other form of hoisting facility that the contractor has on site at the time the sub-contract works are being carried out.
    A sub-contractor should particularly note that:

  — It only has reasonable and *non-exclusive* use of hoisting facilities. For this reason it is usually sensible to agree with the contractor a hoisting/lifting register on a daily or weekly basis so that the time periods for any hoisting required by the sub-contractor can be planned in advance.

  — It has no control over the type of hoisting facilities that will be provided by the contractor. This point can be vitally important to a sub-contractor and can seriously affect the period over which the sub-contract works will be executed.
      For example, mast climbers can be prevented from use by relatively low wind levels, and, because on a mast climber the hoisting facility and the working platform are normally one and the same thing, this can affect the flexibility that may be required when carrying out the sub-contract works.
      Therefore, it is sensible for the hoisting facilities to be agreed in advance and for this agreement to be recorded on a numbered document referred to in sub-contract particulars item 15 of SBCSub/A (item 16 of SBCSub/D/A).

- The benefit of all reasonable watching to be provided by the contractor under the main contract
The fact that watching (e.g. security guards) may be provided by the contractor under the main contract does not absolve the sub-contractor from its liability for the loss or damage of its unfixed site materials.

## Site clearance

Site clearance is dealt with under clause 3.18 of SBCSub/c and SBCSub/D/C and under item 6.4 of the sub-contract particulars (SBCSub/A and SBCSub/D/A).
    If all rubbish arising from the carrying out of the sub-contract works is to be disposed of off site (by the sub-contractor) then no entry needs to be made against item 6.4 of the sub-contract particulars. If no entry is made, then the default position is that all rubbish resulting from the carrying out of the sub-contract works is to be disposed of off site by the sub-contractor.
    If there are specific requirements for the manner in which rubbish is to be disposed of, then details should be inserted under item 6.4 of the sub-contract particulars.
    These details could be, for example, that the sub-contractor is to dispose of his rubbish to a central skip to be provided by the contractor; or may be that certain waste (for example used fluorescent tubes which contain PCBs (i.e. polychlorinated biphenyls)) must be disposed of in a particular manner.
    To the extent that the sub-contractor has liability for site clearance (as set out above), clause 3.18 of SBCSub/C and SBCSub/D/C states that the sub-contractor must:

- Clear away all rubbish resulting from his carrying out of the sub-contract works so as to keep access to the sub-contract works clear at all times. (Although not expressly stated, it is considered that it would be implied that access to the main contract works generally would also need to be kept clear.)
- Upon practical completion of the sub-contract works or of such works in each section, clear up and leave those works clean and tidy to the reasonable satisfaction of the contractor together with all areas made available to him (and used by him) for the purposes of carrying out the sub-contract works.

## Other attendance items

In addition to the attendance items listed above, the following are some of the more common attendance items that the contractor and the sub-contractor may wish to discuss and/or agree should be provided by the contractor; with any such agreement being recorded on a numbered document referred to in sub-contract particulars item 15 of SBCSub/A (item 16 of SBCSub/D/A).

- Setting out of the sub-contract works
  This is commonly resolved on the basis of the contractor providing main grid lines and site datum levels, and the sub-contractor setting out the sub-contract works from those main grid lines and site datum levels.

- Unloading
  This is normally the responsibility of the sub-contractor, but if materials or goods are being delivered before a sub-contractor has a presence on site, it is sometimes agreed that the contractor will unload (but will accept no liability for) those particular materials or goods.

- Distribution of materials
  This is normally the responsibility of the sub-contractor, but on occasion, depending on the site circumstances, the contractor may agree to distribute the materials to (or near to) the area of work.

- Space for storage
  In addition to the requirement for a space for temporary site accommodation, the sub-contractor may wish to store materials on site, and may, therefore wish to agree upon the time period that the storage space will be required for, the size of the space required and the proximity of the storage space to the working area.

- Covered/secure store
  In certain situations, the contractor may agree to provide a covered/secure store for high risk items (even though the liability of loss or damage to any of the sub-contractor's unfixed materials would, unless expressly agreed otherwise, remain with the sub-contractor).

- Steps/trestles/towers
  For the avoidance of any possible future disagreement it is wise for the sub-contractor and the contractor to agree which party will provide, erect and remove, steps, trestles and towers (for work that is not more than 3.30 metres high) that are necessary for the execution of the sub-contract works. If this

matter is not discussed then the default position is that it is the sub-contractor's liability to provide, erect and remove, the steps, trestles and towers.

- Personal protective equipment (PPE)
  It would normally be the responsibility of the sub-contractor to ensure that his operatives have adequate and suitable PPE (e.g. boots, safety helmets, goggles, ear defenders, masks, etc.). However, if very special PPE is required, the contractor may agree to provide this.

- Protection
  It would normally be the responsibility of the sub-contractor to provide protection to any of his goods, materials or works that are not fully and finally incorporated into the main contract works.
  If the contractor agrees to provide any protection materials or carry out any protection work, this needs to be confirmed as an additional attendance item.

- Control of or removal of water
  This problem can often be overlooked, but should be discussed and agreed to avoid any future disagreements. This can be a particular problem in situations such as:

  — diamond drilling (where water is used to cool the cutting tool, etc.), particularly where the diamond drilling is carried out in an otherwise finished area of the works;
  — grit or sand blasting (with a water base) where the excess water needs to be removed (without the grit or sand blocking up the drain runs);
  — temporary roofs, where consideration is often not given to the discharge of rainwater into gutters, rainwater down-pipes and drains.

- Materials, fuel, power and/or water for commissioning and testing
  Item 6.1 of the sub-contract particulars specifically notes that materials, fuel, power and/or water for commissioning and testing are excluded as an attendance item that a contractor would normally be expected to provide to a sub-contractor free of charge.
  The sub-contractor may therefore need to allow for the necessary materials, fuel, power and/or water for commissioning and testing within its tender.
  A point that is sometimes overlooked is that three-phase electricity is sometimes required for testing and commissioning (e.g. for lift installations, etc.) *before* the permanent supply has been connected. This factor can have major programme implications.

# Health and safety and CDM requirements

## Health and safety

Clause 3.19 of SBCSub/C and SBCSub/D/C requires the sub-contractor to comply, at no cost to the contractor, with:

- all health and safety legislation relevant to the sub-contract works and the manner in which they are carried out;

- all reasonable directions of the contractor to the extent necessary for compliance by the contractor and the sub-contractor with such legislation as it affects the sub-contract works; and
- within the time reasonably required, any written request by the contractor for information reasonably necessary to demonstrate compliance by the sub-contractor with clause 3.19 of SBCSub/C and SBCSub/D/C.

### What is the relevant health and safety legislation?

Generally, the laws which govern health and safety are not industry specific, but they do relate to many construction activities (including design, where appropriate). There are several Acts of Parliament and regulations that apply.

Some of the principal Acts that deal with health, safety and welfare in construction are:

- Mines and Quarries Act 1954
- Factories Act 1961
- Offices, Shops and Railway Premises Act 1963
- Health and Safety at Work Act 1974
- Control of Pollution Act 1974
- Highways Act 1980
- New Roads and Streetworks Act 1991

The fundamental Act governing health and safety in construction is the Health and Safety at Work Act 1974. This Act has over sixty regulations, but the principal regulations of this Act which affect design and construction are:

- Health and Safety (First Aid) Regulations 1981
- Noise at Work Regulations 1989
- Electricity at Work Regulations 1989
- The Health and Safety Information for Employees Regulations 1989
- Manual Handling Operations Regulations 1992
- Personal Protective Equipment Regulations 1992
- Gas Safety (Installation and Use) Regulations 1994
- Construction (Design and Management) Regulations 1994 (known as the CDM Regulations)
- Reporting of Injuries, Diseases and Dangerous Occurrences Regulations 1995 (known as RIDDOR)
- Construction Health Safety and Welfare Regulations 1996
- Provision and Use of Work Equipment Regulations 1998 (known as PUWER 98)
- The Lifting Operations and Lifting Equipment Regulations 1998 (known as LOLER 98)
- The Control of Major Accident Hazards Regulations 1999 (known as COMAH)
- Management of Health and Safety at Work Regulations 1999
  The original Management of Health and Safety at Work Regulations 1992, were part of the so-called 'six-pack' of regulations that came into force (for the most part) on 1 January 1993 and which had a major impact on the implementation of health and safety in the construction industry.

The other five regulations that formed part of the 'six-pack' were:

— the Workplace Health Safety and Welfare Regulations 1992
— the Display Screen Equipment Regulations 1992
— the Personal Protective Equipment Regulations 1992
— the Manual Handling Regulations 1992
— the Provision and Use of Work Equipment Regulations 1992

- The Chemicals (Hazards Information and Packaging for Supply) Regulation 2002 (known as CHIP)
- The Control of Substances Hazardous to Health Regulations 2002 (known as COSHH)

Some of the more important elements of the above Acts and regulations are described below.

### The Health and Safety at Work Act 1974

The intention of this Act was to make further provision for securing the health, safety and welfare of persons at work, for protecting others against risks to health or safety in connection with the activities of persons at work, for controlling the keeping and use, and preventing the unlawful acquisition, possession and use of dangerous substances, and for controlling certain emissions into the atmosphere.

The first part of section 2 of the Act contains a general statement of the duties of employers (which would include contractors and sub-contractors) to their own employees while at work, and is qualified in subsection 2 which instances particular obligations to:

(1) Provide and maintain plant and systems of work that are safe and without risks to health. Plant covers any machinery, equipment or appliances including portable power tools and hand tools.
(2) Ensure that the use, handling, storage and transport of articles and substances is safe and without risk.
(3) Provide such information, instruction, training and supervision to ensure that employees can carry out their jobs safely.
(4) Ensure that any workshop under his control is safe and healthy and that proper means of access and egress are maintained, particularly in respect of high standards of housekeeping, cleanliness, disposal of rubbish and the stacking of goods in the proper place.
(5) Keep the workplace environment safe and healthy so that the atmosphere is such as not to give rise to poisoning, gassing or the encouragement of the development of diseases. Adequate welfare facilities should be provided.

In respect of the above, and in all of the following regulations, an employer would be a contractor and/or a sub-contractor, if and as appropriate.

Further duties are placed on an employer by:

(1) Section 2(3) of the Act, which requires the employer to prepare and keep an up-to-date written safety policy supported by information on the organization

and arrangements for carrying out the policy. The safety policy has to be brought to the notice of employees; however, where there are five or fewer employees this section does not apply.

The safety policy should consist of three parts:

(a) a general statement of intent;
(b) details of the organization (people and their duties);
(c) details of the practical arrangements (i.e. systems and procedures), for example:

  (i)     safety training;
  (ii)    safe systems of work;
  (iii)   environmental control;
  (iv)    safe place of work;
  (v)     safe plant and equipment;
  (vi)    noise control;
  (vii)   use of toxic materials;
  (viii)  utilization of safety committee(s) and safety representatives;
  (ix)    fire safety and prevention;
  (x)     medical facilities and welfare;
  (xi)    maintenance of records;
  (xii)   accident reporting and investigation;
  (xiii)  emergency procedures.

(2) Section 2(6) of the Act, which requires an employer to consult with any safety representatives appointed by recognized trade unions to enlist their cooperation in establishing and maintaining high standards of safety.

(3) Section 2(7) of the Act, which requires an employer to establish a safety committee if requested by two or more safety representatives.

The employer is not allowed by section 9 of the Act to charge any employee for anything done or provided to meet statutory requirements.

The employees' duties are laid down in sections 7 and 8 of the Act, and these sections state that, whilst at work, every employee must take care for the health and safety of himself and of other persons who may be affected by his acts or omissions.

Also, employees should cooperate with their employer to meet legal obligations, and they must not, either intentionally or recklessly, interfere with or misuse anything, whether plant equipment or methods of work, provided by their employer to meet the obligations under this or any other related Act.

### Health and Safety (First Aid) Regulations 1981

This regulation deals with the requirements for first aid.

### Noise at Work Regulations 1989

This regulation requires employers to take action to protect employees from hearing damage.

### Electricity at Work Regulations 1989

This regulation requires people in control of electrical systems to ensure that they are safe to use and are maintained in a safe condition.

### The Health and Safety Information for Employees Regulations 1989

This regulation requires (amongst other things) that employers display a health and safety information poster.

### The Manual Handling Operations Regulations 1992

These regulations place an obligation on employers to carry out an assessment of manual handling activities undertaken and to reduce any identified risks. Manual handling includes lifting, pulling, pushing, carrying, lowering and turning.

Once the risks have been identified, the appropriate control measures must be identified and used.

### The Personal Protective Equipment Regulations 1992

This regulation places an obligation on employers to assess and review the provision and suitability of personal protective equipment at work; as noted above, personal protective equipment is usually simply referred to as PPE.

An employer must assess the need to provide PPE, but PPE should be provided only when other control measures have been examined and either implemented or dismissed as not being reasonably practicable.

It is the responsibility of an employer to provide PPE and the employer should not charge its employees for the PPE issued.

### Gas Safety (Installation and Use) Regulations 1994

This regulation covers safe installation, maintenance and use of gas systems and appliances in domestic and commercial premises.

### The Construction (Design and Management) Regulations 1994 (known as the CDM Regulations)

These regulations are dealt with in more detail in the next section of this chapter.

### Reporting of Injuries, Diseases and Dangerous Occurrences Regulations 1995 (known as RIDDOR)

These regulations require that certain accidents must be reported to the Health and Safety Executive (usually referred to as the HSE).

These accidents are:

- an accident that is fatal or involves a major injury such as a fracture, amputation, or loss of sight;
- a work-related accident which results in more than three days off work;
- an accident on, or related to, the site that results in a member of the public being killed or sent to hospital.

### Construction Health Safety and Welfare Regulations 1996

These regulations cover a wide range of basic health, safety and welfare issues such as precautions against falls, falling objects, excavations, access routes, emergency procedures, ventilation, heating, lighting, seating and welfare facilities.

### Provision and Use of Work Equipment Regulations, 1998 (known as PUWER 98)

These regulations place an obligation on all employers to carry out an assessment of work equipment being used within a business and to reduce risks, so far as is reasonably practicable.

### The Lifting Operations and Lifting Equipment Regulations 1998 (known as LOLER 98)

These regulations require that lifting equipment provided for use at work is strong and stable enough for the particular use, that it is positioned and installed to minimize any risks, that it is used safely and is subject to on-going inspection.

### The Control of Major Accident Hazards Regulations 1999 (known as COMAH)

This regulation requires those who manufacture, store or transport dangerous chemicals or explosives in certain quantities to notify the relevant authority.

### Management of Health and Safety at Work Regulations 1999

Regulation 3 of the Management of Health and Safety at Work Regulations 1999 places a legal duty on employers to carry out a risk assessment.

### Risk assessment

A risk assessment is simply a careful examination of what, in the works to be carried out, could cause harm to people (including employees, other operatives, and the public).

There are five basic steps to producing a risk assessment:

- look for the hazards;
- decide who might be harmed, and how they might be harmed;
- evaluate the risks arising from the hazards identified and decide whether existing precautions are adequate or if more should be done;
- record the findings;
- review the assessment from time to time and revise it if necessary.

### Method statement

A method statement is a method of control that is usually used after a risk assessment of an operation is carried out.

The method statement is used to control the operation and to ensure that all concerned are aware of the hazards associated with the work and the safety precautions to be taken.

Method statements should be 'project specific' and one of the main criticisms of sub-contractors (and sometimes of contractors) is that they attempt to use 'generic' method statements (rather than project specific method statements).

### The Chemicals (Hazards Information and Packaging for Supply) Regulation 2000 (known as CHIP)

This regulation requires suppliers to classify, label and package dangerous chemicals and provide safety data sheets for them.

### The Control of Substances Hazardous to Health Regulations 2002 (known as COSHH)

This regulation requires employers to control exposure to hazardous substances to prevent illness; both employees and others who may be exposed must be protected.

These hazardous substances include things such as adhesives, paints, cleaning agents, fumes from soldering and welding, biological agents, and carcinogenic chemicals.

To comply with COSHH, there are eight basic steps to follow:

* assess the risks;
* decide what precautions are needed;
* prevent or adequately control exposure;
* ensure that control measures are used and maintained;
* monitor the exposure;
* carry out appropriate health surveillance;
* prepare plans and procedures to deal with accidents, incidents and emergencies;
* ensure employees are properly informed, trained and supervised.

## The CDM Regulations (i.e. the Construction (Design and Management) Regulations 1994)[3]

Clause 3.20 of SBCSub/C and SBCSub/D/C deals with the matter of the CDM Regulations as they apply to the sub-contractor.

Clause 3.20 is only applicable where *all* of the CDM Regulations apply.

Therefore, if the main contract particulars as annexed to the schedule of information in SBCSub/A and SBCSub/D/A state that only regulation 7 (notification of a project) and/or regulation 13 (requirement on designer) apply, then clause 3.20 of SBCSub/C and SBCSub/D/C will not apply.

---

[3] Revised Construction (Design and Management) Regulations are due to be implemented in 2007. The comments in this section relate to the 1994 CDM Regulations only.

However, assuming that *all* of the CDM Regulations apply, then clause 3.20.3 of SBCSub/C and SBCSub/D/C states that the sub-contractor must comply at no cost to the employer or the contractor with all reasonable requirements of the principal contractor relating to compliance by the sub-contractor with the CDM Regulations, and further notes that no extension of time will be given for such compliance.

The CDM Regulations require that health and safety is taken into account and managed throughout all stages of a project, from conception, design and planning, through to site work and subsequently to maintenance and repair of the building.

The CDM Regulations affect everyone who takes part in the construction process: the employer, the designers, the contractors and the sub-contractors.

The regulations introduced two new roles:

- the planning supervisor;[4]
- the principal contractor.

The regulations also introduced:

- the health and safety plan;
- the health and safety file.

The CDM Regulations apply to most common building, civil engineering and engineering construction work, but there are certain exceptions. For example, the CDM Regulations do not apply where the local authority is the enforcing authority for health and safety purposes (i.e. where the project is not reportable to the Health and Safety Executive).

The CDM Regulations apply to all design work carried out for construction purposes (including demolition and dismantling).

As far as contractors and sub-contractors are concerned, the CDM Regulations apply to all demolition and dismantling work. They also apply to other construction work unless:

- the work will last 30 days or less and involve fewer than five people on site at any one time; or
- the work is being done for a domestic client (that is someone who lives, or will live, in the premises where the work is being done).[5] In this case only the duties to notify HSE (regulation 7) and those placed on designers (regulation 13) apply.

The CDM Regulations require that:

---

[4] Revised CDM Regulations, due to be implemented in 2007, are expected to abolish the role of planning supervisor. The planning supervisor is expected to be replaced by a coordinator, whose function will be to advise the employer, designers and contractors on health and safety issues throughout the project, and who must be appointed by the employer before design work begins.

[5] In some instances domestic clients may enter into an arrangement with a developer who carries on a trade, business or other activity. For example, a developer may sell domestic premises before the project is complete. The domestic client then owns the incomplete property, but the developer still arranges for the construction work to be carried out. In this case the CDM requirements apply to the developer.

- The employer should:
  - appoint a planning supervisor and a principal contractor for each project;
  - take reasonable steps to satisfy themselves that the planning supervisor, principal contractor, project designers and any contractors they appoint directly, are competent and adequately resourced to deal with health and safety problems associated with the project;
  - pass on relevant information reasonably available to them about health and safety matters which relate to the project to those who are planning the project (if there is a health and safety file already available, relevant sections of this file should be provided);
  - ensure that construction work does not start unless a suitable health and safety plan has been prepared.

- The designer (which, in certain circumstances, could include a sub-contractor) should:
  - give adequate regard to risk control of health and safety hazards when designing; eliminate hazards where feasible (e.g. if specifying roof lights, specify non-fragile materials); or reduce risks from those hazards that cannot be eliminated;
  - give adequate regard to how the structure can be maintained and repaired safely once built (e.g. specify 'maintenance free' coatings for materials at height to reduce the need/frequency for repainting);
  - ensure the design includes adequate information about health and safety (e.g. if a set sequence of assembly or demolition is required to maintain structural stability then this should be stated);
  - cooperate with the planning supervisor and any other designer involved in the project;
  - provide design information for inclusion in the health and safety plan and the health and safety file.

The term 'designer' includes everyone preparing drawings and specifications for the project, including variations and temporary works designs, and also producing detailed design work to satisfy a performance specification. Where appropriate, designers include architects, consulting engineers (e.g. structural engineers, building services engineers), contractors and sub-contractors.

CDM Regulation 13 (requirements on designers) applies to all construction projects whether or not the exceptions to the regulations apply as detailed above.

Therefore, if a sub-contractor is responsible for design (either in whole or in part) of any construction project, then he will be expected to comply with the duties of a designer under the CDM Regulations as outlined above, and as detailed under CDM Regulation 13.

- The planning supervisor
  The planning supervisor is appointed by the employer. The role of planning supervisor may be taken on as part of a company or as an individual, although, as noted above, the employer must take reasonable steps to satisfy himself that the planning supervisor is adequately resourced to deal with health and safety problems associated with the project.

The identity of the planning supervisor will be provided in the main contract particulars as annexed to the schedule of information in SBCSub/A and SBCSub/D/A.

Clause 3.20.1 of SBCSub/C and SBCSub/D/C notes that if under clause 3.26 of the main contract conditions the employer notifies the contractor of a new appointee as planning supervisor, the contractor must immediately copy that notification to the sub-contractor.

The role of the planning supervisor is to:

— coordinate health and safety during the design and planning phase of the project;
— ensure that the pre-construction stage health and safety plan for the project is produced in time for it to be provided to bidding contractors as part of the selection process;
— give advice about health and safety competence and resources needed for the project;
— ensure (where appropriate) that written notice of the project is given to the Health and Safety Executive;
— collect information for inclusion in the health and safety file which the planning supervisor is to ensure is prepared, before passing the file to the employer on completion of the project.

The CDM Regulations do not require planning supervisors to visit the site or to assess the performance of the principal contractor once construction work has begun.

• The principal contractor
The principal contractor is appointed by the employer to plan, manage and control health and safety during the construction phase of the project.

Site work should not start until the principal contractor has developed a construction phase health and safety plan based upon information provided in the pre-construction health and safety plan.

The plan may need to be developed during the construction phase to take account of changing conditions on site as work progresses or as the design changes.

The principal contractor's health and safety plan should take account of general issues, specific hazards, risk control measures and the general principles of risk assessment.

It should be noted that whilst the contractor would normally be the principal contractor, this may not always be the case.

The identity of the principal contractor will be provided in the main contract particulars as annexed to the schedule of information in SBCSub/A and SBCSub/D/A.

Clause 3.20.1 of SBCSub/C and SBCSub/D/C notes that if, under clause 3.26 of the main contract conditions, the employer notifies the contractor of a new appointee as the principal contractor, the contractor must immediately copy that notification to the sub-contractor.

• Sub-contractors
Sub-contractors are required to help the principal contractor to achieve safe and healthy site conditions. They should cooperate with other sub-contractors

working on the site and provide health and safety information (including risk assessments) to the principal contractor.

- Health and safety competence
  The CDM Regulations require that when sub-letting contracts a contractor should satisfy himself that those who are to do the work:

  — are competent in relevant health and safety issues; and
  — intend to allocate adequate resources, including time, equipment and properly trained workers, to do the job safely and without risks to health.

It is quite common for a sub-contractor to be asked to complete a health and safety competence questionnaire before its tender is considered for acceptance.

### The health and safety plan

The health and safety plan develops with the project and has at least two clear phases:

- The pre-construction stage health and safety plan is associated with design and planning of the project before tendering or contractor selection.
  The purpose of the plan is to ensure information relevant to health and safety is passed on to those who need it. The pre-construction stage health and safety plan needs to be available to possible principal contractors at the start of selection or at the start of the tendering process.
  The plan informs the principal contractor of the health, safety and welfare matters they need to take into account when planning for site work.

- The construction stage health and safety plan is developed by the principal contractor so that it addresses key issues of health, safety and welfare relevant to the project.
  Issues which need to be considered for inclusion in the plan include:

  — How health and safety will be managed during the construction phase.
  — What high risk activities will require risk assessments and method statements to be produced.
  — Information about welfare arrangements.
  — Information on necessary levels of health and safety training for those working on the project and arrangements for project-specific awareness training and refresher training such as toolbox talks.
    A toolbox talk is a method of passing health and safety information on to the site operatives at site level. The toolbox talk is normally given by the site safety representative to a group of operatives on site, and is normally in relation to a particular health and safety issue (for example, the safe use of electric power tools, or the need for leading edge protection, etc.).
  — Arrangements for monitoring compliance with health and safety law.

The construction stage health and safety plan should be developed as far as possible before construction work actually starts, and should then be reviewed as necessary to account for changing project circumstances.

Clause 3.20.2 of SBCSub/C and SBCSub/D/C states that the contractor must ensure that the sub-contractor is immediately supplied with a copy of any development of the health and safety plan by the principal contractor.

### The health and safety file

The health and safety file is a record of information for the employer or the end user.

The planning supervisor ensures that it is produced at the end of the project and that it is then passed to the employer. The file gives details of health and safety risks that will have to be managed during maintenance, repair, renovation or demolition.

Contractors (and sub-contractors) should pass information that may affect the maintenance, repair, renovation or demolition of the finished building to the planning supervisor for inclusion in the file.

The employer should make the file available to those who will work on any future design, construction, maintenance or demolition of the structure (and this may include the contractor or the sub-contractor when they are carrying out defect rectification work).

A health and safety file will often incorporate operating and maintenance manuals and as-built drawings (prepared by the contractor and the sub-contractors), as well as general details of the construction methods and materials used.

Clause 3.20.4 of SBCSub/C and SBCSub/D/C requires the sub-contractor to provide to the contractor, within the time reasonably required by the contractor, such information in respect of the sub-contract works as is reasonably necessary to enable the contractor to comply with clause 3.25.3 of the main contract conditions, which relates to the information reasonably required by the planning supervisor for inclusion in the health and safety file.

To ensure that sub-contractors provide the health and safety file information timeously, contractors may consider amending SBCSub/C and SBCSub/D/C standard conditions to the effect that the first moiety of retention will not be released in respect of the sub-contract works or for a relevant section, until the required health and safety file information has been provided by the sub-contractor for the sub-contract works or for a section of it, as appropriate.

## Suspension of main contract by contractor

### Notification by contractor

Clause 3.21 of SBCSub/C and SBCSub/D/C states that if the contractor under clause 4.14 of the main contract conditions gives the employer a written notice of his intention to suspend the performance of his obligations under the main contract because of non-payment by the employer, he is to copy that notice to the sub-contractor immediately, and if he then actually suspends his obligations under the main contract, he is to notify the sub-contractor immediately.

It is important for the sub-contractor to note that he is not to suspend his performance in respect of the sub-contract works simply because he has been notified by the contractor that the contractor has suspended his performance in respect of the main contract works. If the sub-contractor suspends his performance in such circumstances, this is likely to be found to be a wrongful suspension by the sub-contractor, and this may amount to a repudiation, which, if accepted by the contractor, may lead to the sub-contractor suffering all of the contractor's damages arising from that repudiation (for commentary on repudiation see Chapter 12).

## Directions following suspension

In fact, following the notice from the contractor advising that the contractor has suspended his performance in respect of the main contract works, the sub-contractor should (where possible) simply proceed regularly and diligently with his sub-contract works, until and/or unless he receives an express direction from the contractor (under clause 3.22.1 of SBCSub/C or SBCSub/D/C) to cease the carrying out of the sub-contract works.

If such a direction is issued, the contractor may issue further directions as may be necessary following such cessation. Although SBCSub/C and SBCSub/D/C does not elaborate upon what those 'further directions' may be, it would be reasonable to assume that they would include things such as a requirement that unfinished works are adequately protected, and that the site working area is left secure.

Clause 3.22.2 of SBCSub/C and SBCSub/D/C states that, if the contractor resumes the performance of his obligations under the main contract, the contractor must, if he has directed the sub-contractor to cease carrying out the sub-contract works, direct the sub-contractor to recommence those works and may issue further directions in regard to such recommencement.

Again, although SBCSub/C and SBCSub/D/C does not elaborate upon what those 'further directions' may be, it would be reasonable to assume that they would allow for such things as the removal of protection from unfinished works, and the cleaning down of work left standing during the period of suspension to allow other trades to proceed.

The sub-contractor's position is safeguarded from any suspension due to the above in respect of:

- time – by the inclusion of clause 2.19.6 of SBCSub/C and SBCSub/D/C as a relevant sub-contract event to be considered by the contractor when giving the sub-contractor an extension of time to the completion date or to a sectional completion date of the sub-contract works (the matter of extensions of time is dealt with in Chapter 5);
- loss and expense – by the inclusion of clause 4.20.4 of SBCSub/C and SBCSub/D/C as a relevant sub-contract matter to be considered by the contractor when agreeing the amount of direct loss and expense due to the sub-contractor (the matter of loss and expense is dealt with in Chapter 9).

## *Other provisions*

### Strikes

Clause 3.23 of SBCSub/C and SBCSub/D/C deals with strikes.

This clause says that if the main contract works are affected by a strike, lockout or local combination of workmen affecting any of the trades employed upon them or any of the trades engaged in the preparation, manufacture or transportation of any of the goods or materials required for the main contract works then:

- neither the contractor nor the sub-contractor will be entitled to make any claim upon the other for any loss and expense resulting from such action;
- the contractor must take all reasonable steps to keep the site open and available for the use of the sub-contractor; and
- the sub-contractor must take all reasonable practical steps to continue with the sub-contract works.

Clause 3.23 of SBCSub/C and SBCSub/D/C also notes that nothing in clause 3.23 will affect any other right of the contractor or the sub-contractor under the sub-contract if such action occurs.

Although not elaborated upon in SBCSub/C or SBCSub/D/C, this caveat would probably, for example, protect the parties' respective positions in respect of the provisions for termination under the sub-contract.

## Benefits under the main contract

Provided that an appropriate indemnity and security for costs is provided by the sub-contractor in line with the reasonable requirements of the contractor, clause 3.24 of SBCSub/C and SBCSub/D/C gives the sub-contractor a right to require the contractor to take certain actions under the main contract which may be relevant in the case of contentious opinions and instructions by the architect/contract administrator under the main contract (provided that those actions are relevant to the sub-contract works and are not inconsistent with any other express terms of the sub-contract.)

Examples of such contentious opinions and instructions by the architect/contract administrator could be:

- those under clause 2.4.3 of SBCSub/C and SBCSub/D/C, where the approval of the standards of the quality of materials or goods or of the standard of workmanship is a matter for the opinion of the architect/contract administrator (and not of the contractor); or
- those under clauses 3.11 to 3.14 of SBCSub/C and SBCSub/D/C in respect of opening up and remedial measures.

Any such action as outlined above is to be undertaken at the expense of the sub-contractor.

In an extreme case, a sub-contractor could ask the contractor to obtain for the sub-contractor the benefit, for example, of the power of an arbitrator to open up, review and revise a decision or certificate of the architect/contract administrator.

Such arbitration (although at the expense of the sub-contractor) would only be between the contractor and the employer, and, if the contractor did not agree with the claim that the sub-contractor wished to make, it is difficult to see how such an approach would prove to be successful.[6]

---

[6] For information, it should be noted that a similar clause to the above in the pre-1980 Green Form of Nominated Sub-Contract was considered in the case of *Gordon Durham & Co. v. Haden Young* (1990) 52 BLR 61.

**Certificates under the main contract**

Clause 3.25 of SBCSub/C and SBCSub/D/C requires the contractor, upon receipt of a written request from the sub-contractor, to notify the sub-contractor of the dates that the following documents are issued under the main contract:

- the fixing or confirmation of the completion date for the main contract works or any section;
- any written statement issued under clause 2.33 of the main contract conditions (i.e. partial possession by the employer);
- any certificate issued under clause 2.35 of the main contract conditions (i.e. confirmation that any defects, shrinkages or other faults in the part taken into possession by the employer have been made good);
- the practical completion certificate or each sectional completion certificate for the main contract works;
- each certificate of making good (defects);
- the final certificate.

The completion date for the main contract works is relevant in respect of any liquidated damages which the contractor may be required to pay under the provisions of the main contract, which in certain circumstances the sub-contractor may be liable for.

It should be noted that the date given in a written statement issued under clause 2.33 of the main contract conditions (i.e. partial possession by the employer) is deemed to be practical completion of that portion of the main contract works.

The practical completion certificate date or each sectional completion certificate date in respect of the main contract works is the deemed latest date for the practical completion and/or the sectional completion of the sub-contract works.

Each certificate of making good (defects) provides the latest date for triggering the release of the balance of any retention to the sub-contractor.

The final certificate under the main contract triggers the release of the final payment to be made by the contractor to the sub-contractor.

# Chapter 8
# Payment

## *Payment – generally*

Payment is the lifeblood of the construction industry; it is a well understood fact that the majority of business failures arise from cash flow problems rather than an imbalance between liabilities and assets.

The question of payment to sub-contractors has traditionally been a major bone of contention between sub-contractors and contractors. The contractor's view was for a long time that the now outlawed 'pay when paid'[1] process was perfectly fair as it simply passed the risk of late or non-payment from the employer to the contractor, down to the sub-contractor. This was seen as being simply a transfer of risk down the contractual chain.

On the other hand, the sub-contractor saw the 'pay when paid' process as being a means of protecting the contractor's cash flow at the expense of the sub-contractor's cash flow. It was also viewed by sub-contractors as being entirely unfair and inequitable, as often payments would not be made to sub-contractors simply because the contractor had not itself received payment for reasons that had absolutely nothing to do with the sub-contractor's work. Sub-contractors considered that, following the rule of privity of contract, payments due under their contracts with contractors should not be delayed because of factors outside the realm of their own sub-contract.

The problems associated with the practice of 'pay when paid' were rife within the construction industry for many years, and it was because of these problems (amongst other things) that the Latham Report (*Constructing the Team*, 1994) was produced.

In terms of payment, the Latham Report recommended that a period should be stipulated within which interim payments must be made to all participants in the process (irrespective of whether or not the payer had first received payment from its own employer), failing which the payee would have an automatic right to compensation, involving payment of interest at a sufficiently heavy rate to deter slow payment.

These recommendations were dealt with and were, in effect, incorporated within the Housing Grants, Construction and Regeneration Act 1996.

---

[1] 'Pay when paid' is the phrase used when a contractor inserts a term into its sub-contract with its sub-contractors to the effect that the sub-contractor will not receive payment for works executed, etc. until after the contractor has first received payment from its own employer for the works in question.

The Housing Grants, Construction and Regeneration Act 1996, sets out certain provisions in respect of payment that must be dealt with by the express terms of all construction contracts as follows (noting that section 104 of Part II of the Act, defines 'construction contracts' as being a contract to carry out construction operations; and the meaning of 'construction operations' (as in section 105 of Part II the Act) has been refined through various court cases – refer for example to *Palmers* v. *ABB*[2] and *Homer Burgess* v. *Chirex*[3]) as follows:

- Part II section 109 relates to the entitlement to stage payments;
- Part II section 110 relates to the dates for payment (and to payment notices);
- Part II section 111 relates to the notice of intention to withhold payment (the withholding notice);
- Part II section 112 relates to the right to suspend performance for non-payment;
- Part II section 113 prohibits conditional payment provisions (i.e. outlaws 'pay when paid') except in the case of insolvency of the party making payment to the contractor if this is expressly stated in the sub-contract.

The Act adds that if the express terms of the construction contract do not deal with the above matters then the Scheme (i.e. the Scheme for Construction Contracts (England and Wales) Regulations, 1998) rules will be supplanted into the contract.

In such a situation, the terms under the Scheme only supplant those *non-compliant* express terms within the construction contract; the *compliant* express payment terms remain in place.

In common with nearly all standard forms of building contract and sub-contract, SBCSub/C and SBCSub/D/C deals with the payment terms adequately and there is no need to import the Scheme terms at all. Of course, if the standard terms are amended by a contractor, then the position could be entirely different.

A significant point to note is that interim payments are simply payments on account leading towards a final payment. The fact that a payment on account is made for a variation or a claim does not mean that this cannot be rescinded at a later date.

Therefore, a payment on account could be made in interim payments against a specific variation claimed or against a particular claim made, but no such payment may be made within the final sub-contract sum because the sub-contract does not make provision for the variation or claim in question.

Often sub-contractors are of the view that once a payment has been made against a variation claimed, or against another claim made, this money cannot be rescinded.

However, this view is incorrect.

The important point is simply, payment will only be made if the money claimed is due to be paid under the terms of the sub-contract.

## Sub-contract sum and sub-contract tender sum

### Taking adjustments into account

Clause 4.5 of SBCSub/C and SBCSub/D/C notes that where the conditions provide that an amount is to be taken into account in the calculation of the final

---

[2] *Palmers* v. *ABB Power Construction* [1999] BLR 426.
[3] *Homer Burgess* v. *Chirex* [2000] BLR 124.

sub-contract sum, then, as soon as the amount is ascertained in whole or in part, the ascertained amount must be taken into account in the next computation of an interim payment.

This clause has been included to prevent contractors from suggesting that an amount that is to be taken into account in the calculation of the final sub-contract sum can only be paid within the final sub-contract sum.

Despite this clause, there is nothing to prevent contractors from making payments in interim payments for variations claimed or other claims made for cash flow or 'political' reasons that are not eventually included within the final sub-contract sum.

Certain contractors may make payments during the currency of the works (for items for which they may have no intention of including a value in the final sub-contract sum) either to prevent a dispute developing with the sub-contractor during the progress of the works, or to assist the sub-contractor's cash flow during the currency of the works, thereby preventing these issues from affecting the progress of the works on site.

The logic being followed is that it is better (from a contractor's perspective) for an argument about money to occur *after* the work has been completed on site, than *before* the work has been completed.

## What methods are used to calculate the final sub-contract sum under SBCSub/C and SBCSub/D/C terms?

The sub-contract can be placed on the basis of either a sub-contract sum (article 3A) or a sub-contract tender sum (article 3B). This is indicated by the footnote to Article 3 which says 'Complete Article 3A or 3B as applicable, and delete the Article not required.'

When the sub-contract sum is used this means that the final sub-contract sum must be calculated on an adjustment basis (refer to clause 4.1.3 of SBCSub/C and SBCSub/D/C), and when the sub-contract tender sum is used this means that the final sub-contract sum must be calculated on a remeasurement basis (refer to clause 4.2.2 of SBCSub/C and SBCSub/D/C).

Clause 4.7.1 of SBCSub/C and SBCSub/D/C makes it clear that both the sub-contract sum and the sub-contract tender sum are exclusive of VAT (value added tax), and adds that in relation to any payment made to the sub-contractor under the sub-contract, the contractor must pay the amount of VAT properly chargeable in respect of that payment.

## Adjustment basis

The adjustment basis is set out under clause 4.1 of SBCSub/C and SBCSub/D/C, and this represents what is generally referred to as a lump sum contract.

In this context, a lump sum contract is a contract to complete specified works for a lump sum price, but payment would still normally be paid if the specified works were not entirely complete. A lump sum contract is different to an 'entire' contract, which rarely applies in any event, where payment is only made if the contract works are *entirely* complete.

Under the adjustment basis, the quality and quantity of work included in the sub-contract sum is that set out in the bills of quantities (and the sub-contractor's design portion documents where SBCSub/D/C applies), or where there are no bills of quantities, the quality and quantity of work is that included in the sub-contract documents taken together.

If the contractor provides quantities and these are contained in the sub-contract documents, then the quantity of work contained in the sub-contract sum must be in line with those quantities (as clause 4.1.1 of SBCSub/C and SBCSub/D/C).

When the adjustment basis applies, the sub-contract sum must not be altered in any way other than in accordance with the express provisions of SBCSub/C and SBCSub/D/C conditions (e.g. by the issue of a variation instruction). This approach is in line with the traditional approach of a lump sum contract, with additions and/or omissions for variations.

Importantly, it must be noted that, in line with clause 4.1.2 of SBCSub/C and SBCSub/D/C, other than for any errors or inadequacies in the bills of quantities (as noted under clause 2.9 of SBCSub/C and SBCSub/D/C), any error, whether an arithmetic error or otherwise shall be deemed to have been accepted by the parties.

Therefore, if there is an arithmetical or other error in the computation of a bill of quantities item, a bill of quantities page total, a bill of quantities section total, or a bill of quantities summary page, these errors are deemed to have been accepted by the parties and no subsequent adjustment will be made for such errors.

Therefore, for example, a bill of quantities page may have been priced as follows:

Item A    £100.00
Item B    £100.00
Item C    £100.00
Item D    £100.00

Total    £300.00 (which should obviously total £400.00)

If, after works start on site, it is discovered that the above bill of quantities page total should have read £400.00 rather than £300.00, then, pursuant to clause 4.1.2 of SBCSub/C and SBCSub/D/C, no adjustment will be made to the sub-contract sum total in respect of this error, and the sub-contractor will be obliged to carry out bill of quantity items A to D inclusive for £300.00 (rather than £400.00).

## Final sub-contract sum – adjustment basis

Under the adjustment basis, the final sub-contract sum is calculated by making the following adjustments to the sub-contract sum:

(1) The amount stated (either an addition or an omission) by the contractor in his acceptance of any schedule 2 quotation.

(2) The amount (either an addition or an omission) of any variation to any schedule 2 quotation (refer to clause 4.3.1 of SBCSub/C and SBCSub/D/C).

(3) The deduction of all provisional sums (as clause 4.3.2.1 of SBCSub/C and SBCSub/D/C), and the addition of the amount of the valuation of the works

executed by or the disbursements made by the sub-contractor in accordance with the directions of the contractor as to the expenditure of those provisional sums (as clause 4.3.3.3 of SBCSub/C and SBCSub/D/C).

(4) The deduction of the value of all work described as provisional included in the sub-contract documents (as clause 4.3.2.1 of SBCSub/C and SBCSub/D/C), and the addition of the amount of the valuation of the works executed by or the disbursements made by the sub-contractor in accordance with the directions of the contractor as to the expenditure against all work described as provisional included in the sub-contract documents (as clause 4.3.3.3 of SBCSub/C and SBCSub/D/C).

(5) The deduction of the value of all approximate quantities included in any bill of quantities (as clause 4.3.2.1 of SBCSub/C and SBCSub/D/C), and the addition of the amount of the valuation of the works executed by or the disbursements made by the sub-contractor in accordance with the directions of the contractor as to the expenditure against all approximate quantities described in any bill of quantities or in the contractor's requirements (as clause 4.3.3.3 of SBCSub/C and SBCSub/D/C).

(6) The adjustment for the amount due (whether this be an addition or an omission) (as clauses 4.3.2.2 and 4.3.3.2 of SBCSub/C and SBCSub/D/C) of each variation valued in line with the valuation rules, together with the adjustment (whether this be an addition or an omission) of the amount included in the sub-contract documents for any other work which has suffered a substantial change in its conditions due to the variation in question (as referred to in clause 5.11 of SBCSub/C and SBCSub/D/C).

(7) The deduction of those items covered by clause 4.3.2.3 of SBCSub/C and SBCSub/D/C. These items relate to:
   (a) clause 2.23 of SBCSub/C and SBCSub/D/C, that is deductions under the main contract conditions; and to
   (b) clause 3.12.2 of SBCSub/C and SBCSub/D/C, that is deductions in respect of non-compliant work.

(8) In line with clause 4.3.3.1 of SBCSub/C and SBCSub/D/C, the addition of any payment due to the sub-contractor as a result of payments made or costs incurred by the sub-contractor under clauses 2.10 (levels and setting out), 2.21 (fees or charges legally demandable) and 2.23 (royalties and patent rights) of the main contract conditions, as referred to in clause 2.5 of SBCSub/C and SBCSub/D/C.

(9) The amount of any valuation under clause 3.14.2 of SBCSub/C and SBCSub/D/C in respect of non-compliant work by others (as clause 4.3.3.5 of SBCSub/C and SBCSub/D/C).

(10) The amount ascertained under clause 4.19 of SBCSub/C and SBCSub/D/C in respect of the sub-contractor's loss and expense (as clause 4.3.3.4 of SBCSub/C and SBCSub/D/C).

(11) Any adjustment (whether this be an addition or an omission) due to the fluctuation option applicable (as clauses 4.3.2.3 and 4.3.3.6 of SBCSub/C and SBCSub/D/C).

(12) The adjustment (whether this be an addition or an omission) of any other amount (as clauses 4.3.2.4 and 4.3.3.7 of SBCSub/C and SBCSub/D/C).

Adjustments of any other amount may, for example, include costs incurred by the contractor associated with the non-compliance of directions by the sub-contractor (refer to clause 3.6 of SBCSub/C or SBCSub/D/C); or associated with insurance taken out on behalf of the sub-contractor (refer to clause 6.5.4 of SBCSub/C or SBCSub/D/C); or associated with a breach of the joint fire code (refer to clause 6.14.4 of SBCSub/C or SBCSub/D/C). Or, alternatively additional costs incurred by the sub-contractor in complying with a joint fire code that is amended or revised after the sub-contract base date (refer to clause 6.15 of SBCSub/C or SBCSub/D/C).

## Remeasurement basis

The remeasurement basis is set out under clause 4.2 of SBCSub/C and SBCSub/D/C, and this represents what is generally referred to as a remeasurement contract.

Under the remeasurement basis, the quality and quantity of work included in the sub-contract tender sum is that set out in the bills of quantities (and the sub-contractor's design portion documents where SBCSub/D/C applies), or where there are no bills of quantities, the quality and quantity of work will be that included in the sub-contract documents taken together. If the contractor provides quantities and these are contained in the sub-contract documents, then the quantity of work contained in the sub-contract tender sum must be in line with those quantities (refer to clause 4.2.1 of SBCSub/C and SBCSub/D/C).

When the remeasurement basis applies the sub-contract works will be subject to complete remeasurement (as clause 4.2.2 of SBCSub/C and SBCSub/D/C).

## Final sub-contract sum – remeasurement basis

When the remeasurement basis applies, the final sub-contract sum is based upon a valuation under Clause 5.2.2 of SBCSub/C and SBCSub/D/C of a complete remeasurement of the works rather than being based on additions to or omissions from the sub-contract tender sum (refer to clause 4.4 of SBCSub/C and SBCSub/D/C).

This amount is then adjusted by:

(1) The amount stated (either an addition or an omission) by the contractor in his acceptance of any schedule 2 quotation (as clause 4.4.2 of SBCSub/C and SBCSub/D/C).

(2) The amount (either an addition or an omission) of any variation to any schedule 2 quotation (as clause 4.4.2 of SBCSub/C and SBCSub/D/C).

(3) The addition of any payment due to the sub-contractor as a result of payments made or costs incurred by the sub-contractor under clauses 2.10 (levels and setting out), 2.21 (fees or charges legally demandable) and 2.23 (royalties and patent rights) of the main contract conditions as referred to in clause 2.5 of

SBCSub/C and SBCSub/D/C (refer to clause 4.4.1 of SBCSub/C and SBCSub/D/C).

(4) The amount of any valuation under clause 3.14.2 of SBCSub/C and SBCSub/D/C in respect of non-compliant work by others (as clause 4.4.4 of SBCSub/C and SBCSub/D/C).

(5) The deduction of those items covered by clause 4.4.5 of SBCSub/C and SBCSub/D/C; these items relate to:

    (a) clause 2.23 of SBCSub/C and SBCSub/D/C, i.e. deductions under the main contract conditions; and to
    (b) clause 3.12.2 of SBCSub/C and SBCSub/D/C, i.e. deductions in respect of non-compliant work.

(6) The amount ascertained under clause 4.19 of SBCSub/C and SBCSub/D/C in respect of the sub-contractor's loss and expense (as clause 4.4.3 of SBCSub/C and SBCSub/D/C).

(7) Any adjustment (whether this be an addition or an omission) due to the fluctuation option applicable (as clause 4.4.5 of SBCSub/C and SBCSub/D/C).

(8) The adjustment (whether this be an addition or an omission) of any other amount (as clause 4.4.6 of SBCSub/C and SBCSub/D/C).

Adjustments of any other amount may, for example, include costs incurred by the contractor associated with the non-compliance of directions by the sub-contractor (refer to clause 3.6 of SBCSub/C or SBCSub/D/C); or associated with insurance taken out on behalf of the sub-contractor (refer to clause 6.5.4 of SBCSub/C or SBCSub/D/C); or associated with a breach of the joint fire code (refer to clause 6.14.4 of SBCSub/C or SBCSub/D/C), or, alternatively additional costs incurred by the sub-contractor in complying with a joint fire code that are amended or revised after the sub-contract base date (refer to clause 6.15 of SBCSub/C or SBCSub/D/C).

## Calculation of final sub-contract sum

In line with clause 4.6.1 of SBCSub/C and SBCSub/D/C, the sub-contractor is to send to the contractor all documents necessary for the purposes of calculating the final sub-contract sum not later than four months after practical completion of the sub-contract works.

Because of the nature of building works, the 'documents necessary' needs to be judged on a project-by-project basis and cannot be more precisely defined; and, therefore, there may be disputes regarding what documents are actually necessary.

Although this is a concern, and sub-contractors may on occasion feel that more and more documents are being requested by the contractor under the 'documents necessary' requirement, simply to delay agreement of the final sub-contract sum and to delay payment being made, in the modern world of contracting (where partnering in the wider sense and repeat business are to the fore) certainty and budgets are all important.

Therefore it is usually in the best interest of the sub-contractor to provide the 'documents necessary' as requested by the contractor (within reason, of course) as quickly as possible, since by doing so the likelihood of the sub-contractor's account being settled at an early date is increased.

In fact, SBCSub/G makes it clear under paragraph 139 that the dates noted in the sub-contract are 'latest' dates and that it is beneficial if the relevant information is provided as early as possible.

The SBCSub/G also adds the very sensible advice that where there are discrete sections and the sub-contract programme is lengthy, the relevant information in respect of a particular section should be prepared and provided on completion of the sub-contract works in that section, rather than being left to the completion of the sub-contract works as a whole.

In line with clause 4.6.2 of SBCSub/C and SBCSub/D/C, the contractor is to prepare and send to the sub-contractor a statement of the calculation of the final sub-contract sum:

(1) not later than eight months after receipt by the contractor from the sub-contractor of the documents referred to in clause 4.6.1 of SBCSub/C and SBCSub/D/C; and
(2) before the architect/contract administrator issues the final certificate under clause 4.15 of the main contract conditions.

Therefore, in effect, a long-stop period of eight months after receipt by the contractor of the documents referred to in clause 4.6.1 of SBCSub/C and SBCSub/D/C is set for the contractor to send to the sub-contractor a statement of the calculation of the final sub-contract sum.

This long-stop period may of course be much shorter, and *must* be shorter if the period between the date when the sub-contractor sends to the contractor all documents necessary for the purpose of calculating the final sub-contract sum and the date that the architect/contract administrator issues the final certificate under clause 4.15 of the main contract conditions is less than eight months.

Depending upon the terms and/or circumstances under the main contract, the statement of the calculation of the final sub-contract sum may be issued some considerable period (perhaps many months) before the issue of the final certificate under clause 4.15 of the main contract conditions.

The statement of the calculation of the final sub-contract sum does not trigger the release of the final payment (see further comments later in this chapter) but is still an important document.

If there is any dispute regarding the time of issue or the content of the statement of the calculation of the final sub-contract sum, this dispute may be taken to adjudication, for example, and the amount decided within any ensuing adjudication action may have an effect upon the amount actually paid under the final payment.

# Payments

## VAT

VAT is dealt with under clauses 4.7.1 and 4.7.2 of SBCSub/C and SBCSub/D/C.

Clause 4.7.1 of SBCSub/C makes it clear that both the sub-contract sum and the sub-contract tender sum are exclusive of VAT, and adds that in relation to any payment made to the sub-contractor under the sub-contract, the contractor must pay the amount of VAT properly chargeable in respect of that payment.

If, after the sub-contract base date, the supply of any goods and services to the contractor by the sub-contractor becomes exempt from VAT and the VAT therefore cannot be recovered by the sub-contractor, the contractor is to pay the sub-contractor an amount equal to the amount of the VAT input tax on the supply to the sub-contractor of the goods and services which contribute to the appropriate part of the sub-contract works but which, because of the exemption, the sub-contractor cannot recover.

This rule only applies where the supply of any goods and services to the contractor by the sub-contractor becomes exempt from VAT *after* the sub-contract base date. The sub-contract base date is given under item 3 of the sub-contract particulars.

The rules on VAT are governed by the Finance Act 1972, and are operated by the commissioners of Customs and Excise. It is beyond the scope of this book to deal with the VAT rules in detail, but it should be noted that there could be situations whereby the VAT chargeable by a sub-contractor is different to the VAT chargeable by a contractor on the same project.

## Construction Industry Scheme (CIS)

Clause 4.8 of SBCSub/C and SBCSub/D/C notes that the obligation of the contractor to make any payment under the sub-contract is subject to the provisions of the Construction Industry Scheme (CIS).

The CIS sets out special rules for contractors and sub-contractors in the construction industry and how they may be affected when dealing with income tax and national insurance matters.

Most businesses working in construction, whether they are companies, partnerships or self-employed individuals, will be affected by the CIS.

The CIS was originally introduced to combat the problems of perceived tax evasion by 'lump labour' and/or labour-only sub-contractors. Lump labour is a term that has traditionally been used for self-employed casual workers that were paid in lump sums, on the basis that the workers would themselves pay their own income tax liability. Unfortunately, it was apparent that many of these workers did not pay their own income tax liability, and the CIS was introduced to address this issue.

Under the CIS, before sub-contractors can get paid they must hold either a registration card or a sub-contractor's tax certificate. To get a sub-contractor's registration card or a sub-contractor's tax certificate, a sub-contractor must first be registered with the Inland Revenue.

Sub-contractors, who meet certain qualifying conditions, will be issued with sub-contractor's tax certificates by the Inland Revenue. Those who do not meet the qualifying conditions will be issued with registration cards.

Where a sub-contractor holds a registration card only, the contractor must make a deduction from all payments for labour of an amount on account of the sub-contractor's tax and national insurance contribution liability. However, where the sub-contractor holds a sub-contractor's tax certificate, the contractor will pay him, or her, gross (i.e. with no amounts taken off).

There are currently five types of registration cards and certificates used under the CIS:

- CIS4(P) is the *permanent registration card* issued to most sub-contractors. It entitles the holder to be paid with a deduction on account of tax and national insurance contributions. It doesn't have an expiry date but it shows the photograph and signature of the authorized holder, along with their national insurance number.

- CIS4(T) is the *temporary registration card* issued to sub-contractors who don't hold or do not know their national insurance number. It enables the holder to be paid with a deduction on account of tax and national insurance contributions while they obtain a valid national insurance number. This card carries an expiry date instead of the national insurance number, but in other respects is the same as CIS4(P).

- CIS6 is the *sub-contractor certificate* issued to individuals, partners in firms and directors of most companies that meet the required turnover, business and compliance requirements. The certificate shows the photograph and signature of the holder and entitles them to be paid gross (i.e. with no deduction).

- CIS5 is the *sub-contractor certificate* issued to companies that cannot be issued with a CIS6. There is no photograph on the certificate but it bears the company secretary's signature. It entitles the sub-contracting company to be paid gross. (i.e. with no deduction).

- CIS5 (Partner) is the *sub-contractor certificate* issued to one partner in business partnerships that cannot be issued with a CIS6. There is no photograph on the certificate but it bears the signature of the partner nominated to hold the certificate by the firm. It entitles the partnership to be paid gross (i.e. with no deduction).

The CIS is currently under review, and a new version of the scheme is due to come into effect in 2007.

## Issue of interim payments

### When is the first interim payment due?

Clause 4.9.1 of SBCSub/C and SBCSub/D/C notes that the first interim payment will be due on the date for issue of the interim payment certificate under the main contract conditions immediately following the commencement of the sub-contract works.

This date is often referred to as the payment due date.

The payment due date is not the date when the sub-contractor receives payment, but is the date from which the final date of payment (the date when the sub-contractor should actually receive payment) is established. Also the payment due date is not necessarily the date when the sub-contractor's work is valued, since clause 4.13 of of SBCSub/C and SBCSub/D/C makes it clear that the valuation date could be not more than seven days before the payment due date.

Clause 4.9.1 then adds that, in the situation where no date for the issue of interim payment certificates under the main contract is stated in the main contract particulars, the first interim payment to the sub-contractor will become due not later than one month after the date of commencement of the sub-contract works on site.

Therefore, in summary, it would appear that the intention is that the first interim payment to a sub-contractor is due either:

(1) On the date for issue of the interim payment certificate under the main contract immediately following the commencement of the sub-contract works, assuming that such a date has been inserted into the main contract particulars.

It could be argued that if no date for the issue of the interim payment certificates under the main contract has been inserted into the main contract particulars, then that date could be calculated by adding one month periods onto the date of possession (of the main contract works) until a date is arrived at that is immediately after (i.e. within one month of) the commencement of the sub-contract works.

(2) Alternatively, as a general default position, the first interim payment to a sub-contractor must, in any event, become due not later than one month after the date of commencement of the sub-contract works *on site.*

Contrary to the above, it could be argued that the default position only comes into effect when no date for interim payment certificates under the main contract has been inserted into the main contract particulars. Therefore if, contrary to the intention of the un-amended main contract, the parties to the main contract agreed interim payment certificate dates that were more than one month apart, and inserted those dates into the main contract particulars, then the default position (arguably) would not come into effect and the sub-contractor would (arguably) be bound by the 'extended' dates inserted in the main contract particulars.

It is worthy of note that option 1 above relates to the commencement of the sub-contract works generally, whilst the default position under option 2 relates to the commencement of the sub-contract works *on site.*

Of course, sub-contract works may commence with off-site manufacture or design, and therefore under option 1 above, the first payment to a sub-contractor would be linked to the commencement of the sub-contract works (whether this be an on-site or an off-site commencement) whereas the default position is linked to a commencement of the sub-contract works *on site* only.

This factor could have major implications for any sub-contractors that have a large off-site involvement in terms of manufacture, design, or other activities.

### When are subsequent interim payments due?

The payment due dates for subsequent interim payments are on the same date in each month following the date of the payment due date for the first interim payment (refer to clause 4.9.2 of SBCSub/C and SBCSub/D/C). If any of the dates so calculated fall on a non-business day (i.e. a Saturday, Sunday or public holiday) then the payment due date for that month must be the nearest business day to the date in question in that month. It should be noted that the wording of clause 4.9.2 refers simply to the 'nearest business day' rather than to the next business day. Therefore, for example, if the payment due date for any particular month fell on a

Saturday, then it could be argued that the nearest business day would be the Friday before rather than the Monday following that date.

In line with clause 4.9.2 of SBCSub/C and SBCSub/D/C, interim payments continue up to and including the month following the date of practical completion of the sub-contract works as a whole. Therefore, if practical completion of the sub-contract works was achieved in May, for example, then the last 'monthly' interim payment would be the appropriate date in the following month (i.e. June).

That is not to say that interim payments will not continue after that date, but they will not necessarily continue on a monthly basis.

Clause 4.9.2 notes that after the interim payment in the month following practical completion of the sub-contract works, interim payments are only issued as and when further amounts are ascertained as due and payable.

When these interim payments are due, the payment due date will be on the same date of the month (or, as appropriate, the nearest business day to that date) as the pre-practical completion interim payment dates.

Clause 4.9.2 also notes that where the first interim payment becomes due on a date which does not recur in a subsequent month, then the interim payment for that subsequent month will be due on the last day of that month. Therefore, for example, if the first interim payment became due on 31 October, then (as there are not 31 days in November) the next interim payment would become due on 30 November (assuming of course that 30 November is a business day; if it is not. then the date applicable would be the nearest business day to the 30 November).

### When is the final date for payment in respect of interim payments?

As noted above, the payment due date is not the date when the sub-contractor receives payment, but is the date from which the final date of payment is established. The latest date that the sub-contractor can actually receive payment is referred to as the final date for payment.

Under SBCSub/C and SBCSub/D/C (clause 4.9.3) the final date for payment of interim payments is to be 21 days after the date on which those payments become due.

Under the main contract the final date for payment of interim payment certificates is '14 days from the date of that interim certificate' (clause 4.13.1 of the main contract). Whilst there is no specific relationship between amounts due in interim certificates under the main contract and those due under interim payments under the sub-contract, the seven-day differential between the main contract final date for payment and the sub-contract final date for payment should help to ensure (in the normal course of events) that the contractor is in funds from the employer before the payments to sub-contractors need to be made.

## Interim payments – amount due

In line with clause 4.10.1 of SBCSub/C and SBCSub/D/C, the amount due in interim payments is based either on:

- the value of any stage payments agreed between the contractor and the sub-contractor; or, if no stage payments have been agreed

- the contractor's gross valuation as referred to in clause 4.13 of SBCSub/C and SBCSub/D/C.

Irrespective of which of the above options is used, the following deductions are to be made:

(1) in line with clause 4.10.1.1 of SBCSub/C and SBCSub/D/C, any amount which may be deducted and retained as retention in respect of the sub-contract works in accordance with clause 4.15 of SBCSub/C and SBCSub/D/C; and
(2) in line with clause 4.10.1.2 of SBCSub/C and SBCSub/D/C, the total amount previously due as interim payments under the sub-contract.

## Stage payments

Stage payments relate to a list of payments that become due upon the completion of various stages of the sub-contract works. Therefore, for example, a sub-contract sum for £200,000 may be broken down into stage payments as follows:

| | |
|---|---|
| First fix works basement | £20,000 |
| Second fix works basement | £15,000 |
| Final fix works to basement | £10,000 |
| Test and commission basement | £5,000 |
| First fix works ground floor | £20,000 |
| Second fix works ground floor | £15,000 |
| Final fix works to ground floor | £10,000 |
| Test and commission ground floor | £5,000 |
| First fix works first floor | £20,000 |
| Second fix works first floor | £15,000 |
| Final fix works to first floor | £10,000 |
| Test and commission first floor | £5,000 |
| First fix works second floor | £20,000 |
| Second fix works second floor | £15,000 |
| Final fix works to second floor | £10,000 |
| Test and commission second floor | £5,000 |
| TOTAL | £200,000 |

This list of stage payments would need to be incorporated into the sub-contract by being a numbered document listed under and attached to sub-contract particulars item 15 of SBCSub/A (item 16 of SBCSub/D/A).

When the amount due for an interim payment is being considered the contractor only takes into account those items from the list of stage payments that have been entirely completed.

Those items that have not been completed, or those items that have only been partly completed, are discounted in the valuation of the amount due for an interim payment.

Therefore, based upon the above list of stage payments, if, at the time of an interim payment, a sub-contractor had completed all of the first fix works, had carried out 50% of all second fix works, and had carried out 15% of all final fix works, the value of the interim payment (excluding any deductions for retention and/or previous payments, and excluding any VAT adjustment) would be as shown below.

| | |
|---|---|
| First fix works basement | £20,000 |
| Second fix works basement | £NIL |
| Final fix works to basement | £NIL |
| Test and commission basement | £NIL |
| First fix works ground floor | £20,000 |
| Second fix works ground floor | £NIL |
| Final fix works to ground floor | £NIL |
| Test and commission ground floor | £NIL |
| First fix works first floor | £20,000 |
| Second fix works first floor | £NIL |
| Final fix works to first floor | £NIL |
| Test and commission first floor | £NIL |
| First fix works second floor | £20,000 |
| Second fix works second floor | £NIL |
| Final fix works to second floor | £NIL |
| Test and commission second floor | £NIL |
| TOTAL | £80,000 |

It should be noted that the above calculation excludes any second fix and final fix works because none of the second fix works or the final fix works were 100% complete, and therefore no value *at all* is included for these elements of the sub-contract works.

There is no limit as to what stages the parties may agree upon in respect of the breakdown of the sub-contract sum; therefore a breakdown simply stating the stages as being first fix, second fix, final fix, and testing and commissioning, would be equally as valid as a breakdown that was on a room by room, or an area by area basis.

## Gross valuation

### Ascertainment

As noted above, the contractor's gross valuation in respect of an interim payment, must be made no more than seven days before the relevant payment due date.

The gross valuation to be made by the contractor must be the total of the amounts referred to in clauses 4.13.1 to 4.13.6 of SBCSub/C and SBCSub/D/C, less the total of the amounts referred to in clause 4.13.7 of SBCSub/C and SBCSub/D/C.

These clauses are reproduced below:

'4.13.1 The total of the sub-contract work on-site properly executed by the Sub-Contractor (including any work so executed to which clause 5.2 refers, any work so executed for which a Schedule 2 Quotation has been accepted by the Contractor and any variations thereto to which clause 5.3.3 applies), together, where applicable, with any adjustment of that total under Fluctuations Option C, but excluding any restoration, replacement or repair of loss or damage and removal and disposal of debris which under clause 6.7.4 are treated as a Variation. Where a priced Activity Schedule is included in the Numbered Documents, the value to be included in respect of the sub-contract work in each activity to which it relates shall be a proportion of the price stated for the sub-contract work in that activity equal to the proportion of the sub-contract work in that activity that has been properly executed;

4.13.2 The total value of the materials and goods delivered to or adjacent to the Main Contract Works for incorporation therein by the Sub-Contractor but not so incorporated, provided that the value of such materials and goods shall only be included as and from the times that they are reasonably, properly and not prematurely so delivered and have been adequately protected against weather and other casualties;

4.13.3 The total value of any Listed items, provided that the conditions set out in clause 4.14 have been fulfilled;

4.13.4 Any amounts to be included in interim payments in accordance with clause 4.5 as a result of payments made or costs incurred by the Sub-Contractor under clauses 2.10, 2.21 or 2.23 of the Main Contract Conditions, as referred to in clause 2.5 of these Conditions;

4.13.5 Any amounts ascertained under clause 4.19 of these Conditions or any amounts in respect of any restoration, replacement or repair of loss or damage and removal and disposal of debris which under clause 6.7.4 are treated as a Variation;

3.13.6 Any amount payable to the Sub-Contractor under Fluctuations Options A or B, if applicable; and

3.13.7 Any amounts deductible under clause 2.23 or 3.12.2 and any amount allowable by the Sub-Contractor to the Contractor under Fluctuations Options A or B, if applicable.'

Each of these subclauses is commented on further, below.

### Clause 4.13.1. The value of the works

Under this subclause the gross valuation to be made for each interim payment should include:

- The total value of the sub-contract works on site properly executed by the sub-contractor, at the appropriate date.

- The value of any work executed (at the appropriate date) to which clause 5.2 of SBCSub/C and SBCSub/D/C refers.
  This relates to the valuation of variations in line with the terms of the sub-contract (for commentary on the valuation of variations, refer to Chapter 10).

- The value of any work executed (at the appropriate date) for which a schedule 2 quotation has been accepted by the contractor. A schedule 2 quotation relates to the valuation of variations (for commentary on schedule 2 quotations, refer to Chapter 10).

- The value of any work executed (at the appropriate date) in respect of variations to a schedule 2 quotation.

- Where applicable, the adjustment of the total value of the foregoing four items (excluding any restoration, replacement or repair or loss or damage and removal and disposal of debris which under insurance clause 6.7.4 of SBCSub/C and SBCSub/D/C are treated as a variation) under fluctuations option C.

  See further commentary on the various fluctuation options later in this chapter.

  The exclusion of the restoration, replacement, repair, loss, or damage and removal and disposal of debris, which, under insurance clause 6.7.4 of SBCSub/C and SBCSub/D/C, are treated as a variation from the total amount to which fluctuations option C applies, is because the valuation of the work under clause 6.7.4 is on the basis of a current day cost, whilst fluctuation option C applies an increased cost factor on a formula basis based upon a base month stated in the sub-contract. It would not be correct for an increased costs formula to be applied to a current day cost figure.

- Where a priced activity schedule is included in the numbered documents, the value to be included in respect of the sub-contract work in each activity to which it relates must be the price stated for the sub-contract work in respect of that activity, multiplied by the proportion of the sub-contract work in that activity that has been properly executed.

In SBCSub/A, the third recital of SBCSub/A (the seventh recital in SBCSub/D/A) refers to the option of using a priced schedule of activities unless article 3B applies. (Article 3B is used where a sub-contract tender sum applies and the works are subject to complete remeasurement. In such a situation, a priced schedule of activities would not be applicable.)

When a priced schedule of activities is provided, it is to be used in calculating the gross valuation; it generally makes that calculation simpler and quicker than one based on measurement.

The footnote to the third recital of SBCSub/A (the seventh recital in SBCSub/D/A) notes that in the priced schedule of activities each activity should be priced, so that the sum of those prices equals the sub-contract sum excluding provisional sums and excluding the value of work for which approximate quantities are included in any bills of quantities.

An example of a priced activity schedule is given below.

| | **Mechanical services** | £ |
|---|---|---|
| A | Heating installation | 150,000.00 |
| B | Hot water installation | 4,000.00 |
| C | Cold water installation | 10,000.00 |
| D | Gas installation | 1,500.00 |
| E | Mechanical ventilation | 18,000.00 |
| F | Controls and wiring | 6,000.00 |
| G | Testing and commissioning | 3,000.00 |
| H | As installed drawings and O & M manuals | 2,500.00 |
| | **Electrical services** | |
| A | Distribution boards and sub-main cabling | 20,000.00 |
| B | Power installation | 55,000.00 |

| C | Lighting installation including fitting | 70,000.00 |
|---|---|---|
| D | Fire alarm system | 10,000.00 |
| E | Security system | 4,000.00 |
| F | Earthing and bonding | 3,000.00 |
| G | Testing and certification | 1,500.00 |
| H | As installed drawings and O & M manuals | 2,500.00 |
| | **Total** | **361,000.00** |

NB: sub-contract sum £411,000.00 less
provisional sums £50,000.00 = £361,000.00

### Clause 4.13.2 of SBCSub/C and SBCSub/D/C (unfixed materials and goods delivered to or adjacent to the main contract works)

Under this subclause the gross valuation to be made for each interim payment should include the total value of the materials and goods delivered to or adjacent to the main contract works for incorporation by the sub-contractor, provided that the value of such materials and goods will only be included as and from the time that they are reasonably, properly and not prematurely delivered and have been adequately protected against weather and other casualties.

Although this subclause refers to 'materials and goods delivered to *or adjacent* to the Main Contract Works' (italics added), it is considered that this still means that the materials and goods must be on site, or at the very least on land next to the site which the employer owns or has control over. This is because clause 4.13.3 of SBCSub/C and SBCSub/D/C relates specifically to materials and goods stored off site; therefore it is logical that clause 4.13.2 must relate to materials and goods stored on site.

Further, the materials and goods need only be paid for if they are reasonably, properly and not prematurely delivered to site. Therefore, if an electrical sub-contractor (for example) had a ten-month sub-contract period, and light fittings were due to be installed during the tenth month, if the sub-contractor delivered the light fittings to site during its first month on site, it may be considered that the light fittings had been delivered prematurely. If this view was taken by the contractor, then it would not be necessary for the contractor to make payment for the light fittings.

Of course, if the sub-contractor could show that the materials or goods were needed on site at an earlier date because of some legitimate reason (perhaps some design, builder's work, or coordination reason) then the contractor would be liable to make payment for the materials or goods.

In respect of this matter, the sub-contractor is to ensure that the delivered materials or goods are adequately protected against weather and other casualties. Despite this stipulation, it should be noted that the sub-contractor remains responsible for any of his materials and goods until (at the earliest) they are fully, finally and properly incorporated into the main contract works (refer to clause 6.7.2 of SBCSub/C and SBCSub/D/C). This matter is discussed in more depth in Chapter 11.

It should be noted that even if payment is not made for any materials or goods delivered to or adjacent to the main contract works (which are intended for the main contract works) by the sub-contractor, these materials or goods must not be removed from the site unless the contractor with the agreement of the architect/

contractor administrator consents in writing to such removal. This is confirmed under clause 2.15.1 of SBCSub/C and SBCSub/D/C which also makes it clear that such consent shall not be unreasonably delayed or withheld.

Clause 2.15.2 of SBCSub/C and SBCSub/D/C notes that when the value of any of the sub-contractor's unfixed materials or goods on site have been included in any interim payment certificate (under the main contract) under which the amount properly due to the contractor has been paid to him by the employer, those materials or goods will become the property of the employer, and the sub-contractor must not deny that they are and have become the property of the employer. Therefore, in such a scenario, the sub-contractor's unfixed materials and goods would become the property of the employer even though the sub-contractor may not have been paid for them.

Further, clause 2.15.3 of SBCSub/C and SBCSub/D/C notes that if the contractor pays the sub-contractor for any unfixed materials or goods on site, in advance of a value for such materials and goods being included in any interim payment certificate (under the main contract), such materials or goods will become the property of the contractor upon payment being made by him to the sub-contractor.

The above clauses are designed to protect the employer and the contractor from right of title claims by the sub-contractor in respect of unfixed materials and goods on site.

However, do these clauses protect the employer and the contractor from a right of title claim by the sub-contractor's suppliers?

In recent years it has become a fairly standard practice for suppliers of goods to include in their conditions of sale a retention of title or a '*Romalpa*' clause (which is named after the first case where their use was considered).[4]

The purpose of these retention of title clauses is to reserve property in the goods with the seller until the buyer has made payment in full for them, even if they have already been delivered.

The particular importance of these clauses is to protect the seller in those cases where the buyer becomes insolvent before payment has been made.

To be effective, it is necessary for such a clause to be expressly inserted as, otherwise, the assumption would be that property in the goods would pass on delivery at the latest.

This is stated in section 19(1) of the Sale of Goods Act 1979, as follows:

'Where there is a contract for the sale of specific goods or where goods are subsequently appropriated to the contract, the seller may, by the terms of the contract or appropriation, reserve the right of disposal of the goods until certain conditions are fulfilled: and in such a case, notwithstanding the delivery of the goods to the buyer, or to a carrier or other bailee or custodier for the purpose of transmission to the buyer, the property in the goods does not pass to the buyer until the conditions imposed by the seller are fulfilled.'

In respect of the payment of listed items (see below) the sub-contractor needs to provide to the contractor reasonable proof that the property in such listed items is vested in the sub-contractor before payment is made by the contractor. However, no such requirement seems to apply in respect of materials and goods delivered

---

[4] *Aluminium Industries BV v. Romalpa Aluminium Limited* [1976] 1 WLR 676.

to or adjacent to the main contract works. Therefore, it is conceivable that a sub-contractor's supplier could seek to recover unfixed materials and goods delivered to or adjacent to the main contract works that it had not been paid for.

It is for this reason that contractors may wish to receive reasonable proof from the sub-contractor that the property in unfixed materials and goods delivered to or adjacent to the main contract works is vested in the sub-contractor before payment is made for those materials or goods by the contractor.

The above concern only exists whilst the materials or goods are unfixed. Materials and goods which are to be incorporated into the works in almost all construction projects are bound at some stage to become 'attached to the soil', and thus become 'fixtures' to the land.

Once the materials and goods become fixtures they become the property of the owner of the freeholder (normally the employer) in any event. It is sometimes difficult to determine whether materials or goods are fixtures and this is a mixed question of law and fact which is to be determined by a judge when considering all of the circumstances.[5]

### Clause 4.13.3 of SBCSub/C and SBCSub/D/C (unfixed materials and goods off-site (listed items))

Under this subclause the gross valuation to be made for each interim payment should include the total value of any listed items, provided that the conditions set out in clause 4.14 of SBCSub/C and SBCSub/D/C have been fulfilled.

Listed items are defined under clause 1.1 of SBCSub/C and SBCSub/D/C as the items (if any) listed as such under item 7 of the sub-contract particulars SBCSub/A and SBCSub/D/A, as referred to in clause 4.14 of SBCSub/C and SBCSub/D/C.

If there are listed items inserted under item 7 of the sub-contract particulars, these will be those materials and goods for which, subject to clause 4.14 of SBCSub/C and SBCSub/D/C, the contractor will pay the sub-contractor before delivery to the site.

These items are what are commonly known as 'off-site' materials, and the important point to note is that, unless listed items are inserted under item 7 of the sub-contract particulars, the sub-contractor will not get paid for materials stored off site at all.

This factor could have serious consequences for sub-contractors. If, for example, a lift installation sub-contractor needed to purchase and store a lift off site many months prior to its installation on site (perhaps because of some manufacturing time slot issue), but that lift is not included as a listed item, then the sub-contractor would not get paid for the lift until after it has been delivered to site.

In such a situation, the sub-contractor would be financing perhaps 80% or more of the sub-contract value for many months, unless he can come to some arrangement with his lift supplier to defer payment of the lift until the lift is delivered to site.

On a similar basis it should of course be noted that if the contractor wishes off-site materials and goods to be included in an interim payment certificate (in respect of the main contract works) he is required, by the definition of listed items in clause 1.1 of the main contract conditions, to ensure that they are listed as such items by

---

[5] *Holland v. Hodgson* (1872) LR 7 CP 328.

the employer in a list supplied by the contractor and annexed to the contract bills and/or to the specification/work schedules, as appropriate.

The contractor is also to show that the pre-conditions in clause 4.17 of the main contract conditions are satisfied, and this includes providing reasonable proof that the property in such off-site materials and goods is vested in him.

Because of the realities of contracting life, it is unlikely that a contractor will agree to a sub-contractor including listed items in the sub-contract that are not already listed items in the main contract. Therefore, at tender stage it would be sensible for the sub-contractor to determine what materials and goods are listed items under the main contract.

So the first point to note is that, unless a material or goods are included as a listed item in the sub-contract, the contractor will not be liable to pay for the materials or goods when they are stored off site.

However, even if materials or goods are listed items, the sub-contractor is only entitled to be paid if the conditions set out in clause 4.14 of SBCSub/C and SBCSub/D/C have been fulfilled.

These conditions are:

(1) That the listed items are in accordance with the sub-contract. This effectively means that the item must be in line with the specification requirements. However, a listed item could also not be in accordance with the sub-contract if, because of a variation, the work requiring the listed item had been omitted or altered.

(2) That the sub-contractor has provided the contractor with reasonable proof that the property in such listed items is vested in the sub-contractor.

This requirement is designed to quell any right of title claims on the materials or goods by the sub-contractor's suppliers. This matter is dealt with in more detail under clause 4.13.2 of SBCSub/C and SBCSub/D/C above.

The actual requirement under this clause is that after the listed items in question have been included in a gross valuation as being properly due to the sub-contractor, and this amount has been paid to the sub-contractor, the appropriate listed items become the property of the contractor.

(3) That the listed items are insured for their full value against loss or damage under a policy of insurance protecting the rights of the employer, the contractor and the sub-contractor. The insurance needs to be in force from the period when the property (right of title) of the materials or goods passes to the sub-contractor up until the time that those materials or goods are delivered to, or adjacent to, the main contract works. Also, the insurance needs to provide cover for the specified perils, which are:

    (a) fire;
    (b) lightning;
    (c) explosion;
    (d) storm;
    (e) tempest (i.e. a violent storm);
    (f) flood;
    (g) escape of water from any water tank, apparatus or pipes;
    (h) earthquake;

(i)  aircraft and other aerial devices or articles dropped therefrom;
(j)  riot and civil commotion.

These specified perils, however, do not include for the excepted risks.
The excepted risks are defined under clause 6.1 of SBCSub/C and BCSub/D/C as being:

'ionizing radiations or contamination by radioactivity from any nuclear fuel or from any nuclear waste from the combustion of nuclear fuel, radioactive toxic explosive or other hazardous properties of any nuclear assembly or nuclear component thereof, pressure waves caused by aircraft or other aerial devices travelling at sonic or supersonic speeds.'

(4)  That, at the premises where the listed items have been manufactured or assembled, or are stored, the materials or goods are set apart or have been clearly and visibly marked (individually or in sets) by letters or figures or by reference to a predetermined code, and there is in relation to such items clear identification of:

(a)  the contractor and the employer to whose order they are held; and
(b)  their destination as the main contract works.

(5)  That, if item 7 of the sub-contract particulars in SBCSub/A and SBCSub/D/A says that such a bond is required, the sub-contractor has provided from a surety approved by the contractor a bond in favour of the contractor or the employer, as the contractor directs, in the amount set out in the sub-contract particulars and in the terms set out in part 1 of schedule 3 attached to SBCSub/C and SBCSub/D/C.

It is important to note that a bond can only be asked for (in respect of listed items) if this has been noted as being required under item 7 of the sub-contract particulars in SBCSub/A and SBCSub/D/A.

When the sub-contract particulars are completed it is important that in respect of a bond under item 7 (listed items) the wording 'is required' or 'is not required' is deleted as appropriate; and if the wording 'is not required' is deleted, that the surety's maximum liability under the bond is stated as a sum in pounds.

The form of the bond (which has been agreed between the JCT and the British Bankers' Association) is set out under schedule 3 part 1 of SBCSub/C and SBCSub/D/C.

SBCSub/G makes the note that if a sub-contractor encounters difficulties over the terms of the bond with the branch of a British bank from which he seeks the bond, he should request that the branch refer the matter to its head office who should be able to confirm that, as the terms are as agreed by the British Bankers' Association with the JCT, they are acceptable.

The form of bond in question is what is known as an 'on demand' bond.

An on demand bond is a bond that can be called upon without any condition (i.e. purely 'on demand') and the sub-contractor needs to be aware of this fact before entering into the bond.

On demand bonds have been criticized by many commentators as being quite inappropriate for the UK construction industry, and many contractors and sub-contractors refuse to use them.

One of the recommendations of the Latham Report (*Constructing the Team*, 1994) was that bonds 'should not be on demand and unconditional but should have clearly defined circumstances set out in them for being called' (ref: paragraph 5.10.2 of the Latham Report).

Of course a sub-contractor may be in a difficult position in that if he does not agree to enter into the on demand bond the probability is that (if a bond is stated as being required under item 7 of the sub-contract particulars) he will not get paid for off-site materials and may therefore need to finance the listed items until they are delivered to or adjacent to the main contract works. Alternatively, if he does agree to enter into the on demand bond he will be obliged to pay the premium for the bond (which is non-recoverable, but which should be allowed for within the sub-contractor's tender) and may need to suffer the vagaries of claims made against the on demand bond.

When completing the bond as set out in part 1 of schedule 3, the information that needs to be inserted includes:

- The name and registered office address of the surety (i.e. the party that guarantees payment of the bond amount).
- The name of the contractor.
- The name of the sub-contractor
- The location of the project/site.
- The surety's maximum aggregate liability amount under the bond stated as a sum in pounds. This should be the same sum as stated in the sub-contract under item 7 of the sub-contract particulars in SBCSub/A and SBCSub/D/A.
- A 'longstop' date at which time the obligations of the surety under the bond will cease.

The bond then needs to be signed as a deed by or on behalf of the surety.

When making a claim against the bond, the contractor must complete the schedule to the bond, and provide the following information:

- The date of the notice making a claim against the bond.
- The date of the bond.
- The name of the contractor.
- Details of the surety.
- The amount demanded in pounds (which obviously cannot exceed the maximum aggregate liability stated on the bond).
- The address where payment should be sent.

The schedule must be signed by those persons who are authorized by the contractor to act for and on his behalf, and those signatures must be authenticated by the contractor's bankers; then the schedule must be sent to the surety at its registered office.

### Clause 4.13.4 of SBCSub/C and SBCSub/D/C

Under this subclause the gross valuation to be made for each interim payment should include any amounts to be included in interim payments in accordance with clause 4.5 of SBCSub/C and SBCSub/D/C, as a result of payments made or costs incurred by the sub-contractor under clauses 2.10, 2.21 or 2.23 of the main contract conditions, as referred to in clause 2.5 of SBCSub/C and SBCSub/D/C.

The above items relate to payments made or costs incurred by the sub-contractor in respect of:

- clause 2.10 of the main contract conditions relating to levels and setting out;
- clause 2.21 of the main contract conditions relating to fees or charges legally demandable;
- clause 2.23 of the main contract conditions relating to royalties and patent rights.

### Clause 4.13.5 of SBCSub/C and SBCSub/D/C

Under this subclause the gross valuation to be made for each interim payment should include any amounts ascertained under clause 4.19 of SBCSub/C and SBCSub/D/C, or any amounts in respect of any restoration, replacement or repair of loss or damage and removal and disposal of debris which under clause 6.7.4 SBCSub/C and SBCSub/D/C are treated as a variation.

The above relates to:

- any loss and expense payments (which are considered in detail in Chapter 9); and
- the restoration of the sub-contract works and the replacement or repair of any sub-contract site materials that are lost or damaged, and the removal and disposal of debris which under insurance clause 6.7.4 of SBCSub/C and SBCSub/D/C are treated as a variation.

As noted above, under clause 4.13.1 of SBCSub/C and SBCSub/D/C, this variation is treated on a current day cost basis.

The insurance provisions under SBCSub/C and SBCSub/D/C (which includes clause 6.7.4) are dealt with in Chapter 11.

### Clause 4.13.6 of SBCSub/C and SBCSub/D/C

Under this subclause the gross valuation to be made for each interim payment should include any amount payable to the sub-contractor under fluctuations options A or B, if applicable.

This relates to any 'increased costs' payable to the sub-contractor as a result of either fluctuation option A or B.

Fluctuation options A and B are considered in more detail later in this chapter.

### Clause 4.13.7 of SBCSub/C and SBCSub/D/C

Under this subclause the gross valuation to be made for each interim payment should include any amounts deductible under clause 2.23 or 3.12.2 of SBCSub/C and SBCSub/D/C, and any amount allowable by the sub-contractor to the contractor under fluctuations options A or B, if applicable.

This subclause relates to a deduction in respect of inaccurate setting out by the sub-contractor or to defects or other faults in the sub-contract works (refer to clause 2.23 of SBCSub/C and SBCSub/D/C).

This deduction can only be made if a comparable deduction has been made under clause 2.10 or 2.38 of the main contract conditions.

Clause 4.13.7 also relates to any deduction applicable where the architect/contract administrator under clause 3.18.2 of the main contract conditions allows all or any non-compliant work carried out by the sub-contractor to remain in place (refer to clause 3.12.2 of SBCSub/C and SBCSub/D/C).

Clause 4.13.7 finally relates to any decrease in value that is payable to the sub-contractor (which equates to a deduction of the amount that would otherwise be due) as a result of either fluctuation option A or B.

The sub-contractor needs to be aware that when fluctuation options A or B apply this can result in either an increase in prices or a decrease in prices (as, indeed, it can if fluctuation option C applies).

Fluctuation options A and B (and C) are considered in more detail later in this chapter.

## Can a sub-contractor make a payment application, and what is its effect?

Clause 4.13 of SBCSub/C and SBCSub/D/C states that not later than seven days before an interim payment is due the sub-contractor may submit an application setting out what he considers to be the amount of the gross valuation.

If the sub-contractor does submit such an application, and if the contractor agrees with the gross valuation in the sub-contractor's application, he can adopt that valuation for the purposes of determining the gross valuation due in the appropriate interim payment.

If the contractor does not agree with the gross valuation in the sub-contractor's application, the conditions do not require him to advise the sub-contractor why he does not agree with the sub-contractor's application nor does he need to identify the reasons for the difference between his own valuation and that submitted by the sub-contractor.

Although the conditions do not require the contractor to advise the sub-contractor why he does not agree with the sub-contractor's application, or to identify the reasons for the difference between his own valuation and that submitted by the sub-contractor, if the contractor does not provide this information then a dispute may unnecessarily arise.

If a dispute did arise, in such circumstances, the contractor would need to show that he had properly valued the works; and by providing an adequate breakdown at the time of the valuation, the chances of a dispute would obviously be reduced.

It must be noted that the conditions do not require the sub-contractor to submit an application for payment *at all*, the conditions simply state that the sub-contractor *may* submit an application for payment.

If the sub-contractor did not submit an application for payment, *the contractor would still be liable to make payments to the sub-contractor in line with the agreed payment terms.*

However, on a commercial basis it is obviously sensible for the sub-contractor to make an application for payment for (at least) four very good reasons:

(1) It reminds the contractor that the sub-contractor is seeking payment. When a sub-contractor has not been paid because it has not submitted an application

for payment, a contractor will frequently say something along the lines of 'as you did not ask for any money, I assumed you did not want to be paid'.

(2) It advises the contractor of the amount that the sub-contractor expects to be paid. If a sub-contractor does not make an application for payment, a contractor will often take the opportunity to value the works conservatively and (sometimes because of a lack of knowledge) may not include all of the items that the sub-contractor is entitled to be paid for. If the sub-contractor makes an application for payment, the likelihood of this happening is reduced.

(3) If there is a dispute regarding the valuation of the works, the sub-contractor's position is more easily established.

(4) It acts as a regular milestone document for the sub-contractor to summarize its payment entitlement in terms of works executed, variations, dayworks, materials, on or off site, and loss and expense, etc. and also provides a yardstick against which costs can be allocated by the sub-contractor for internal cost control purposes.

## Retention

### Rules governing retentions

Retention is a sum of money held against the value of works executed by the sub-contractor in respect of latent defects on those works. Retention should only apply to latent defects (i.e. defects that are not apparent), since any patent defects (i.e. obvious defects) should be dealt with by way of a reduction (or abatement) in the valuation of the works.

Retention is usually released in two parts (normally two halves); the first part is normally released at the practical completion stage and the second part is normally released when all defects have been cleared at the end of the rectification or defects liability period.

The withholding of retention monies in the construction industry has caused difficulties for many years.

Considerable delay in the release of retention (particularly the second part of retention) is commonplace, and in the event of the insolvency of any party in the payment chain, at any time, retention may be lost altogether.

In the case where retention is held by a contractor against a sub-contractor and the contractor becomes insolvent and ceases trading, the likelihood is that the sub-contractor would not recover any of the retention monies at all. (Although the sub-contractor would be an unsecured creditor, it is highly unlikely that there will be sufficient funds left after the liquidator/receiver has balanced the contractor's total assets with the contractor's total liabilities for the sub-contractor to recover anything other than a very minor percentage of the retention monies held. When speaking about the likely recovery of monies in such a situation, the colloquial phrase in the construction industry is 'ten pence in the pound, in ten years' time' – and that may well be optimistic.)

It was to counter such a situation that the retention was traditionally held in trust. In such a situation, the party holding the retention only had a fiduciary interest in the retention (but with no obligation to invest the retention monies). The effect of such a provision was that the party from whom the retention was being held was entitled to insist on the retention money being paid into a separate bank account, to protect it from the advances of any creditors of the party that was holding the retention.

However, there is no such provision within the JCT Standard Building Sub-Contract.

Another very typical example is where a sub-contractor's final retention release is not due until after the certificate of making good (defects) under the main contract works has been issued. In such a situation there may be a mechanical and electrical sub-contractor (for example) with a sub-contract value of £1 million, whose final release of retention (at say 1.5%) equals £15,000.00. There may also be another sub-contractor (e.g. a mastic pointing sub-contractor) whose sub-contract value is £10,000.00, and whose final release of retention (at, for example, 1.5%) equals £150.00. The mechanical and electrical sub-contractor has cleared all of its defects at the end of the rectification/defects liability period, and therefore, in the normal course of events, would expect to be paid the retention monies held. However, because the mastic pointing sub-contractor has not cleared all of its defects (possibly because the release of £150.00 to it is not of great importance) the certificate of making good (defects) under the main contract works has not been issued; and because that certificate has not been issued, the mechanical and electrical sub-contractor cannot recover the £15,000.00 (final release) of retention held against its account.

This situation is what is frequently referred to as 'pay when certified' which many commentators consider is so closely akin to 'pay when paid' that it should be outlawed by way of the Housing Grants, Construction and Regeneration Act 1996, but which the courts appear to consider is a legitimate mechanism.

Because of this pay when certified mechanism (particularly in respect of the final release of retention), sub-contractors can be kept from their retention monies sometimes for many months after they have cleared all of their defects, simply because another, totally unrelated sub-contractor, has not cleared its own defects.

In the example given above, the mechanical and electrical sub-contractor could have £15,000.00 withheld perhaps for many months, and would obviously continue pursuing the release of the retention monies. However, if, a smaller sub-contract package only had £150.00 final retention held, the sub-contractor may sometimes effectively 'write-off' the retention rather than spend excessive monies pursuing the release of the retention for month after month. Anecdotal evidence suggests that it is not unknown for sub-contractors of smaller sub-contract package sizes to effectively build into their price an assumption that the final retention release will not be made, since making this allowance in its tender is more cost effective than spending disproportionate costs in pursuing a relatively small amount of money.

Against the above background, SBCSub/C and SBCSub/D/C deals with retention either by way of a retention percentage deducted or retained (but not held in trust, as noted above), or by way of a retention bond.

### Retention percentage deducted or retained

Clause 4.10.1.1 of SBCSub/C and SBCSub/D/C states 'any amount which *may* be deducted and retained as Retention' (italics added).

The operative word above is 'may' (rather than 'shall') and therefore, although it may seem to be an inane point to make, the option is for the contractor not to deduct or retain any retention monies at all. However, in the harsh reality of contracting, the probability is that, even if the contractor was not having any retention monies at all deducted from its own payments, it would still hold retention monies against its sub-contractors.

As mentioned above, it should be particularly noted that any retention that is retained is *not* required to be held in trust under SBCSub/C or SBCSub/D/C.

In line with Clause 4.15.1 of SBCSub/C, the retention which may be deducted is to be calculated as follows:

(1) Where the sub-contract works or such works in any section have not reached practical completion, the retention which the contractor may deduct and retain must be the percentage stated under item 8 of the sub-contract particulars applied to the part of the gross valuation (appropriate at that time) covered by clauses 4.13.1 to 4.13.3 inclusive (refer to clause 4.15.1.1 of SBCSub/C and SBCSub/D/C). The retention percentage may be inserted by the parties to the sub-contract under item 8 of the sub-contract particulars; however, if no rate is inserted the default percentage is 3% (which is the same as the default position under the main contract).

(2) Where the sub-contract works or such works in any section have reached practical completion, the retention which the contractor may deduct and retain will be one half of the amount calculated under item 1 above (refer to clause 4.15.1.2 of SBCSub/C and SBCSub/D/C).

If there are no defects in the sub-contract works (or such works in a section) on the date of the expiry of the rectification period (previously known as the defects liability period) of the main contract for the main contract works (or relevant section) the balance of any retention deducted and retained by the contractor must be included in the next interim payment following the expiry of the applicable rectification period (refer to clause 4.15.2 of SBCSub/C and SBCSub/D/C).

The rectification period of the main contract for the main contract works is the period inserted into the main contract particulars (against the reference to clause 2.38 of the main contract) and relates to the period after practical completion of the main contract works (or a section) where the contractor is liable for any defects, shrinkages or other faults in the works. The rectification period entered into the main contract particulars is whatever is agreed between the parties to the main contract. This period is frequently entered as twelve months, but the default position in the main contract (i.e. the period that is assumed to apply if no entry is made against clause 2.38 in the main contract particulars) is six months.

Therefore, if at the end of the applicable rectification period the sub-contractor has no defects, then its retention release is not tied to the release of the retention under the main contract works, but is dealt with entirely independently.

However, if at the end of the applicable rectification period, the sub-contractor has defects outstanding in respect of its works, then its retention release is dealt with as follows.

First, the applicable defects are to be stated in a list issued by the contractor to the sub-contractor (as clause 4.15.3 of SBCSub/C and SBCSub/D/C).

When the contractor and the sub-contractor agree that all of the defects on the list of defects have been cleared, the contractor is to write to the sub-contractor to confirm the date when the defects were all cleared (as clause 4.15.3 of SBCSub/C and SBCSub/D/C).

If the contractor and the sub-contractor are unable to agree that the defects have all been cleared (which would normally be the situation where the sub-contractor considers that the defects have been cleared and the contractor considers that they have not been cleared) then the dispute regarding this matter may be referred to adjudication for a decision.

Once the contractor and the sub-contractor have agreed that all of the defects have been cleared, or once the clearance of all defects has been determined by adjudication, the balance of any retention monies deducted and retained by the contractor must be included in the next interim payment following the date (either as agreed or as determined by adjudication) that the defects have all been cleared. Therefore, if the first interim payment was due on 12 January 2004, and the date when the defects were cleared was 9 July 2005, then the next interim payment following the clearance of the defects would be 12 July 2005.

In the event that the contractor and the sub-contractor are unable to agree that the defects have all been cleared and the dispute regarding this matter is not referred to adjudication for determination, then the retention monies deducted and retained by the contractor must be included in the next interim payment following the issue of the certificate of making good (defects) under the main contract for the main contract works or a relevant section of it.

Under the above provisions, the sub-contractors' hands are no longer tied in respect of obtaining the release of their own retention monies independently from the completion of all other defects in relation to the main contract works or a section of it. However, given that a party cannot (normally) recover its own costs in an adjudication action, the effectiveness of this particular provision, where relatively small amounts of retention monies have been deducted and retained by a contractor against a particular sub-contractor, remains to be seen.

## Retention bond

As an alternative to retention monies being held, SBCSub/C and SBCSub/D/C also has provision for a retention bond.

A retention bond only applies where item 9 of the sub-contract particulars in SBCSub/A and SBCSub/D/A states that a retention bond is to apply.

The form of the bond (which has been agreed between the JCT and the British Bankers' Association) is set out under schedule 3 part 2 of SBCSub/C and SBCSub/D/C.

SBCSub/G makes the note that if a sub-contractor encounters difficulties over the terms of the bond with the branch of a British bank from which he seeks the bond, he should request that the branch refer the matter to its head office who should be able to confirm that, as the terms are as agreed by the British Bankers' Association with the JCT, they are acceptable.

When completing the bond as set out in part 2 of schedule 3, the information that needs to be inserted includes:

- The date of the bond.
- The name and address of the surety (i.e. the party that guarantees payment of the bond amount).
- The name and address of the contractor.
- The name and address of the sub-contractor.
- The surety's maximum aggregate liability amount under the bond stated as a sum in pounds. This will be the same sum as stated in the sub-contract under item 9 of the sub-contract particulars.
- The surety's address where a demand under the bond should be sent.
- The surety's address where a copy of the written notice to the sub-contractor of his liability for the amount demanded should be sent.
- The (default) expiry date of the bond. This will be the date as stated in the sub-contract under item 9 of the sub-contract particulars.

The bond then needs to be signed as a deed by or on behalf of the surety.

The bond operates in conjunction with clause 4.16 of SBCSub/C and SBCSub/D/C.

When a retention bond is in place, the contractor (at the date of each interim payment) is to prepare a statement which specifies the deduction in respect of retention that would have been made had the retention bond not been in place (refer to clause 4.16.1 of SBCSub/C and SBCSub/D/C).

The retention bond needs to be in place and needs to be provided by the sub-contractor to the contractor on or before the date of commencement of the sub-contract works, and the retention bond must be maintained by the sub-contractor until the expiry date of the bond. In line with clause 4.16.2 of SBCSub/C and SBCSub/D/C, the proposed surety must be approved by the contractor.

If the sub-contractor does not provide or maintain the retention bond as required, then the contractor is entitled to deduct and retain retention monies in line with the provisions of clauses 4.10.1.1 and 4.15 of SBCSub/C and SBCSub/D/C in the next applicable interim payment to the sub-contractor. If the sub-contractor subsequently provides and thereafter maintains the required retention bond the contractor must (not may), in the next interim payment after such compliance, release to the sub-contractor the retention deducted during the period of the breach (refer to clause 4.16.3 of SBCSub/C and SBCSub/D/C).

The surety's maximum aggregate liability amount under the bond is normally based upon the retention percentage applied to the original scope of the sub-contract works. Clearly, if variations are issued, the value of the sub-contract works to which the retention percentage would apply may increase, and in such a case the maximum retention monies may exceed the maximum aggregate liability amount under the bond.

If such circumstances apply, clause 4.16.4 of SBCSub/C and SBCSub/D/C makes it clear that the sub-contractor can either arrange with the surety for the aggregate liability amount to be increased so that it equates to such an increased retention sum, or, alternatively, the contractor may deduct and retain the retention monies due that are in excess of the maximum aggregate liability amount on the retention bond.

Item 2 of the notes section to the retention bond (which will not appear on the bond issued by the surety) states that it is understood that a surety will, at additional cost to the sub-contractor and possibly subject to other terms and conditions of the surety, provide for a greater sum than that originally inserted as the maximum aggregate liability amount inserted in the bond.

The purpose of the retention bond is to allow the contractor to recover from the surety:

(1) The costs actually incurred by the contractor by reason of the failure of the sub-contractor to comply with the directions of the contractor under the sub-contract (as clause 4.3.1 of the retention bond).

(2) Any expenses or any direct loss and/or damage caused to the contractor as a result of the termination of the sub-contractor's employment by the contractor (as clause 4.3.2 of the retention bond).

(3) Any costs, other than the amounts referred to in items 1 and 2 above, which the contractor has actually incurred and which, under the sub-contract, he is entitled to deduct from monies otherwise due or to become due to the sub-contractor (as clause 4.3.3 of the retention bond).

The second paragraph to item 3 of the notes to the retention bond states that any demand under clause 4 of the retention bond must not exceed the costs actually incurred by the contractor, therefore (and accordingly) the contractor cannot add profit to any costs incurred when making a claim under the retention bond.

Before making a claim under the retention bond, the contractor is to give written notice to the sub-contractor of his liability for the amount demanded, and is to request the sub-contractor to discharge his liability. At the time when such a notice is sent to the sub-contractor a copy is to be sent to the surety at the address noted under clause 4.4 of the retention bond. A claim under the retention bond cannot be made until 14 days after this written notice has been provided to the sub-contractor, and can only be made if the sub-contractor has not complied with the requirements of the written notice.

When making a claim against the bond, the contractor must:

- provide details of the retention bond;
- provide the date of the demand;
- state the amount of the retention that would have been held by the contractor at the date of the demand had retention been deductible (refer to clause 4.2 of the retention bond);
- state the amount demanded, and identify which one or more of clauses 4.3.1, 4.3.2 or 4.3.3 of the retention bond is being relied upon for the demand being made;
- incorporate a certification that the sub-contractor has been given 14 days' written notice of his liability for the amount demanded by the contractor and that the sub-contractor has not discharged that liability;
- provide evidence that a copy of the written notice to the sub-contractor was simultaneously sent to the surety at the address noted under clause 4.4 of the retention bond.

The demand must be signed by persons who are authorized by the contractor to act for and on his behalf, and those signatures must be authenticated by the contractor's bankers; then the demand must be sent to the surety at the address noted under clause 4.1 of the retention bond.

The last paragraph of clause 4 of the retention bond states:

'Such demand as above shall, for the purposes of this Bond but not further or otherwise, be conclusive evidence (and admissible as such) that the amount demanded is properly due and payable to the Contractor by the Sub-Contractor.'

Item 3 of the notes section to the retention bond (which will not appear on the bond issued by the surety) states that the inclusion in the above quoted paragraph of the words *'but not further or otherwise'* is to make clear that the sub-contractor would not be prevented by the terms of clause 4 (of the retention bond) from alleging, under the sub-contract, that the sub-contractor was not in breach of any of the matters stated in clauses 4.3.1 to 4.3.3 of the bond.

In other words, a demand made under clauses 4.3.1 to 4.3.3 of the retention bond will be taken as being conclusive evidence to the surety that the amount demanded is properly due and payable (therefore, the retention bond is, in effect, an on-demand bond) but any such demand will not prevent the sub-contractor from pursuing its claim under the terms of the sub-contract.

Finally, clause 4.16.5 of SBCSub/C and SBCSub/D/C notes that where the contractor has required the sub-contractor to provide a performance bond, then, in respect of any default to which that performance bond refers which is also a matter for which the contractor could make under the terms of the retention bond, the contractor must first have recourse to the retention bond.

# Discount

## What discount can be deducted by the main contractor?

There are two types of discount commonly used in the construction industry:

(1) trade discount;
(2) cash discount.

Neither trade discount nor cash discount are referred to in SBCSub/C or SBCSub/D/C, and the contractor does not have the entitlement to deduct either discount under the terms of SBCSub/C or SBCSub/D/C.

Despite this fact, discounts are an extremely common part of the contract/sub-contract relationship, and many contractors may attempt to incorporate some form of discount provision by way of a numbered document (refer to item 15 of the sub-contract particulars SBCSub/A (item 16 of the sub-contract particulars SBCSub/D/A)).

Therefore, a brief commentary is given on the question of discount below.

As noted above, the construction industry is familiar with two types of discount, namely: trade discounts and discounts for prompt payment (normally known as cash discounts).

Trade discounts are usually granted to specific contractors by virtue of their commercial standing with the sub-contractor, or alternatively may be given in respect of a particular project to secure work. A trade discount is, in effect, simply a reduction of price and normally applies (without condition) to all quoted prices.

Therefore, a sub-contractor may have provided a quotation for £100,000 but to secure the works in question needs to reduce his price by £5,000. Rather than re-price the works entirely, the sub-contractor will often offer a trade discount of 5% (i.e. £5,000 on £100,000) thereby effectively reducing his price to £95,000.

Trade discounts do not (normally) have any conditions attached to them, and are given purely in an effort to secure sub-contract works.

On the other hand when cash discounts are offered these are normally conditional upon an improvement upon the existing payment terms. Therefore, a sub-contractor may offer a 2.5% cash discount if the payment period between the payment due date is reduced from, for example, 21 days as SBCSub/C and SBCSub/D/C to, for example, seven days. Alternatively, a sub-contractor may offer a 2.5% cash discount if the period between interim payments is increased in frequency from, for example, monthly as SBCSub/C and SBCSub/D/C to, for example, every two weeks. The cash discount is therefore offered in exchange for better payment terms, and the cash discount can, in such a situation, only be taken by the contractor if the improved payment terms are met.

When dealing with cash discount, the case of *Team Services* v. *Kier Management*[6] is of particular interest. In that case Mr Justice Bowsher found the cash discount mechanism on the terms being considered (which were fairly typical terms used in the industry) were of only temporary effect. Therefore, although it is often assumed that if a contractor fails to pay an instalment in due time to the sub-contractor he permanently loses the right to deduct the cash discount on that payment; the wording of a typical cash discount clause makes it clear that the calculation of cash discount falls to be made on the whole of the sum due, including previous instalments, on every occasion. Accordingly, if the contractor fails to make one payment on time, he can substantially recoup his position by making the next payment on time; all he loses is the interest on the cash discount for one month.

Therefore, the converse of this point is that the contractor should take special care to pay the final instalment on time, otherwise he will lose his cash discount not just on that final instalment, but on the whole amount that has fallen due.

## The final payment

The final payment is simply the final sub-contract sum, which must be calculated by the contractor in accordance with whichever of clause 4.3 or 4.4 of SBCSub/C or SBCSub/D/C applies (see commentary earlier in this chapter), less only the total amount previously due as interim payments under the sub-contract (refer to clause 4.12.1 of SBCSub/C and SBCSub/D/C).

In line with clause 4.12.2 of SBCSub/C and SBCSub/D/C, the final payment is due not later than seven days after the date of issue (not the date of receipt) of the final certificate under clause 4.15 of the main contract condition. The final date for payment of the final payment must be 28 days after the date that it becomes due (as clause 4.12.2 of SBCSub/C and SBCSub/D/C).

The above position is effectively a 'pay when certified' situation which, as noted earlier in this chapter, many commentators consider is so closely akin to 'pay when

[6] *Team Services plc* v. *Kier Management and Design Limited* (1992) CILL 786.

paid' that it should be outlawed by way of the Housing Grants, Construction and Regeneration Act 1996, but which the courts appear to consider is a legitimate mechanism.

In this situation the sub-contractor's final payment (under the sub-contract) is directly linked to the date of issue of the final certificate (under the main contract), something that the sub-contractor has absolutely no control over.

Although the contractor must advise the sub-contractor, upon receipt of a written request from the sub-contractor, of the date of the final certificate (in line with clause 3.25.5 of SBCSub/C and SBCSub/D/C), this still does not give the sub-contractor any control over this issue.

However, even though the above situation may not be entirely to the liking of a sub-contractor, the impact of the clause is lessened by the fact that in line with clause 4.9.2 of SBCSub/C and SBCSub/D/C, interim payments to a sub-contractor continue to become due after practical completion of the sub-contract works as and when further amounts are ascertained as due and payable, and a sub-contractor's retention may be released irrespective of the fact that defects on other parts of the main contract works have not been cleared (as clauses 4.15.2 and 4.15.3 of SBCSub/C and SBCSub/D/C).

## Counterclaim, set-off and abatement

It is important to understand the difference between counterclaim, set-off and abatement.

This is because case law (whilst not entirely conclusive) suggests that while an effective withholding notice is necessary to be able to set-off against monies otherwise payable, the common law defence of abatement survives the absence of such a notice.

Where an account between the parties of several debits and credits are closely related to the same underlying transaction, no question of set-off arises because the debt is regarded to be the balance of the account.

As Lord Mansfield said:[7] 'Where the nature of the employment, transaction, or dealings, necessarily constitutes an account consisting of receipts and payments, debts and credits, it is certain that only the balance can be the debt.'

### Counterclaim

In respect of counterclaims, Lord Denning said:[8]

'The word "counterclaim" is not defined in section 28, but, following Halisbury again, 3rd edition, volume 34, page 411, I think it is:

"any claim that could be the subject of an independent action. It is not confined to money claims . . . and . . . it need not relate to or be connected with the original subject of the cause or matter."'

---

[7] *Green* v. *Farmer* (1768) 4 Burr 2214.
[8] *Henriksens A/S* v. *Rolimpex THZ* [1974] 1 QB 233, CA.

Therefore, a counterclaim must be capable of an existence independent of the other party's claim being made, although it can be used to set-off if it meets the necessary requirements.

## Set-off

A set-off acts as a mechanism to absolve a party from actually paying another party's claim. There are three types of set-off:

- A set-off at common law is available in respect of opposing claims for readily ascertainable liquidated debts or monetary demands (but not damages), which do not need to be related to the same transaction.
- Equitable set-off applies in the case of a cross-claim that is so closely associated with the claim that justice requires it to be taken into account. Unlike set-off at common law, it can include a claim for damages but not a separate or independent counterclaim.
- Contractual set-off is made in accordance with the express machinery of the contract.

## Abatement

The principle of abatement was described by Lord Morris in *Gilbert Ash* v. *Modern Engineering*[9] as follows:

'It has long been an established principle of law that if one man does work for another, the latter, when sued, may defend himself by showing that the work was badly done and that the claim made in respect of it should be dismissed.'

### Value of work

Most contracts provide that payment is to be only for work properly carried out and, even if adequate payment terms are not incorporated into the contract, the fallback provision under the Scheme[10] is for payment in respect of work 'performed in accordance with [the] contract'.

Therefore, it can readily be argued that the 'value' of work is only arrived at after discounting that part of the work that is not in accordance with the contract (i.e. after abatement) by reason of breach.

If a sub-contractor applies for more money than he is entitled to because he has not yet carried out part of the work then there is no set-off or abatement; he is simply not entitled to all he has claimed (i.e. the whole sum is not due).

Assuming that there is no contractual provision under which the sum claimed is the amount due, no withholding notice is required to reduce the sub-contractor's entitlement.

---

[9] *Gilbert Ash Ltd* v. *Modern Engineering (Bristol) Ltd* [1974] AC 689.
[10] The Scheme for Construction Contracts (England and Wales) Regulations, 1998.

### Quantification of abatement

The measure of any abatement is the amount by which the worth of the subject matter is less because of the breach of contract. However, this creates some difficulties. In particular, should the cost of remedial works to be taken into account?

In practice it may be that a realistic measure of abatement is dependent upon the circumstances, in particular whether the sub-contractor has a right to remedy defects and whether he has failed to exercise or has been denied that right.

## The payment notice

Section 110 (2) of the Housing Grants, Construction and Regeneration Act 1996, states:

> 'Every construction contract shall provide for the giving of a notice by a party not later than five days after the date on which a payment becomes due from him under the contract, or would have become due if:
>
> (a) the other party had carried out his obligations under the contract; and
> (b) no set-off or abatement was permitted by reference to any sum claimed to be due under one or more contracts;
>
> specifying the amount (if any) of the payment made or proposed to be made, and the basis on which that amount was calculated.'

This notice is commonly referred to as a payment notice.

Unfortunately, the Act is silent on what happens if a payment notice as provided in the construction contract is not issued.

The failure of a party to serve a payment notice does not mean that the sum applied for automatically becomes payable, unless the contract *expressly provides* for that consequence. SBCSub/C and SBCSub/D/C does not make such an express provision.

In the absence of such an express provision, an adjudicator may abate a claim, as explained by Judge Bowsher QC in *Whiteways Contractors v. Impresa Castellie*:[11]

> 'It is common for a party to a building contract to make deductions from sums claimed on the Final Account (or on earlier interim applications) on account of overpayment on previous applications and it makes no difference whether those deductions are by way of set-off or abatement. The scheme of the HGCRA is to provide that, for the temporary purposes of the Act, notice of such deductions is to be made in a manner complying with the requirements of the Act. In making that requirement, the Act sees no distinction between set-offs and abatements. I see no reason why it should have done so, and I am not tempted to try to strain the language of the Act to find some fine distinction between its applicability to abatements as opposed to set-offs. Of course, in considering a dispute, an Adjudicator will make his own valuation of the claim before him and in doing so, he may abate the claim in respects not mentioned in the notice of intention to withhold payment.'

---

[11] *Whiteways Contractors (Sussex) Ltd* v. *Impressa Castelli Construction UK Ltd* [2000] CILL 1664.

However, even with the power to abate it would be unwise to ignore section 110 (2) of the Act and not serve a payment notice. This is because, in practice, an adjudicator may wrongly take the view that the sum applied for in the absence of a payment notice is automatically payable, or, more possibly, take the incorrect view that the absence of a payment notice in effect shifts the burden of proof so that it is for the paying party to justify in the adjudication why the sum applied for should not be paid in full. Whilst this is obviously an incorrect view because, despite the absence of a payment notice, the basic principle remains that 'he who asserts must prove', it is a risk that should be avoided if possible.

SBCSub/C and SBCSub/D/C addresses the issue of payment notices under clause 4.10.2 of SBCSub/C and SBCSub/D/C (for interim payments) and under clause 4.12.2 of SBCSub/C and SBCSub/D/C (for the final payment).

## The payment notice – interim payments

Clause 4.10.2 of SBCSub/C and SBCSub/D/C notes that not later than five days after the date on which an interim payment becomes due the contractor must give a written notice to the sub-contractor specifying the amount of the payment which is proposed to be made in respect of the sub-contract works, to what the amount of the payment relates, and the basis on which that amount was calculated. An example of a payment notice in respect of interim payments is given in Appendix B of this book.

As detailed above, the amount to be included in the gross valuation for each interim payment is set out in clause 4.13 of SBCSub/C and SBCSub/D/C.

Clearly, the contractor should set out his calculation of the gross valuation for each interim payment in sufficient detail to enable the sub-contractor to understand how the total gross valuation has been arrived at, and what the basis of the calculation of the individual parts of the gross valuation was. As a guide, the level of detail provided by the contractor should be at least to the level provided by the sub-contractor in respect of any relevant payment application made.

If insufficient detail is provided by the contractor, this may cause uncertainty, and this uncertainty may lead to a (possibly unnecessary) dispute developing between the sub-contractor and the contractor.

If a dispute did develop, the contractor would need to show that it had operated the provisions of clause 4.13 of SBCSub/C and SBCSub/D/C correctly. Clearly, if an adequate breakdown of the gross valuation is provided by the contractor in respect of interim payments, within the timescales set out in the sub-contract (i.e. not later than five days after the date on which an interim payment becomes due), the likelihood of a dispute developing is considerably reduced.

### What is the effect of not giving a payment notice in respect of interim payments?

Many sub-contractors are of the view that if a payment notice is not given in respect of an application submitted by a sub-contractor, the sub-contractor will automatically be entitled to receive payment in line with the application submitted. This is not the case.

A series of court cases have shown that the lack of a payment notice has very little practical effect. All the sub-contractor is entitled to be paid, irrespective of whether or not a payment notice is given or not, is the amount that is actually due.

The amount that is actually due is not determined by the value claimed by the sub-contractor within a payment application submitted, nor, indeed, by the value indicated in any payment notice issued by the contractor, but is based on the correct gross valuation of the works (in line with clause 4.13 of SBCSub/C and SBCSub/D/C) at the appropriate payment due date, less any retention applicable and/or any previous payments made; all as detailed under clause 4.10.1 of SBCSub/C and SBCSub/D/C.

Of course, if a payment notice was not issued, and if there was a dispute regarding the amount paid in respect of a particular interim payment, and if that dispute was referred to adjudication, then the adjudicator may take into account the lack of a payment notice (as a background issue), but this should not have any impact upon his decision as to the correct amount due.

The correct amount due should be determined simply by applying the provisions of the sub-contract to arrive at the gross valuation, and then deducting any retention and previous payments applicable.

The above position is amplified by clause 4.10.4 of SBCSub/C and SBCSub/D/C which states:

'Subject to any notice given under clause 4.10.3 [i.e. a withholding notice, which is discussed further below], the Contractor shall no later than the final date for payment pay the amount specified in his notice given under clause 4.10.2 [i.e. a payment notice] or, *in the absence of a notice under clause 4.10.2, the amount calculated in accordance with clause 4.10.1.*' (italics added)

In other words, if a payment notice under clause 4.10.2 is not issued, the sub-contractor is *only* entitled to be paid the amount due calculated in accordance with clause 4.10.1. (This ignores the effect of any withholding notices, which are discussed later in this chapter.)

Despite the above comments, and as noted above, the contractor would be well advised to submit a payment notice in any event to minimize the possibility of an unnecessary dispute developing.

## The payment notice – final payment

Clause 4.12.2 notes that not later than five days after the date on which the final payment becomes due the contractor must send the final payment notice in writing by special or recorded delivery, notifying the sub-contractor of the amount of the final payment to be made to the sub-contractor, to what the amount of the payment relates, and the basis on which it was calculated. An example of a final payment notice is given in Appendix B of this book.

As detailed earlier in this chapter, the amount to be included in the final payment is the final sub-contract sum, which must be calculated by the contractor in accordance with whichever of clause 4.3 or 4.4 of SBCSub/C and SBCSub/D/C applies.

If insufficient detail is provided by the contractor in his final payment notice, this may cause uncertainty, and this uncertainty may lead to a (possibly unnecessary) dispute developing between the sub-contractor and the contractor.

Clearly, the contractor should set out his calculation of the final sub-contract sum in sufficient detail to enable the sub-contractor to understand how the final sub-contract sum has been arrived at, and also to show the basis of the calculation of the individual parts of the final sub-contract sum.

If a dispute did develop, the contractor would need to show that it had operated the provisions of clause 4.3 or 4.4 of SBCSub/C and SBCSub/D/C correctly. Clearly, if an adequate breakdown of the final sub-contract sum is provided by the contractor within the timescales set out in the sub-contract (i.e. not later than five days after the date on which the final payment becomes due), the likelihood of a dispute developing is considerably reduced.

### What is the effect of not giving a final payment notice in respect of the final payment?

Many sub-contractors may be of the view that if a final payment notice is not given in respect of an application (or a final account) submitted by a sub-contractor (even though the sub-contract conditions do not actually require the sub-contractor to submit a final account), the sub-contractor will automatically be entitled to receive payment in line with the application (or final account) submitted.

As noted above in respect of interim payments, this is not the case, and a series of court cases have shown that the lack of a final payment notice will, in reality, have very little practical effect.

Exactly as for interim payments, all the sub-contractor is entitled to, irrespective of whether or not a final payment notice is issued, is the amount that is actually due.

The amount that is actually due is not determined by the value claimed by the sub-contractor, nor, indeed, by the value indicated in any final payment notice submitted by the contractor, but is based on the correct gross valuation of the works (according to clause 4.3 or 4.4 of SBCSub/C and SBCSub/D/C), less any previous payments made, all as detailed under clause 4.12.1 of SBCSub/C and SBCSub/D/C.

Of course, if a final payment notice was not issued, and if there was a dispute regarding the amount paid that was referred to adjudication, then the adjudicator may take into account the lack of a final payment notice (again as a background issue only), but this should not have any impact upon his decision as to the correct amount due.

The correct amount due should be determined simply by applying the provisions of the contract to arrive at the final sub-contract sum and then deducting any previous payments made to arrive at the amount due.

This position is further clarified by paragraph 155 of SBCSub/G, which says:

> 'If the Contractor fails to give the Sub-Contractor a Final Payment Notice, then, unless he gives a withholding notice under clause 4.12.3, *he is required to pay the Sub-Contractor the amount calculated in accordance with clause 4.12.1.*' (italics added) (Refer also to clause 4.12.3 of SBCSub/C and SBCSub/D/C.)

In other words, if a final payment notice under clause 4.12.2 of SBCSub/C or SBCSub/D/C is not issued, the sub-contractor is *only* entitled to be paid the amount due calculated in accordance with clause 4.12.1 of SBCSub/C and SBCSub/D/C.

Despite the above comments, the contractor would be well advised to submit a final payment notice in any event, for two very good reasons:

(1) to minimize the possibility of an unnecessary dispute developing; and
(2) because if the final payment notice is not issued, there will be no finality of the matters referred to in clause 1.9 of SBCSub/C and SBCSub/D/C (refer to Chapters 3 and 13 of this book).

It may be considered by some that because the final payment notice must be sent by special or recorded delivery (according to clause 4.12.2 of SBCSub/C and SBCSub/D/C), whereas a payment notice for an interim payment does not need to be sent in such a way, there must be some particular significance in respect of the payment due to a sub-contractor as a result of a final payment notice not being issued.

However, the terms of the sub-contract do not support this proposition at all, and, in terms of the amount of payment actually due, SBCSub/C and SBCSub/D/C do not make the lack of a final payment notice in respect of the final payment any more influential than the lack of a payment notice in respect of interim payments.

It appears clear that the reason why the final payment notice must be sent by special or recorded delivery is simply because of the effect that the final payment notice has regarding the finality of the matters referred to in clause 1.9 of SBCSub/C and SBCSub/D/C (refer to Chapters 3 and 13 of this book), and not because of the effect that the final payment notice has upon the amount actually due to the sub-contractor.

## The withholding notice

Section 111 of the Housing Grants, Construction and Regeneration Act 1996 states:

'(1) A party to a construction contract may not withhold payment after the final date for the payment of a sum due under the contract unless he has given an effective notice of intention to withhold payment.

(2) To be effective such a notice must specify:

(a) the amount proposed to be withheld and the ground for withholding payment; or
(b) if there is more than ground, each ground and the amount attributable to it;

and must be given not later than the prescribed period before the final date for payment.'

This is commonly referred to as the withholding notice.

Section 111 of the Act was considered by Judge Hicks QC in *VHE Construction v. RBSTB*.[12]

In that case the defendant sought to rely upon rights of set-off notwithstanding the absence of any section 111 notice. However, Judge Hicks decided:

---

[12] *VHE Construction plc v. RBSTB Trust Co. Ltd* [2000] CILL 1592.

'The first subject of dispute as to the effect of Section 111 is whether Section 111(1) excludes the right to deduct money in exercise of a claim to set off in the absence of an effective notice of intention to withhold payment. Mr Thomas, for RBSTB, submits that it does not. I am quite clear, not only that it does, but that that is one of its principal purposes. I was not taken to the reports or other preparatory material leading to the introduction of this part of the Act, nor to anything said in Parliament but the see-saw of judicial decision, drafting, fashion and editorial commentary in this area is familiar to anyone acquainted with construction law, and in my judgment, Section 111 is directed to providing a definitive resolution of the debate. The words "may not withhold payment" are in my view ample in which to have the effect of excluding set-offs and there is no reason why they should not mean what they say.'

Therefore, the above decision, confirms that a party wishing to set-off against sums due must provide a compliant section 111 notice (a withholding notice).

There are important provisions as to the timing and content of a withholding notice. It is clear that for a withholding notice to be effective it must be given within the prescribed period, and if it is given even one day outside the prescribed period it will not be effective.

It has also been found through court actions that it is not acceptable for a party to rely upon a notice provided in advance of any application for payment. This point was confirmed by the Scottish Authority of *Strathmore Building Services* v. *Colin Scott Grieg*.[13]

In that case, the defending party tried to argue that a letter detailing certain claims sent to the claimant in advance of the particular application for payment forming the subject matter of the adjudication constituted an effective withholding notice. In respect of that claim, Lord Hamilton found as follows:

'The purpose of Section 111 is to provide a statutory mechanism on compliance with which, but only compliance with which a party otherwise due to make payment may withhold such payment. It clearly, in my view, envisages a notice given under it being a considered response to the application for payment, in which response it is specified how much of the sum applied for it is proposed to withhold and the ground or grounds for withholding that amount. Such a response cannot, in my view, effectively be made prior to the application itself being made.'

In addition to the timing of the withholding notice, there are also considerations as to its content. For a withholding notice to be effective the specific wording of section 111 of the Act must be complied with in terms of detailing the ground or grounds for withholding payment.

Also, any notice needs to be in writing, needs to be addressed to the other party, must be sent from or on behalf of the withholding party, and must state an intention to withhold monies.

Therefore, vague letters intimating a possible future claim in respect of loss and expense etc. will almost certainly not be sufficient.

SBCSub/C and SBCSub/D/C addresses the issue of withholding notice under clause 4.10.3 of SBCSub/C and SBCSub/D/C (for interim payments) and under clause 4.12.3 of SBCSub/C and SBCSub/D/C (for the final payment).

---

[13] *Strathmore Building Services Ltd* v. *Colin Scott Greig* (2001) 17 Const LJ 72.

## Withholding notice – interim payments

Clause 4.10.3 of SBCSub/C and SBCSub/D/C notes that not later than five days before the final date for payment of an interim payment the contractor may give a written notice to the sub-contractor which shall specify any amount proposed to be withheld and/or deducted from the amount notified under clause 4.10.2 of SBCSub/C and SBCSub/D/C (i.e. the payment notice), the ground or grounds for such withholding and/or deduction, and the amount of withholding and/or deduction attributable to each ground.

An example of a withholding notice is given in Appendix B of this book.

It should be noted that the Act does not specify a period before the final date for payment when the withholding notice must be issued, but simply says: 'A party to a construction contract may not withhold payment after the final date for the payment of a sum due under the contract unless he has given an effective notice of intention to withhold payment.' Therefore, clause 4.10.3 of SBCSub/C and SBCSub/D/C may be amended to say that a withholding notice may be issued not later than *one day* before the final date for payment, and this would still be compliant with the Act.

Alternatively, as a withholding notice complying with clause 4.10.3 of SBCSub/C and SBCSub/D/C is required to be given by the contractor not later than five days before the final date for payment, the contractor could, if he wishes, give the sub-contractor a withholding notice at the same time as he gives a payment notice.

If a withholding notice is combined with a payment notice in this way, then for that 'combined' notice to be effective it would need to be issued not later than five days after the applicable payment due date, and the combined notice would need to satisfy the requirement of both a payment notice and a withholding notice.

However, as noted above, to be compliant on its own (in line with the un-amended sub-contract terms), a withholding notice must be issued by no later than five days before the final date for payment of an interim payment, must detail any amount proposed to be withheld and/or deducted and must specify the ground or grounds for such withholding and/or deduction.

Therefore, a case in point may be where a contractor wishes to withhold from a groundworks sub-contractor's interim payment, the cost associated with the rebuilding of a brick wall that the groundworker had damaged with his excavator whilst excavating drain trenches nearby.

For such a withholding notice to be valid, the withholding notice would need to be issued timeously (i.e. no later than five days before the final date for payment of the interim payment in question) and would need to say that the amount proposed to be withheld was, for example, £950 and the ground for the withholding was because a brick wall (as detailed) needed to be rebuilt (on a certain date as recorded) because it had been demolished by the groundworker's excavator (on a certain date as recorded).

The sub-contract conditions do not specifically require that details of the costs are provided. However, to avoid any future unnecessary dispute (and to defeat any possible argument that may be raised in a possible future adjudication action that the cost information being provided in an adjudication action had not previously been seen by the other party and was therefore new information that could, arguably, be disallowed), it is more sensible if details of the costs are provided at the time when the withholding notice is issued.

Therefore, rather than simply saying that the costs are £950 (as the example above), a build-up to the figure of £950 should be given, along the lines of:

| | |
|---|---|
| Clear bricks from site | |
| One labourer for half a day @ £80.00 per day | = £ 40.00 |
| One skip @ £125.00 per skip | = £125.00 |
| | |
| Rebuild brick wall | |
| Two bricklayers for one day @ £150.00 per bricklayer per day | = £300.00 |
| One labourer for one day @ £100.00 per day | = £100.00 |
| Scaffold tower for one day @ £50.00 per day | = £ 50.00 |
| 500 No. facing bricks @ £500.00 per thousand | = £250.00 |
| 0.4 tonnes of cement @ £100.00 per tonne | = £ 40.00 |
| 0.4 tonnes of sand @ £50.00 per tonne | = £ 20.00 |
| One cement mixer for one day @ £25.00 per day | = £ 25.00 |
| | |
| Total | = £950.00 |

If insufficient detail is provided by the contractor, this may cause uncertainty, and this uncertainty may lead to a (possibly unnecessary) dispute developing between the sub-contractor and the contractor.

If a dispute did develop, the contractor would need to show that it had operated the provisions of clause 4.10.3 of SBCSub/C and SBCSub/D/C correctly, and the more detail that is provided, the easier it will be to show that the contractor's actions were in line with the terms of the sub-contract.

### What is the effect of not giving a withholding notice in respect of an interim payment?

Unlike the courts' view of payment notices, it is now quite clear, following a series of court cases, that if a timeous withholding notice is not issued, no withholding or deduction will be permitted within an adjudication action. This provision is strictly operated, and if a withholding notice is not issued within the timescale detailed within the sub-contract (i.e. no later than five days before the final date for payment, of the payment in question), the proposed withholding and/or deduction will almost certainly be disallowed and the merits of the proposed withholding and/or deduction will not be considered at all.

## Withholding notice – final payment

Clause 4.12.3 of SBCSub/C and SBCSub/D/C notes that not later than five days before the final date for payment of the final payment the contractor may give a written notice to the sub-contractor which must specify any amount to be proposed to be withheld and/or deducted from the amount notified under clause 4.12.2 (the payment notice), the ground or grounds for such withholding and/or deduction, and the amount of withholding and/or deduction attributable to each ground.

An example of a withholding notice is given in Appendix B of this book.

The general commentary above regarding a withholding notice for interim payments equally applies to a withholding notice in respect of the final payment.

*What is the effect of not giving a withholding notice in respect of the final payment?*

Clause 4.12.3 of SBCSub/C and SBCSub/D/C states:

> 'If no such notice of withholding and/or deduction is given, the Contractor shall, not later than the final date for payment, pay the amount notified to the Sub-Contractor under clause 4.12.2 or (in the absence of any notice under clause 4.12.2) the amount calculated in accordance with clause 4.12.1.'

As noted above, in respect of a withholding notice for interim payments, it is now quite clear following various court cases that if a timeous withholding notice is not issued, no withholding or deduction will be permitted.

This provision is strictly operated, and if a withholding notice is not issued within the timescale detailed within the sub-contract (i.e. no later than five days before the final date for payment of the final payment), the proposed withholding and/or deduction will almost certainly be disallowed within an adjudication action and the merits of the proposed withholding and/or deduction will not be considered at all.

## Timing of payments, payment notices and withholding notices

For ease of reference, a brief summary is provided below in respect of the timing of payments, payment notices and withholding notices for both interim payments and the final payment.

### Interim payments

- For the date when an interim payment becomes due refer to clause 4.9 of SBCSub/C and SBCSub/D/C.
- Within five days of when an interim payment becomes due, a written notice of the amount proposed to be paid under clause 4.10.2 of SBCSub/C and SBCSub/D/C (a payment notice) is required to be given.
- Within 16 days of when an interim payment becomes due (or, put another way, not later than five days before the final date of payment of the interim payment), any withholding notice under clause 4.10.3 of SBCSub/C and SBCSub/D/C that is required to be given, must be given.
- Within 21 days of when an interim payment becomes due, the interim payment is required to be paid (i.e. this is the final date for payment), as clause 4.9.3 of SBCSub/C and SBCSub/D/C.

### Final payment

- Within seven days of the date of issue of the final certificate under the main contract the final payment becomes due (clause 4.12.2 of SBCSub/C and SBCSub/D/C).

- Within five days of when the final payment becomes due, the 'Final payment notice' is required to be sent (clause 4.12.2 of SBCSub/C and SBCSub/D/C).
- Within 23 days of when the final payment becomes due (or, put another way, not later than five days before the final date of payment of the final payment), any withholding notice under clause 4.12.3 of SBCSub/C and SBCSub/D/C that is required to be given, must be given.
- Within 28 days of when the final payment becomes due, the final payment is required to be paid (i.e. this is the final date for payment), as clause 4.12.2 of SBCSub/C and SBCSub/D/C.

# Interest

Interest means a financial charge that is made (in this case by a sub-contractor) against another party (in this case the contractor) in respect of outstanding monies owed (which normally relates to overdue or late payments in part or in full).

Until fairly recently, a party had no automatic right or implied right to interest, unless there was an express term in the contract which dealt with the question of interest.

The position has, however, now changed by way of the introduction of the Late Payment of Commercial Debts (Interest) Act 1998.

There is now an implication in most construction contracts/sub-contracts (amongst many other contracts, with a very few exceptions only) that debts carry interest and a right to compensation.

Subsections 1(1) and (2) of the Late Payment of Commercial Debts (Interest) Act 1998 provide as follows:

'(1) It is an implied term in a contract to which this Act applies that any qualifying debt created by the contract carries simple interest subject to and in accordance with this Part.

(2) Interest carried under that implied term (in this Act referred to as 'statutory interest') shall be treated, for the purposes of any rule of law or enactment (other than this Act) relating to interest on debts, in the same way as interest carried under an express contract term.'

The right to compensation arises under section 5A[14] of the Act for contracts entered into from 7 August 2002, and this provides as follows:

'5A. (1) Once statutory interest begins to run in relation to a qualifying debt, the supplier shall be entitled to a fixed sum (in addition to the statutory interest on the debt).

(2) That sum shall be:

(a) for a debt less than £1,000, the sum of £40;
(b) for a debt of £1,000 or more, but less than £10,000, the sum of £70;
(c) for a debt of £10,000 or more, the sum of £100.

(3) The obligation to pay an additional fixed sum under this section in respect of a qualifying debt shall be treated as part of the term implied by section 1(1) in the contract creating the debt.'

---

[14] Introduced by The Late Payment of Commercial Debts Regulations 2002.

The rate of interest under the Act is 8% over the official dealing rate (i.e. the Bank of England Base Rate); and interest runs from 30 days after the date of the supply or the date of the invoice, whichever is later.

It is envisaged by the Act that some parties may agree a specific interest rate or compensation in their contract. If so, the Act will not apply and the agreed express terms will apply.

However, to avoid larger companies abusing their bargaining power, any agreed interest rate or compensation for late payment must be 'substantial'; if not, the Act will apply instead.

SBCSub/C and SBCSub/D/C deals with the matter of interest under clauses 4.10.5 (for interim payments) and 4.12.4 (for the final payment).

## Interim payments – clause 4.10.5

In respect of interim payments, clause 4.10.5 states that if the contractor fails properly to pay the amount due, or any part of it, to the sub-contractor under SBCSub/C or SBCSub/D/C, by the final date of payment, the contractor must pay to the sub-contractor, in addition to the amount not properly paid, simple interest thereon at the interest rate for the period until such payment is made. Clause 4.10.5 adds that the non-payment of such interest will be treated as a debt due to the sub-contractor by the contractor.

If this clause is considered part by part, it is clear that:

(1) Interest is only payable on amounts that are due to the sub-contractor under SBCSub/C or SBCSub/D/C. The question of the amounts due to a sub-contractor under SBCSub/C and SBCSub/D/C is dealt with earlier in this chapter. Interest is not payable on the basis of a sub-contractor's application for payment (unless, of course, this equals the amount that is due to the sub-contractor under SBCSub/C and SBCSub/D/C).

(2) Interest is payable from the date of the final date for payment.

(3) Interest stops accruing when the amount due to a sub-contractor under SBCSub/C or SBCSub/D/C conditions has been paid.

(4) Interest can only be claimed on the basis of simple interest, not compound interest. (Simple interest means interest paid on the principle sum only; compound interest means interest paid on the principle sum and on interest accumulated in respect of the principle sum.)

(5) The interest rate that applies (as defined under clause 1.1 of SBCSub/C and SBCSub/D/C) is 'a rate 5% per annum above the official dealing rate of the Bank of England correct at the date that a payment due under this contract becomes overdue'.

From the above, it appears that the interest rate applicable to a particular overdue payment is fixed (i.e. at 5% per annum above the official dealing rate of the Bank of England) at the date that the payment became due; and does not, as one might expect, change during the period of time that the debt remains outstanding to suit any fluctuations (whether these be up or down) to the official dealing rate of the Bank of England.

(6) The non-payment of any interest due shall be treated as a debt due to the sub-contractor by the contractor, and may be pursued by the sub-contractor in the same way as any other debt due under the sub-contract.

The other points of significance are:

(1) The sub-contractor does not need to claim interest to be entitled to be paid interest. Clause 4.10.5 of SBCSub/C and SBCSub/D/C says plainly:

'If the Contractor fails properly to pay the amount, or any part of it, due to the Sub-Contractor under these Conditions, by the final date for its payment, the Contractor shall pay to the Sub-Contractor in addition to the amount not properly paid simple interest thereon at the Interest Rate for the period until such payment is made.'

Therefore, there is no prerequisite for a sub-contractor to apply for interest before the payment of interest is made, and, because the word 'shall' is used (which is clearly directional rather than discretionary) the contractor is obliged to pay interest (on overdue payments) irrespective of whether or not any application for interest payment has been made by a sub-contractor.

Of course, irrespective of the above, it is sensible for a sub-contractor to make an application for interest payments, since it is highly unlikely that a contractor will make a payment for interest to a sub-contractor unless he has first been prompted by the sub-contractor.

(2) Clause 4.10.5 of SBCSub/C and SBCSub/D/C makes it clear that the acceptance of any payment of interest by the sub-contractor must not in any circumstances be construed as a waiver by the sub-contractor of his right:

   (a) to proper payment of the principal amounts due from the contractor to the sub-contractor in accordance with SBCSub/C and SBCSub/D/C conditions;
   (b) to suspend performance of his obligations under clause 4.11 of SBCSub/C and SBCSub/D/C (see commentary on this matter below);
   (c) to terminate his employment under section 7 of SBCSub/C (refer to Chapter 12).

(3) One particular question currently remains unresolved. The Late Payment of Commercial Debts (Interest) Act 1998 makes it clear that parties may only agree interest rates as express terms of their contract if such rates are 'substantial'. Given that the interest rate in SBCSub/C and SBCSub/D/C is 3% lower than the interest rate in the Act, and given also that SBCSub/C and SBCSub/D/C do not allow for any 'compensation' payments as detailed under section 5A of the Act, there must be some question as to whether or not the interest rate in SBCSub/C and SBCSub/D/C would be considered to be 'substantial' if this were tested in court. If this matter was tested in court, and if SBCSub/C and SBCSub/D/C interest rate was found not to be 'substantial', then the terms of the Act (with a higher interest rate) would be imported into SBCSub/C and SBCSub/D/C as implied terms.

**Final payment – clause 4.12.4**

In respect of the final payment, clause 4.12.4 states that if the contractor fails properly to pay the amount, or any part of it, due to the sub-contractor under SBCSub/C or SBCSub/D/C by the final date for its payment, the contractor must pay to the sub-contractor, in addition to the amount not properly paid, simple interest thereon at the interest rate for the period until such payment is made. Clause 4.12.4 of SBCSub/C and SBCSub/D/C adds that the non-payment of such interest must be treated as a debt due to the sub-contractor by the contractor.

The general commentary in respect of interest under clause 4.10.5 of SBCSub/C and SBCSub/D/C above applies (where applicable), and similar to the above comments, clause 4.12.4 of SBCSub/C and SBCSub/D/C makes it clear that the acceptance of any payment of interest under clause 4.12.4 by the sub-contractor must not in any circumstances be construed as a waiver by the sub-contractor of his rights to proper payment of the amount due.

## Sub-contractor's right of suspension

Despite the fact that for many years a sub-contractor's threat to suspend performance on site (or indeed to actually suspend performance on site) has been used by many sub-contractors as a lever to force contractors to release overdue payments, there is no common law right to suspend performance where the sum due under the contract has not been paid in full.

Section 112 of the Housing Grants, Construction and Regeneration Act 1996 now gives this right of suspension to parties on a qualifying construction contract, and this therefore introduces a potentially powerful sanction against non-payment.

Section 112 does not expressly prohibit a contractual exclusion or waiver of the right of suspension but it is thought that, in the context of the policy of the Act, it is probably intended that parties cannot exclude this right.

SBCSub/C and SBCSub/D/C deals with this matter under clause 4.11.

Under clause 4.11 a sub-contractor is required to give a seven-day written notice to the contractor of his intention to suspend the performance of his obligations under the sub-contract, and the ground or grounds on which it is intended to suspend such performance, if the contractor fails to make the payment that is due to the sub-contractor in full before the expiry of that seven-day period.

If the contractor does not make payment in full (of the amount due) within the seven-day period, then the sub-contractor can suspend performance until payment in full occurs.

This seems very simple, but a few important points to note are:

(1) A sub-contractor cannot (under the conditions of SBCSub/C and SBCSub/D/C) suspend performance until after he has given a seven-day written notice, and until after that seven-day notice period has expired.

(2) The sub-contractor's right to suspend performance does not affect any other rights and remedies of the sub-contractor. Therefore the sub-contractor would be entitled, for example, to give notice of termination under the contract (which is dealt with in Chapter 12) as well as suspend performance.

(3) If the sub-contractor does (correctly) suspend performance, he is entitled to an extension of time to the period for completion under clause 2.19.5 of SBCSub/C and SBCSub/D/C.

The extension of time period is only for the period when the suspension starts (after the seven-day notice has been given) to the time when payment in full occurs.

There is no guidance as to when 'payment in full' occurs; however, given that it is a well-settled principle of law that a cheque is to be treated as cash, and thus is not susceptible to set-off,[15] it is considered that the courts are likely to decide that payment in full is made when the sub-contractor receives a cheque rather than (the later date) of when the cheque had been cleared through the sub-contractor's account. It should be noted, however, that in rare cases the courts have accepted that an employer could cancel an issued cheque to a contractor.[16]

Irrespective of the above, it is submitted that no 'lead-in' time would be permitted for the sub-contractor to return to site after payment in full has occurred.

Therefore, if the final date for payment was a Monday, but payment was not made, then a sub-contractor may send to the contractor a seven-day written notice on the next day i.e. Tuesday (although, he can, if he wishes, send it at a later date).

Clause 4.11 of SBCSub/C and SBCSub/D/C states that a sub-contractor may suspend performance if payment has still not been made within seven days after the written notice has been given.

Assuming that the written notice was issued on the Tuesday (and assuming that there were no public holidays in the period concerned), to prevent the sub-contractor suspending performance, the contractor would therefore need to make payment in full by no later than the following Tuesday (i.e. seven days after the date of the written notice).

If the contractor did not make payment by that following Tuesday, the sub-contractor would be entitled to suspend performance as from the following day (i.e. the Wednesday).

However, if the contractor did not make payment on the Tuesday or the Wednesday, but made payment in full on the following day (i.e. the Thursday) the sub-contractor would be obliged to return to site on the day following that (i.e. the Friday). The sub-contractor would not (it is submitted) be entitled to an extension of the period for completion under clause 2.19.5 of SBCSub/C and SBCSub/D/C for any delay in the return to work after payment in full has been made.

Therefore, on a practical level it is submitted that not only would the sub-contractor not be entitled to an extension of time to the period for completion under clause 2.19.5 of SBCSub/C and SBCSub/D/C for any extended lead-in periods, but he may also be considered to be in default under clause 7.4.1.1 of SBCSub/C and SBCSub/D/C in the following situations:

(a) the sub-contractor may experience difficulty in returning operatives to site on a Friday and therefore defers returning the operatives to site until the following Monday; or alternatively

---

[15] *Nova (Jersey) Knit Ltd* v. *Kammgarn Spinnerei* [1977] 1 WLR 713, HL; *Isovel Contracts Ltd* (in administration) v. *ABB Building Technologies Ltd* (formerly ABB Steward Ltd) [2002] 1 BCLC 390.
[16] *Willment Brothers Ltd* v. *North West Thames Regional Health Authority* (1984) 26 BLR.

(b) the sub-contractor may have sent the operatives suspended because of non-payment to another site and cannot extract them from that other site without reasonable notice being given, and therefore cannot restart work on the site where performance had been suspended for a further week.

Clause 7.4.1.1 of SBCSub/C and SBCSub/D/C (which is dealt with in Chapter 12) states that a default is where (before practical completion of the sub-contract works) the sub-contractor 'without reasonable cause wholly or substantially suspends the carrying out of the Sub-Contract Works', and if the sub-contractor is considered to be in default in this way this may lead to a notice being issued by the contractor in respect of this default, and, if the default was not corrected by the sub-contractor, this default may lead on to termination by the contractor.

In fact, in an even more extreme case, the contractor may attempt to say that the sub-contractor's 'extended' suspension of performance actually constitutes a repudiation of the sub-contract, which the contractor may accept (the consequences of repudiation are dealt with in Chapter 12).

(4) Further, if the sub-contractor does (correctly) suspend performance, he is entitled, pursuant to clause 4.20.3 of SBCSub/C and SBCSub/D/C, to recover the loss and expense associated with the period of such suspension.

However, somewhat curiously, this right of recovery has the caveat that the suspension by the sub-contractor must not have been 'frivolous or vexatious'.

Frivolous in this context probably means 'paltry' or 'trifling'; and vexatious means 'seeking to annoy'. The intention of this caveat appears to be to prevent the sub-contractor from recovering loss and expense because of suspension where paltry or trifling amounts only are outstanding, or where the suspension is made by the sub-contractor simply to annoy the contractor.

It is unclear how it will be determined what frivolous or vexatious is in respect of any particular sub-contract, and it remains to be seen how often the above caveats will be used, and how effective the caveats are when tested.

(5) The amount to be paid is the amount due, in line with clause 4.10.1 of SBCSub/C and SBCSub/D/C. In addition, the amount to be paid could include:

(a) any VAT properly chargeable in respect of such payment;
(b) any interest payment due under SBCSub/C.

(6) As a sub-contractor can only suspend performance if the amount that is due has not been made at the correct time, a sub-contractor needs to be certain that the amount is actually due before taking this step. The amount actually due may be affected by things such as:

(a) abatement (e.g. a reduction in the value of the works because of defective or incomplete works); and
(b) set-off or a withholding.

A sub-contractor may be at risk in operating this section where the contractor relies in any future action upon an abatement, a set-off or a withholding.

As noted above, a wrongful suspension by the sub-contractor may amount to a repudiation, which, if accepted by the contractor, may lead to the sub-contractor suffering all of the contractor's damages arising from that repudiation (the consequences of repudiation are dealt with in Chapter 12).

(7) The sub-contractor needs to be aware that section 112(4) of the Housing Grants, Construction and Regeneration Act 1996 says:

> 'Any period during which performance is suspended in pursuance of the right conferred by this section shall be disregarded in computing for the purposes of any contractual time limit the time taken, by the party exercising the right *or by a third party*, to complete any work directly or indirectly affected by the exercise of the right.' (italics added)

Also and/or because of the express terms of his sub-subcontract with his own sub-contractors, the sub-subcontractor may also be able to rely upon this provision in respect of its own time period and/or in respect of a loss and expense claim (as appropriate).

The question of suspension of the main contract by the contractor is dealt with in Chapter 7.

## Fluctuation options

Under Clause 4.17 of SBCSub/C and SBCSub/D/C, three fluctuation (i.e. increased cost) provisions are referred to.

Fluctuations option A: contribution, levy and tax fluctuations.
Fluctuations option B: labour and materials cost and tax fluctuations.
Fluctuations option C: formula adjustment.

Which fluctuation option is applicable to any particular sub-contract is indicated under item 10 of the sub-contract particulars in SBCSub/A or SBCSub/D/A.

Footnote 14 to item 10 of the sub-contract particulars states that all but one fluctuations option is to be deleted. Therefore, the intention is that only one fluctuations option will apply to every sub-contract.

Footnote 13 to item 10 of the sub-contract particulars states that where fluctuations option A or B applies in the main contract it is a requirement of that contract that the same option applies to any sub-contract.

Clause 4.18 of SBCSub/C and SBCSub/D/C notes that, irrespective of which fluctuation option applies to the sub-contract works generally, none of the fluctuation options will apply in respect of work for which a schedule 2 quotation has been accepted by the contractor or in respect of a variation to an accepted schedule 2 quotation.

Details of the three fluctuation options are set out in schedule 4 to SBCSub/C and SBCSub/D/C.

### Fluctuations option A

Fluctuations option A relates to 'Contribution, levy and tax fluctuations'.
Under this option, fluctuations are recovered as described below.

## *Labour (covered under paragraph A.1 of fluctuations option A)*

In respect of labour the starting point is that the sub-contract sum or the 'Sub-contract tender sum' is deemed to be based upon the types and rates of contribution, levy and tax (as defined in option A) payable by a person in his capacity as an employer which at the sub-contract base date was payable by the sub-contractor; on the types and rates of refund of the contributions, levies and taxes payable (as defined in option A); and upon the types and rates of premiums receivable (as defined in option A), by a person in his capacity as an employer which at the sub-contract base date was refundable to or receivable by the sub-contractor.

A type and rate so payable are known as a 'tender type' and a 'tender rate' respectively.

The sub-contract base date is indicated under item 3 of the sub-contract particulars (SBCSub/A and SBCSub/D/A).

If, after the sub-contract base date, any of the following apply then the net amount of the difference between what the sub-contractor actually pays (or will pay) in respect of the above and the amount deemed to have been allowed at the sub-contract base date must, as the case may be, be paid to or be allowed by the sub-contractor:

(1) any of the tender rates (except for a rate or levy payable by virtue of the Industrial Training Act 1982) is increased or decreased; or
(2) if a tender type ceases to be payable; or
(3) if a new type of contribution, levy or tax which is payable by a person in his capacity as an employer becomes payable.

The above increases or decreases relate only to workpeople engaged upon or in connection with the sub-contract works (either on or adjacent to the site) and to workpeople directly employed by the sub-contractor whilst they are engaged upon the production of materials and goods for use in or in connection with the sub-contract works (assuming that these workpeople are not on or adjacent to the site).

In respect of the above, 'workpeople' are defined as being 'persons whose rates of wages and other emoluments are governed by the rules or decisions or agreements of the Construction Industry Joint Council or some other recognized wage fixing body for trades associated with the building industry'.

In addition to the above, a sub-contractor is entitled to be paid increased costs in respect of persons employed by the sub-contractor who are engaged upon or in connection with the sub-contract works (either on or adjacent to the site) but who are not covered by the above definition of 'workpeople'. This may cover for people such as foremen, quantity surveyors and contracts managers (when on or adjacent to the site).

When this latter category of persons is considered, the amount of any increase or decrease is to be the same as the rate applicable for a craftsman.

Further, when making calculations for this latter category of persons, it should be noted that the time spent on site is to be measured in whole working days only; no period of less than two whole working days in any week must be taken into account; and periods of less than a whole working day shall not be aggregated to amount to a whole working day.

Paragraph A.12 of fluctuations option A notes that there must be added to the amount paid to or allowed by the sub-contractor in respect of the above labour items, the percentage stated under item 10 of the sub-contract particulars (against the heading 'Percentage addition for fluctuations option A'). The percentage addition inserted is required by the main contract to be the same as that stated in the main contract particulars against the entry for clause 4.21 and schedule 7.

### *Materials (covered under paragraph A.2 of fluctuations option A)*

In respect of materials, the starting point is that the sub-contract sum or the sub-contract tender sum is deemed to be based upon the types and rates of duty and tax (as defined in option A) by whomsoever payable, which at the sub-contract base date was payable on the import, purchase, sale, appropriation, processing, use or disposal of the materials, goods (as defined in option A), electricity, fuels, materials taken from the site as waste or any other solid, liquid or gas necessary for the execution of the sub-contract works by virtue of any Act of Parliament.

A type and rate so payable are known as a 'tender type' and a 'tender rate' respectively.

The sub-contract base date is indicated under item 3 of the sub-contract particulars (SBCSub/A and SBCSub/D/A).

If, after the sub-contract base date, any of the following apply then the net amount of the difference between what the sub-contractor actually pays (or will pay) in respect of the above and the amount deemed to have been allowed at the sub-contract base date shall, as the case may be, be paid to or be allowed by the sub-contractor:

(1) any of the tender rates are increased or decreased; or
(2) a tender type ceases to be payable; or
(3) a new type of duty or tax becomes payable.

Paragraph A.12 of fluctuations option A notes that the percentage stated under item 10 of the sub-contract particulars (against the heading 'Percentage addition for fluctuations option A') must be added to the amount paid to or allowed by the sub-contractor in respect of the above material, etc. items. The percentage addition inserted is required by the main contract to be the same as that stated in the main contract particulars against the entry for clause 4.21 and schedule 7.

### *Sub-let work (covered under paragraph A.3 of fluctuations option A)*

If the sub-contractor sub-lets any portion of the sub-contract works to a sub-subcontractor he must incorporate in the sub-subcontract provisions like the provisions of fluctuations option A, including the percentage stated in the sub-contract particulars (item 10) pursuant to paragraph A.12.

If the price payable under such a sub-subcontract is increased above or decreased below the price in such a sub-subcontract by reason of the operation of the incorporated provisions, then the net amount of such increase or decrease must, as the case may be, be paid to or allowed by the sub-contractor under the sub-contract. It should be noted that the above increase or decrease is recoverable on a net basis only; the sub-contractor cannot apply a further percentage to any net increases or decreases of its sub-subcontractor. This position is consistent with paragraph A.8 of fluctuations option A, as noted below.

SBCSub/C and SBCSub/D/C does not explain what happens if a sub-contractor sub-lets any of the sub-contract works without incorporation of provisions to like effect as noted above. Despite this it would be sensible for a sub-contractor to ensure that such provisions are incorporated into any sub-subcontract that it lets, so that the provisions of paragraph A.3 can be applied.

### Written notice by sub-contractor (covered under paragraph A.4 of fluctuations option A)

Paragraph A.4.1 makes it clear that the sub-contractor must give a written notice to the contractor within a reasonable time after the occurrence of an event which may result in increased (or decreased) costs, as identified above.

Paragraph A.4.2 states that a written notice regarding any particular event, is a condition precedent to any payment being made to the sub-contractor in respect of the event in question.

A 'condition precedent' means that one party to a contract must do something (e.g. issue a notice) before the other party to the contract is obliged to perform his part of the agreement (e.g. make payment). Paragraph A.4.2 appears to be a fairly clear example of a condition precedent.

### Agreement – contractor and sub-contractor (covered under paragraph A.5 of fluctuations option A)

Paragraph A.5 of the fluctuations option A permits the contractor and the sub-contractor to agree what will be deemed for all of the purposes of the sub-contract to be the net amount payable to (or allowable by) the sub-contractor as a result of an event which may result in increased (or decreased) costs, as identified above.

This clause may be sensibly used where the anticipated increase or decrease in value is less than the anticipated administration costs in ascertaining the actual increase or decrease in value. Alternatively, where the firming up of an approximated fluctuation amount will result in an anticipated increase or decrease in value which is less than the anticipated administration costs in ascertaining the actual increase or decrease in value.

### Fluctuations – final sub-contract sum (covered under paragraph A.6 of fluctuations option A)

Although paragraph A.6 only refers to any fluctuation amounts which become payable or allowable being taken into account in calculating the final sub-contract sum or when the sub-contractor prepares and submits an account to the contractor as a consequence of termination under clauses 7.8 to 7.10 of SBCSub/C or SBCSub/D/C (as detailed under clause 7.11.3 of SBCSub/C and SBCSub/D/C), in fact fluctuation amounts should also be included within the gross valuation for interim payments as noted under clause 4.13.6 of SBCSub/C and SBCSub/D/C.

### Evidence and computations by sub-contractor (covered under paragraph A.7 of fluctuations option A)

Under this paragraph, the sub-contractor, as soon as is reasonably practicable, is to provide such evidence and computations as the contractor may reasonably require to enable the fluctuations amount to be ascertained.

In respect of persons other than workpeople (e.g. foremen, quantity surveyors, etc. whether these are the sub-contractor's own employees or the employees of a sub-subcontractor) the sub-contractor is to provide a signed certificate each week certifying the validity of the evidence reasonably required to ascertain such amounts. Such a weekly certificate would need to provide at least some reference (by date or by number) of the week being considered, the name of the person involved, their position (e.g. foreman), and the hours and days that they were engaged upon or in connection with the sub-contract works either on or adjacent to the site in the week in question.

### No alteration to sub-contractor's profit (covered under paragraph A.8 of fluctuations option A)

This paragraph makes it clear that no addition to or deduction from the sub-contract sum as a result of any fluctuations will alter in any way the amount of profit that the sub-contractor includes in the sub-contract sum. Therefore, in respect of both increased and decreased costs the sub-contractor is to take the net increases and the net decreases only. This is consistent with the position in respect of sub-subcontractors (see commentary on paragraph A.3 above).

### Position where sub-contractor is in default over completion (covered under paragraph A.9 of fluctuations option A)

If the sub-contractor has failed to complete on time (as notified by the contractor under clause 2.21 of SBCSub/C or SBCSub/D/C) then fluctuations will not be taken into account past the date when the sub-contractor should have completed its works (which will be in line with either the sub-contractor's original period or periods for completion, or with any extended period or periods for completion as granted by the contractor under clause 2.18.1 of SBCSub/C or SBCSub/D/C, as appropriate).

The above restriction on fluctuations will only apply if:

(1) the printed text of clauses 2.16 to 2.19 of SBCSub/C and SBCSub/D/C is unamended and forms part of the sub-contract conditions; and
(2) the contractor has, in respect of every written notification by the sub-contractor under clause 2.18 of SBCSub/C or SBCSub/D/C given his decision in writing to such revision, if any, of the period or periods for completion of the sub-contract works as he considers to be in accordance with that clause.

### Work, etc. to which paragraphs A.1 to A.3 not applicable (covered under paragraph A.10 of fluctuations option A)

Paragraph A.10 of fluctuations option A makes it clear that fluctuations will not apply to:

(1) work for which the sub-contractor is allowed daywork rates under clause 5.9 of SBCSub/C or SBCSub/D/C; or
(2) changes in the rates of VAT charged on the supply of goods or services by the sub-contractor to the contractor under the sub-contract.

## Fluctuations option B

Fluctuations option B relates to 'Labour and materials cost and tax fluctuations'. Under this option, fluctuations are recovered as described below.

### *Labour (covered under paragraphs B.1 and B.2 of fluctuations option B)*

In respect of labour, the starting point is that the sub-contract sum or the sub-contract tender sum is deemed to be based upon the rates of wages and other emoluments (as defined) payable by the sub-contractor at the sub-contract base date; the transport charges, etc. (as listed under sub-contract particulars item 10 of SBCSub/A or SBCSub/D/A), the types and rates of contribution, levy and tax payable by a person in his capacity as an employer (as defined in option B) which at the sub-contract base date was payable by the sub-contractor, and on the types and rates of refund of the contributions, levies and taxes payable (as defined under option B), and upon the types and rates of premiums receivable (as defined under option B), by a person in his capacity as an employer which at the sub-contract base date was refundable to or receivable by the sub-contractor.

A type and rate so payable is known as a 'tender type' and a 'tender rate' respectively.

The sub-contract base date is indicated under item 3 of the sub-contract particulars (SBCSub/A and SBCSub/D/A).

If, after the sub-contract base date any of the following apply, then the net amount of the difference between what the sub-contractor actually pays (or will pay) in respect of the above and the amount deemed to have been allowed at the sub-contract base date must as the case may be, be paid to or be allowed by the sub-contractor:

(1) any of the tender rates (including wages and emoluments, and transport charges, etc.) are increased or decreased; or
(2) a tender type ceases to be payable; or
(3) a new type of contribution, levy or tax which is payable by a person in his capacity as an employer becomes payable.

The above increases or decreases relate only to workpeople engaged upon or in connection with the sub-contract works (either on or adjacent to the site) and to workpeople directly employed by the sub-contractor whilst they are engaged upon the production of materials and goods for use in or in connection with the sub-contract works (assuming that those workpeople are not on or adjacent to the site).

In respect of the above, 'workpeople' are defined as being 'Persons whose rates of wages and other emoluments are governed by the rules or decisions or agreements of the Construction Industry Joint Council or some other recognized wage fixing body for trades associated with the building industry'.

In addition to the above, a sub-contractor is entitled to be paid increased costs in respect of persons employed by the sub-contractor who is engaged upon or in connection with the sub-contract works (either on or adjacent to the site) but is not covered by the above definition of workpeople. This may cover for people such as foremen, quantity surveyors and contracts managers (when on or adjacent to the site).

When this latter category of persons is considered, the amount of any increase or decrease is to be the same as the rate applicable for a craftsman.

When making calculations for this latter category of persons, it should be noted that the time spent on site is to be measured in whole working days only; no period of less than two whole working days in any week must be taken into account and periods of less than a whole working day shall not be aggregated to amount to a whole working day.

Paragraph B.13 of fluctuations option B notes that the percentage stated under item 10 of the sub-contract particulars (against the heading 'Percentage addition for fluctuations option B') must be added to the amount paid to or allowed by the sub-contractor in respect of the above labour items. The percentage addition inserted is required by the main contract to be the same as that stated in the main contract particulars against the entry for clause 4.21 and schedule 7.

### Materials, goods, electricity and fuels (covered under paragraph B.3 of fluctuations option B)

In respect of materials, goods, electricity and fuels, the starting point is that the sub-contract sum or the sub-contract tender sum is deemed to be based upon the market prices which were current at the sub-contract base date, (including the types and rates of duty and tax) by whomsoever payable which at the sub-contract base date was payable on the import, purchase, sale, appropriation, processing, use or disposal of the materials, goods (as defined), electricity, fuels, materials taken from the site as waste or any other solid, liquid or gas necessary for the execution of the sub-contract works by virtue of any Act of Parliament.

A type and rate so payable are known as a 'tender type' and a 'tender rate' respectively.

The sub-contract base date is indicated under item 3 of the sub-contract particulars (SBCSub/A and SBCSub/D/A).

If, after the sub-contract base date, the market prices increase or decrease, and/or any of the tender rates increase or decrease, and/or a tender type ceases to be payable, and/or a new type of duty or tax becomes payable; then the net amount of the difference between what the sub-contractor actually pays (or will pay) in respect of the above and the amount deemed to have been allowed at the sub-contract base date (as noted above) must, as the case may be, be paid to or be allowed by the sub-contractor.

Paragraph B.13 of fluctuations option B notes that there must be added to the amount paid to or allowed by the sub-contractor in respect of the above materials, items, etc., the percentage stated under item 10 of the sub-contract particulars (against the heading 'Percentage addition for fluctuations option B'). The, addition inserted is required by the main contract to be the same as that stated in the main contract particulars against the entry for clause 4.21 and schedule 7.

### Sub-let work (covered under paragraph B.4 of fluctuations option B)

If the sub-contractor sub-lets any portion of the sub-contract works to a sub-subcontractor he must incorporate in the sub-subcontract provisions the like provisions of fluctuations option B, including the percentage stated in the sub-contract particulars (item 10) pursuant to paragraph B.13.

If the price payable under such a sub-subcontract is increased above or decreased below the price in such sub-subcontract by reason of the operation of the incorporated provisions, then the net amount of such increase or decrease must, as the case may be, be paid to or be allowed by the sub-contractor under the sub-contract. It should be noted that the above increase or decrease is recoverable on a net basis only; the sub-contractor cannot apply a further percentage to any net increases or decreases of its sub-subcontractor. This position is consistent with paragraph B.9 of fluctuations option B, as noted below.

SBCSub/C and SBCSub/D/C does not explain what happens if a sub-contractor sub-lets any of the sub-contract works without incorporation of provisions to like effect. Despite this it would be sensible for a sub-contractor to ensure that such provisions are incorporated into any sub-subcontract that it lets, so that the provisions of paragraph B.4 can be applied.

### Written notice by sub-contractor (covered under paragraph B.5 of fluctuations option B)

Paragraph B.5.1 makes it clear that the sub-contractor must give a written notice to the contractor within a reasonable time after the occurrence of an event which may result in increased (or decreased) costs, as identified above.

Paragraph B.5.2 states that a written notice regarding any particular event, is a condition precedent to any payment being made to the sub-contractor in respect of the event in question.

A 'condition precedent' means that one party to a contract must do something (e.g. issue a notice) before the other party to the contract is obliged to perform his part of the agreement (e.g. make payment). Paragraph B.5.2 appears to be a fairly clear example of a condition precedent.

### Agreement – contractor and sub-contractor (covered under paragraph B.6 of fluctuations option B)

Paragraph B.6 of the fluctuations option B permits the contractor and the sub-contractor to agree what shall be deemed for all of the purposes of the sub-contract to be the net amount payable to (or allowable by) the sub-contractor as a result of an event which may result in increased (or decreased) costs, as identified above.

This clause may be sensibly used where the anticipated increase or decrease in value is less than the anticipated administration costs in ascertaining the actual increase or decrease in value. Alternatively, it can be used where the firming up of an approximated fluctuation amount will result in an anticipated increase or decrease in value which is less than the anticipated administration costs in ascertaining the actual increase or decrease in value.

### Fluctuations – final sub-contract sum (covered under paragraph B.7 of fluctuations option B)

Although paragraph B.7 only refers to any fluctuation amounts which become payable or allowable being taken into account in calculating the final sub-contract sum or when the sub-contractor prepares and submits an account to the contractor as a consequence of termination under clauses 7.8 to 7.10 of SBCSub/C or

SBCSub/D/C (as detailed under clause 7.11.3 of SBCSub/C), in fact fluctuation amounts should also be included within the gross valuation for interim payments as noted under clause 4.13.6 of SBCSub/C and SBCSub/D/C.

### *Evidence and computations by sub-contractor (covered under paragraph B.8 of fluctuations option B)*

Under this paragraph, the sub-contractor, as soon as is reasonably practicable, is to provide such evidence and computations as the contractor may reasonably require to enable the fluctuations amount to be ascertained.

In respect of persons other than workpeople (e.g. foremen, quantity surveyors, etc. whether these are the sub-contractor's own employees or the employees of a sub-subcontractor) the sub-contractor is to provide a signed certificate each week certifying the validity of the evidence reasonably required to ascertain such amounts. Such a weekly certificate would need to provide at least some reference (by date or by number) of the week being considered, the name of the person involved, their position (e.g. foreman), and the hours and days that they were engaged upon or in connection with the sub-contract works either on or adjacent to the site in the week in question.

### *No alteration to sub-contractor's profit (covered under paragraph B.9 of fluctuations option B)*

This paragraph makes it clear that no addition to or deduction from the sub-contract sum as a result of any fluctuations will alter in any way the amount of profit that the sub-contractor includes in the sub-contract sum. Therefore, in respect of both increased and decreased costs the sub-contractor is to take the net increases and the net decreases only. This is consistent with the position in respect of sub-subcontractors (see commentary on paragraph B.4 above).

### *Position where sub-contractor in default over completion (covered under paragraph B.10 of fluctuations option B)*

If the sub-contractor has failed to complete on time (as notified by the contractor under clause 2.21 of SBCSub/C or SBCSub/D/C) then fluctuations will not be taken into account past the date when either the sub-contractor should have completed its works (which will be in line with the sub-contractor's original period or periods for completion, or will be in line with any extended period or periods for completion as granted by the contractor under clause 2.18.1 of SBCSub/C or SBCSub/D/C, as appropriate).

The above restriction on fluctuations will only apply if:

(1) the printed text of clauses 2.16 to 2.19 of SBCSub/C or SBCSub/D/C is unamended and forms part of the sub-contract conditions; and
(2) the contractor has, in respect of every written notification by the sub-contractor under clause 2.18 of SBCSub/C or SBCSub/D/C, given his decision in writing to such revision, if any, of the period or periods for completion of the sub-contract works as he considers to be in accordance with that clause.

***Work, etc. to which paragraphs B.1 to B.4 not applicable (covered under paragraph B.11 of fluctuations option B)***

Paragraph B.11 of fluctuations option B makes it clear that fluctuations will not apply to:

(1) work for which the sub-contractor is allowed daywork rates under clause 5.9 of SBCSub/C or SBCSub/D/C; or
(2) changes in the rates of VAT charged on the supply of goods or services by the sub-contractor to the contractor under the sub-contract.

## Fluctuations option C

Fluctuations option C deals with the recovery of fluctuations on a formula basis.

Paragraph C.1 states that the sub-contract sum or sub-contract tender sum must be adjusted in accordance with the provisions of fluctuations option C and the formula rules issued for use with fluctuations option C by the JCT, current at the sub-contract base date.

The sub-contract base date is indicated under item 3 of the sub-contract particulars (SBCSub/A or SBCSub/D/A).

Using the formula rules, increased costs are calculated by taking the increase in a monthly index figure, calculated from the index figure applicable at the sub-contract base date to the index figure applicable at the date when the work is actually executed.

A non-adjustable element is then applied to the calculation. This non-adjustable element (which is not to exceed 10%) is to be inserted (against clause C.1.4) in the 'Supplemental particulars' forming part of SBCSub/A and SBCSub/D/A.

If fluctuations option C applies to the main contract, the non-adjustable element percentage should be the same percentage as that stated in the schedule of information (of the main contract works) attached as part of SBCSub/A and SBCSub/D/A.

A non-adjustable element is *only* applicable where the employer is a local authority.

The calculation of increased costs on a formula basis is calculated in line with the principle of the following example:

[V] = Value of work executed in interim valuation no. 4:     £10,000
[FIV] = Formula index at date of valuation no. 4:            110
[FIBM] = Formula index at base month:                        100
[NA] = Non-adjustable element:                               10%

Increased costs [IC] equals:

$$\frac{[FIV - FIBM]}{[FIBM]} \times [100 - NA]\% \times [V] = [IC]$$

$$\frac{110 - 100}{100} \times [100 - 10]\% \times £10,000 = £900$$

There are particularly detailed formula rules for specific situations that may be encountered, and information regarding this is to be provided by the parties (where applicable) in the supplemental particulars forming part of SBCSub/A and SBCSub/D/A.

The information to be provided includes:

- the definition of the balance of adjustable work (rule 3);
- the method of dealing with 'fix-only' work (rule 8(i), rule 8(ii), rule 8(iii));
- the works categories applicable to the sub-contract works, and the classification of items as referred to in rule 11a or 11b;
- the weightings between labour and materials for electrical installations; for heating, ventilating and air conditioning installations; and/or for sprinkler installations (as appropriate) (rule 43);
- confirmation as to whether the separate materials index for sprinkler installations will or will not apply (rule 55a);
- confirmation of when the formula adjustment for the lift installation will take effect (rule 61a);
- the average price per tonne of delivered steel and/or of erected steel (rule 64);
- information regarding catering equipment installations (rule 70a).

### Interim valuations

Paragraph C.2 of fluctuations options C notes that the sub-contractor is entitled to make representations to the contractor at the time of each interim valuation as to the value of work to which formula adjustment is to be made and adds that any adjustment under fluctuations option C must be effected in all such payments. Because of the provisional nature of the monthly indices when first published, paragraph C.1.5 allows for a correction (either an addition or a deduction) to the fluctuation amount due (based upon the firm monthly indices when published). This adjustment is to be effected in the appropriate interim valuation.

### Articles manufactured outside the UK

Paragraph C.3 of fluctuations option C deals with the matter of increased costs on articles manufactured outside the United Kingdom.

In such a situation, the sub-contractor is to set out (against clause C.3) in the Supplemental particulars forming part of SBCSub/A and SBCSub/D/A, a list of the applicable articles and the market price of those materials, converted into sterling at the sub-contract base date. Any increases or decreases on those market prices at the time of purchase are recoverable by the sub-contractor.

Paragraph 81 of SBCSub/G notes that if the contractor wishes to recover the cost of the above articles manufactured outside the United Kingdom, then (assuming that fluctuations option C applies to the main contract) the contractor needs to append the same list of articles and prices, etc. to the main contract documents.

### Power to agree – contractor and sub-contractor

Paragraph C.4 of fluctuations option C permits the contractor and the sub-contractor to agree what shall be deemed for all of the purposes of the sub-contract to be the formula adjustment payable.

There are two provisos applicable, as follows:

- any amount agreed should be the same, or approximately the same, as the amount that would be due if the formula calculation was carried out;
- any agreement reached will not bind a sub-subcontractor.

This clause may be sensibly used where the anticipated increase or decrease in value (or the difference between an estimated value and an approximate value) is less than the anticipated administration costs in ascertaining the actual increase or decrease in value.

### Position where monthly bulletins are delayed

Paragraph C.5 details the procedure to be followed if the publication of the monthly bulletins is delayed, or if they cease being published.

### Formula adjustment – failure to complete

The position where a sub-contractor is in default over completion is covered under paragraph C.6 of fluctuations option C.

If the sub-contractor has failed to complete on time (as notified by the contractor under clause 2.21 of SBCSub/C or SBCSub/D/C) then fluctuations will not be taken into account past the date when either the sub-contractor should have completed its works (which will be in line with the sub-contractor's original period or periods for completion, or will be in line with any extended period or periods for completion as granted by the contractor under clause 2.18.1 of SBCSub/C or SBCSub/D/C, as appropriate).

This will be effected by using the monthly index number applicable to the valuation period in which the sub-contractor should have completed its work (as defined above) rather than using the monthly index number current at the interim valuation date.

The above restriction on fluctuations will only apply if:

(1) the printed text of clauses 2.16 to 2.19 of SBCSub/C is unamended and forms part of the sub-contract conditions; and
(2) the contractor has, in respect of every written notification by the sub-contractor under clause 2.18 of SBCSub/C or SBCSub/D/C, given his decision in writing to such revision, if any, of the period or periods for completion of the sub-contract works as he considers to be in accordance with that clause.

# Chapter 9
# Loss and Expense

## Introduction

When considering loss and expense, the first thing that must be noted is that claims in the construction industry generally fall under two differing legal heads. That is:

(1) claims for breach of contract or for other events for which specific provision is made within the conditions of the contract (normally referred to as loss and expense claims); and
(2) claims for breach of contract for which no specific provision has been included in the conditions of contract (normally referred to as common law damages claims).

Loss and expense claims have certain advantages over common law damages claims, and some of these advantages are:

- There is a clear definition of responsibilities and procedures.
- There is a clarification of the events and the circumstances for claims to be made.
- There is a right to interim payments and/or decisions.
- There is the avoidance of the uncertainties involved in claiming at common law.
- Loss and expense claims create a right under the contract to a debt, rather than a claim for damages.

Normally, the inability to recover monies under the express terms of the contract for reasons such as not conforming with the procedures required does not preclude a claim at common law on the same matters, provided there is the entitlement at common law to claim the damages, and provided that the sub-contractor's common law rights have not been specifically excluded by the sub-contract.

If a matter stated in the sub-contract as giving rise to an entitlement to loss and expense is a breach of contract, then the sub-contractor will normally have a right to common law damages. An example of this would be the late issue of information by the contractor.

However, if the matter relied on does not amount to a breach of contract (i.e. issuing variations, etc.) then common law damages would not normally be applicable.

On the other hand, the preservation of the sub-contractor's common law rights may entitle him to make common law damages claims in respect of matters for which there are no express provisions within the loss and expense clauses (for example the breach of any applicable implied terms by the contractor).

The authority on the application of damages arising from the contract both in addition to and, where appropriate, instead of the express provisions of the contract for loss and expense, is clearly stated in the case of *Stanley Hugh Leach* v. *Merton*.[1]

It is important to note that should the sub-contractor need to pursue his claim at common law because he has not followed the procedures under the sub-contract, then it is possible that he may be penalized by the court if his non-conformity with the sub-contract procedure did not allow the contractor to mitigate his loss.

What this means in practice is that if the contractor had been made aware of the pending claim at an earlier stage (in line with the express provisions of the sub-contract) it may have been able to take steps to reduce the effects of the breach. If the contractor was not given the opportunity to mitigate his loss in this way, the sub-contractor may have any future award in damages reduced by a court because of this failing.

Whereas under the main contract the architect/contract administrator cannot, without the express authority of the employer, decide on claims that are not expressly provided for under the contract and therefore cannot decide on common law damages claims, no such restriction exists in the contractor/sub-contractor relationship.

For the avoidance of any doubt, in respect of the above, clause 4.22 of SBCSub/C and SBCSub/D/C preserves the common law damages rights and the remedies of the contractor and the sub-contractor.

## Loss and expense

### Matters affecting regular progress

People in the construction industry often make the mistake (contractually) of linking time and money, in that they assume that financial claim recovery can *only* be secured if an extension of time has been granted and, conversely, that it will *certainly* be due if an extension has been given. In reality, neither of these propositions is accurate.

Extension of time clauses and those for additional financial recovery (i.e. for loss and expense) are separate matters in SBCSub/C and SBCSub/D/C as they are in all other JCT standard forms.

Additional monies may be recovered even if prolongation is not present, or alternatively, the sub-contractor may receive a comprehensive extension of time and still not be entitled to additional financial recovery.

Therefore, it is advisable that delay and money notices are kept separate.

The question of extensions of time has been dealt with in Chapter 5.

In respect of loss and expense, in line with clause 4.19 of SBCSub/C and SBCSub/D/C, if the regular progress of the sub-contract works is materially

---

[1] *Stanley Hugh Leach* v. *London Borough of Merton* (1985) 32 BLR 51.

affected or is likely to be materially affected by any of the relevant sub-contract matters (as detailed below), the sub-contractor may make written application to the contractor for the recovery of the resultant loss and expense incurred.

Any such loss and expense is that for which the sub-contractor would not be reimbursed by a payment under any other provision in SBCSub/C or SBCSub/D/C.

Therefore, for example, where the loss and/or expense is in respect of a variation for which a schedule 2 quotation has been accepted, any applicable loss and expense amount is to be paid as part of the accepted quotation, and clause 4.19 of SBCSub/C and SBCSub/D/C will not apply.

Apart from where the sub-contractor is reimbursed for his loss and expense by a payment under any other provision in SBCSub/C or SBCSub/D/C, if the sub-contractor makes a written application for loss and expense, then the amount of the payable loss and expense agreed between the contractor and the sub-contractor must be taken into account in the calculation of the final sub-contract sum or shall be recoverable by the sub-contractor from the contractor as a debt, provided that:

- the sub-contractor's application is made as soon as it has become, or should reasonably have become apparent to him that the regular progress has been or is likely to be affected (clause 4.19.1 of SBCSub/C and SBCSub/D/C);
- the sub-contractor submits (upon request from the contractor) such information as is reasonably necessary to show that the regular progress has been or is likely to be affected (clause 4.19.2 of SBCSub/C and SBCSub/D/C); and
- the sub-contractor submits details of the loss and expense as reasonably requested by the contractor (clause 4.19.3 of SBCSub/C and SBCSub/D/C).

It is very important to note that loss and expense is only recoverable where it can be shown that the above procedures have been followed and that a relevant sub-contract matter has affected the regular progress of the sub-contract works.

## Relevant sub-contract matters

The relevant sub-contract matters are listed under clause 4.20 of SBCSub/C and SBCSub/D/C, and each of these matters is considered below.

The relevant sub-contract matters, by and large, replicate the relevant sub-contract events detailed under clause 2.19 of SBCSub/C and SBCSub/D/C.

Commentary is given in respect of the relevant sub-contract events in Chapter 5. Therefore, reference should be made to the appropriate relevant sub-contract event (as indicated) for further general commentary on the relevant sub-contract matter being considered.

'Clause 4.20.1 Variations (excluding any for which a Schedule 2 Quotation has been accepted by the Contractor, but including any other matters or directions which under these Conditions are to be treated as, or as requiring, a Variation).'

For further commentary refer to clause 2.19.1 of SBCSub/C and SBCSub/D/C, in Chapter 5.

'Clause 4.20.2 Directions of the Contractor, including those which pass on instructions of the Architect/Contract Administrator:

4.20.2.1 for the expenditure of Provisional Sums included in the Contractor's Requirements or any Bills of Quantities, excluding an instruction for expenditure of a Provisional Sum for defined work;

4.20.2.2 for the opening up for inspection or testing of any work, materials or goods under clause 3.17 of the Main Contract Conditions (including making good), unless the inspection or test shows that the work, materials or goods were not in accordance with the Sub-Contract;

4.20.2.3 for the opening up for inspection or testing of any work, materials or goods under clause 3.10 (including making good), unless the inspection or test shows that the work, materials or goods were not in accordance with this Sub-Contract;

4.20.2.4 in relation to any discrepancy in or divergence between any of the Numbered Documents or any discrepancy in or divergence between any of those documents and the Contract Documents under the Main Contract;

4.20.2.5 for the postponement of any work to be executed under this Sub-Contract (whether in connection with a postponement under the Main Contract or otherwise);

4.20.2.6 with respect to any find of antiquities.'

For further commentary, refer to clause 2.19.2 of SBCSub/C and SBCSub/D/C, in Chapter 5.

In respect of clause 4.20.2.1, the reference to contractor's requirements only relates to SBCSub/D/C.

In respect of clause 4.20.2.2, opening up and testing in respect of clause 3.17 of the main contract conditions is not part of the relevant sub-contract matter, but it is part of the relevant sub-contract event.

'Clause 4.20.3 Suspension by the Sub-contractor under clause 4.11 of the performance of his obligations under this Sub-contract, *provided the suspension was not frivolous or vexatious*' (italics added). For further commentary on this refer to clause 2.19.5 of SBCSub/C and SBCSub/D/C.

The wording, 'provided the suspension was not frivolous or vexatious' is not included as part of the relevant sub-contract event (clause 2.19.5 of SBCSub/C and SBCSub/D/C). It would therefore appear that an extension of time would be granted even if the suspension was frivolous or vexatious, but loss and expense would not be recoverable in such a situation.

Frivolous in this context probably means 'paltry' or 'trifling'; and vexatious means 'seeking to annoy'. The intention of this caveat appears to be to prevent the sub-contractor from recovering loss and expense because of suspension where paltry or trifling amounts only are outstanding, or where the suspension is made by the sub-contractor simply to annoy the contractor.

Therefore it would appear that a sub-contractor will not be penalized in terms of time because of a legitimate suspension (even if, for example, only a trifling amount of monies were outstanding) but he may be penalized in terms of recovering any loss and expense that he incurs associated with such suspension.

It is unclear how it will be determined what frivolous or vexatious is in respect of any particular sub-contract, and it remains to be seen how often the above caveats will be used, and how effective the caveats are when tested.

'Clause 4.20.4 Suspension by the Contractor under clause 4.14 of the Main Contract Conditions of the performance of his obligations under the Main Contract.' For further commentary on this refer to clause 2.19.6 of SBCSub/C and SBCSub/D/C, in Chapter 5.

'Clause 4.20.5 The execution of work for which an Approximate Quantity is included in the Contract Bills which is not a reasonably accurate forecast of the quantity of work required.' For further commentary on this refer to clause 2.19.4 of SBCSub/C and SBCSub/D/C in Chapter 5.

'Clause 4.20.6 Where there are Bills of Quantities, the execution of work for which an Approximate Quantity included in those bills is not a reasonably accurate forecast of the quantity of work required.' For further commentary on this refer to clause 2.19.4 of SBCSub/C and SBCSub/D/C, in Chapter 5.

'Clause 4.20.7 Any impediment, prevention or default, whether by act or by omission, by the Employer, the Architect/Contract Administrator, the Quantity Surveyor or any of the Employer's Persons except to the extent caused or contributed to by any default, whether by act or omission, of the Sub-contractor or of any of the Sub-contractor's Persons.'

For further commentary on this refer to clause 2.19.7 of SBCSub/C and SBCSub/D/C, in Chapter 5.

Clause 4.20.8 'Any impediment, prevention or default, whether by act or by omission, by the Contractor or any of the Contractor's Persons (including, where the Contractor is the Principal Contractor, any default, whether by act or omission, in that capacity) except to the extent caused or contributed to by any default, whether by act or omission, of the Sub-Contractor or of any of the Sub-Contractor's Persons.'

For further commentary on this refer to clause 2.19.7 of SBCSub/C and SBCSub/D/C in Chapter 5.

In respect of the words: 'except to the extent caused or contributed to by any default, whether by act or omission, of the Sub-contractor or of any of the Sub-Contractor's Persons' within clauses 4.20.7 and 4.20.8 of SBCSub/C and SBCSub/D/C, it should be particularly noted that clauses 2.7.3 and 2.7.4 of SBCSub/C and SBCSub/D/C require that:

- such further drawings, details, information and directions referred to in clauses 2.7.1 and 2.7.2 of SBCSub/C and SBCSub/D/C must be provided or given at the time it is reasonably necessary for the recipient party to receive them, having regard to the progress of the sub-contract works and the main contract works; and

- where the recipient party has reason to believe that the other party is not aware of the time by which the recipient needs to receive such further drawings, details, information or directions, he must, so far as is reasonably practicable, advise the other party sufficiently in advance to enable him to comply with the requirements of clause 2.7 of SBCSub/C or SBCSub/D/C.

## Contractor's reimbursement

Clause 4.21 of SBCSub/C and SBCSub/D/C gives the contractor a right to recover from the sub-contractor the agreed amount of any direct loss and expense caused to the contractor where the regular progress of the main contract works is materially affected by any act, omission or default of the sub-contractor, or any of the sub-contractor's persons.

As detailed in Chapter 5, clause 2.21 of SBCSub/C and SBCSub/D/C deals with the recovery by the contractor of any direct loss and/or expense suffered or incurred by the contractor caused by the culpable failure of the sub-contractor to complete the sub-contract works on time.

In respect of clause 4.21, the contractor is to notify the sub-contractor in writing (in line with clause 4.21.1) with reasonable particulars of the effects or likely effects on the regular progress of the main contract works resulting from any act, omission or default of the sub-contractor, or any of the sub-contractor's persons, and must also give details of the resultant loss and expense as the sub-contractor reasonably requests.

Any amount of loss and expense agreed between the contractor and the sub-contractor relating to clause 4.21.1 may be deducted from any monies due to or to become due to the sub-contractor or must be recoverable by the contractor from the sub-contractor as a debt (as clause 4.21.2 of SBCSub/C and SBCSub/D/C).

If the contractor follows the option of deducting the loss and expense from monies due to the sub-contractor, it is important that a withholding notice required by clause 4.10.3 and/or 4.12.3 of SBCSub/C and SBCSub/D/C is issued timeously.

## In pursuing a loss and expense claim, what does a sub-contractor need to prove?

Most claims, whether for time or money, involve establishing what is often called the nexus of cause and effect (i.e. the link between cause and effect). This means that what needs to be proved is that, because of the occurrence of a particular event certain things happened and as a direct result of those events happening one of the parties incurred delays or costs which were not previously contemplated by the parties and which it would not have been reasonable for the parties to have contemplated.

In order to establish this nexus (or link), which is never an easy task in a construction situation, records need to be available which show that the circumstances that existed before an event occurred changed after that event occurred, and that that change in circumstances could only have been as a result of the event in question.

### Burden of proof

Even at an early stage of any claim preparation it is worth noting that if the claim proceeds to formal proceedings, the burden of proof will lie with the party that makes an assertion. This follows the legal maxim 'he who asserts must prove'.

This means, in effect, that it is not for a party defending a claim to disprove the claim; it is for the party pursuing a claim to prove the claim.

However, if the party pursuing a claim adduces sufficient evidence to raise a presumption that what is claimed is true, the burden of proof will pass to the other party, and it is then for that party to adduce sufficient evidence to rebut the presumption.

Therefore, any claim document should be prepared with the above basic principle in mind.

### Standard of proof

Following the matter of burden of proof, the next question is 'what standard of proof needs to be provided?'

In other words, to what degree does the proof need to be made out?

In respect of this matter, again, if the claim proceeds to formal proceedings then the standard of proof that needs to be met will be that required in civil proceedings, which is based upon the 'balance of probabilities' principle.

The balance of probabilities principle means that the court/tribunal makes its decision on the basis that something is more likely to have occurred than not.

Therefore, if a set of scales is imagined as being the balance of probabilities, there only needs to be 51% of probability on one side of the scales to enable that side of the scales to succeed over the other side of the scales. Naturally, it is necessary for the party with the burden of proof at any particular time to push the scales down to (at least) the 51% level.

In contrast to the above, the standard of proof in criminal proceedings is based on the much higher level of 'beyond reasonable doubt'.

It is because of this difference in the standard of proof levels between civil proceedings and criminal proceedings, that a claimant may be successful in civil proceedings, whereas, on the same facts, criminal proceedings may be unsuccessful.

When preparing a claim (in addition to the burden of proof principle outlined above) the required standard of proof needs to be recognized, particularly as, in certain situations, the party receiving the claim (i.e. before proceedings are commenced) appears to expect a standard of proof more akin to beyond reasonable doubt rather than on the balance of probabilities.

Part of this reason may be because of the apparent dichotomy between the standard of proof required for civil proceedings (i.e. the balance of probabilities) and the requirement under clause 4.19.3 of SBCSub/C and SBCSub/D/C which assumes that the contractor will *ascertain* the loss and/or expense value.

The word 'ascertain' is usually understood to mean 'to find out for certain' and this is much closer to the standard of proof for criminal proceedings (i.e. beyond reasonable doubt) than it is to the standard of proof for civil proceedings (i.e. the balance of probabilities).

There have been conflicting court cases regarding this particular matter. For example, in the case of Alfred *Mc Alpine* v. *Property and Land*[2] it was found that when an architect is required to ascertain he is obliged to find out for certain and not merely to make a general assessment. However, in the *How* v. *Lindnor*[3] case, it was held that assessment of loss and expense was akin to assessment of damages,

---

[2] *Alfred McAlpine Homes North Ltd* v. *Property and Land Contractors Ltd* (1995) 76 BLR 59.
[3] *How Engineering Services Ltd* v. *Lindnor Ceilings Ltd* [1999] CILL 1521.

requiring no special standard of proof, and that the exercise of judgment was not only permissible but required where loss had not been proved with absolute certainty.

Further, in the case of *Norwest Holst* v. *Co-op Wholesale Society*[4] Mr Justice Thornton found that under a sub-contract the parties were normally required to agree loss and expense (as is the case in SBCSub/C and SBCSub/D/C) and, in the event that they could not agree, it was in order for an arbitrator to determine such reasonable loss and expense (on a balance of probabilities basis).

Another point to note is that when assessing probabilities, the court or tribunal will have in mind, to whatever extent is appropriate in the particular case, that the more serious the allegation the less likely it is that the event occurred and, hence, the stronger should be the evidence before the court concludes that the allegation is established on the balance of probabilities.

This does not mean that different standards of proof are required where different assertions are made, but it does mean that the inherent probability or improbability of an event is itself a matter to be taken into account when weighing the probabilities and deciding whether, on balance, the event occurred.

The more improbable the event, the stronger must be the evidence that it did occur before, on the balance of probability, its occurrence will be established.[5]

It should be noted that 'more likely than not' is not necessarily a hard test to pass and facts can often be proved to this standard merely by circumstantial evidence, or by one person's word being preferred to another's, although contemporaneous written records are always more influential.

In all circumstances, some admissible evidence must be produced to support everything claimed.

### What needs to be proved in a loss and expense claim document?

Following on from the above, the following initial points need to be proved in a loss and expense claim document:

- that an event actually occurred;
- that the event was one expressly catered for within the sub-contract;
- that the notices required under the sub-contract had been given; and
- what the effect was, in financial terms, of the specified event.

In respect of the above list, what then needs to be shown is the connection between the *cause* and the *effect*, that is showing how the latter item on the above list (the financial effect) resulted directly from the first item on the above list (the event that occurred).

This is the part of the process that normally causes most difficulty.

The above connection is a matter of evidence and it is the part of a claim for which records are absolutely crucial.

Indeed, this is when the quality and extent of the records and documentation comes into its own.

---

[4] *Norwest Holst Construction* v. *Co-operative Wholesale Society* (1997/1998) (unreported) 2 December 1997; 17 February 1998.
[5] Ref Lord Nicholls in Re H (Minors) [1996].

With some of the heads of claim (outlined later in this chapter) such a connection is relatively easy to identify, whilst with others it is rather more problematic.

It is because of the difficulty of establishing the required nexus of cause and effect that it is common for global claims to be produced.

Before considering the individual heads of claim in detail it is worth looking at the background of global claims generally.

# Global claims

As stated above, the assessment of money due to a claimant should be made essentially on the grounds of establishing the connection between actual cause and actual effect.

However, such an approach can often be very difficult, particularly where there are many interrelated events, and in such situations a claimant is tempted to submit a 'global' claim (which is sometimes referred to as 'rolled-up' claim).

These global claims are claims for a financial loss which arises from various different events, where individual sums of money are not claimed for each individual event and instead a single global sum is claimed in respect of the alleged cumulative effect of all of the events.

Quite often claimants will simply base their claims in terms of quantum on the difference between estimated cost and actual cost, normally with an adjustment for any recovery made within the variation account.

The basis of the global claim is the total extra cost incurred, as calculated above, resulting from numerous events whose consequences had a complex interaction, such that it was impossible or impractical to disentangle them.

Global claims have been considered in many court cases, and some of the more important cases (in terms of principle) are considered below.

### Crosby v. Portland UDC (1967)

In the *Crosby* v. *Portland UDC*[6] case, which followed an arbitration, the court recognized that the extra cost incurred in that instance arose from a complex matrix of consequences and that, in such circumstances, it might be difficult or even genuinely impossible to accurately allocate responsibility between the competing causes and that there was no need to make an apportionment between the causes which had no real basis, because such a division would be just as flawed in theory as looking at the consequences in the round.

In the court judgment, Mr Justice Donaldson said:

'I can see no reason why he [the arbitrator] should not recognize the realities of the situation and make individual awards in respect of those parts of individual items of the claim which can be dealt with in isolation and a supplementary award in respect of the remainder of those claims as a composite whole.'

The court therefore took the view that to the extent that it is possible and practicable to attribute individual periods of delay to individual causes and to put a

---

[6] *Crosby and Sons Ltd* v. *Portland Urban District Council* (1967) 5 BLR 121.

monetary value on those individual periods of delay that is what should be done. Only where there is a balance of items where such treatment is not possible or practicable should a composite claim or award be made, and, even then, only on the basis that any monies awarded did not overlap with other parts of the claim.

### London Borough of Merton *v.* Stanley Hugh Leach *(1985)*

In *Merton* v. *Leach*[7] it was found that the contractor may be able to rely on a global approach where it is wholly artificial to apportion the totality of the extra costs incurred between the various events which have contributed towards those costs, provided that the contractor has not unreasonably delayed in making application in regard to each event which is alleged to be compensatory and for which liability has been proven.

In this case Mr Justice Vinelott said:

'I think I should nonetheless say that it is implicit in the reasoning of Donaldson J [in *Crosby* v. *Portland UDC*], first, that a rolled up award can only be made in a case where the loss or expense attributable to each head of claim cannot in reality be separated and secondly that a rolled up award can only be made where apart from that practicable impossibility the conditions which have to be satisfied before an award can be made have been satisfied in relation to each head of claim.'

An important point to note in respect of both the *Crosby* v. *Portland UDC* and the *Merton* v. *Leach* cases is that a global claim approach would appear to be inappropriate where there are concurrent causes of default, one of which does not give rise to an entitlement.

### Wharf *v.* Eric Cumine *(1992)*

Further restrictions on a global claim approach were raised in the *Wharf* v. *Eric Cumine*[8] case. In that case the claim was submitted in pleadings on a simple overall global assessment. The court decided that such an approach had no bearing upon the obligation of a claimant to plead his case with such particularity as is sufficient to alert the opposite party to the case which is going to be made against him at trial. In this case, the court decided there had been a failure even to attempt to specify any discernible nexus between the wrong alleged and the consequent delay.

Lord Oliver of Aylmerton said:

'ECA are concerned at this stage not so much with quantification of the financial consequences – the point with which the two cases [*Crosby* v. *Portland UDC* and *Merton* v. *Leach*] referred to were concerned – but with the specification of the factual consequences of the breaches pleaded in terms of periods of delay. The failure even to attempt to specify any discernible nexus between the wrong alleged and the consequent delay provides, to use Mr Thomas' phrase, "no agenda" for the trial.'

---

[7] *London Borough of Merton* v. *Stanley Hugh Leach* (1985) 32 BLR 51.
[8] *Wharf Properties Ltd* v. *Eric Cumine Associates* (1992) 8 Const LJ 62.

## Mid Glamorgan *v.* Williams *(1992)*

In the *Mid Glamorgan* v. *Williams*[9] case, the court held that:

'Where a claim is made for extra costs incurred through delay as a result of various events whose consequences have a complex interaction that renders specific relation between the event and the time or money consequence impossible or impracticable, it is permissible to maintain a global claim.'

## McAlpine *v.* McDermott *(1992)*

In the *McAlpine* v. *McDermott*[10] case, at a trial lasting 92 days before an official referee, detailed evidence of delay or disruption caused by individual drawing revisions or variations was not given by McAlpine.

As Lord Justice Lloyd put it:

'There are frequent references in Mr Grimmer's evidence and the contemporary minutes to "the vast amount of disruption", the "utter confusion" and the "impossible situation" in which the plaintiffs found themselves. But that was all. There were no particulars.'

McAlpine instead relied on its consultant, who had prepared a claim for indirect costs suffered by McAlpine as a result of the variations, revised drawings and late answers to technical queries. The consultant allocated all the delay which occurred to these matters. However, the way in which he did so suffered from two major defects:

(1) His approach assumed that if one man was working for one day (or on one day) on a particular variation, the whole contract was held up for that day.
(2) In his calculation of additional man-hours and therefore additional labour costs, the consultant assumed that the whole of the workforce planned for a particular activity and was engaged continuously on that activity from the day it started until the day it finished.

Lord Justice Lloyd accordingly stated:

'By reason of these defects, we conclude that the plaintiffs did not prove, or indeed come near to proving, on the evidence which they called at the trial, that the delay . . . was due to the revised drawings, VOs and late response to TQs.'

On the other hand, McDermott gave evidence at trial tracing the impact of every drawing revision, variation and technical query. McDermott's evidence showed, for example, that drawing revisions alleged to disrupt fabrication had had no such effect, because at the relevant time the steel for cutting had not yet arrived in the fabrication shop.

---

[9] *Mid Glamorgan County Council* v. *J. Devonald Williams* (1992) 8 Const LJ 61.
[10] *McAlpine Humberoak Ltd* v. *McDermott International Inc. (No. 1)* (1992) 58 BLR 1.

Lord Justice Lloyd stated:

'The [trial] judge dismissed the defendants' approach to the case as being "a retrospective and dissectional reconstruction by expert evidence of events almost day by day, drawing by drawing, TQ by TQ and weld procedure by weld procedure, designed to show that the spate of additional drawings which descended on McAlpine virtually from the start of the work really had little retarding or disruptive effect on its progress". In our view the defendants' approach is just what the case required.'

McDermott presented a positive case on delay, to the effect that delay was in fact caused by failure of McAlpine to qualify their weld procedures, delay in manufacture, remedial work due to McAlpine's poor workmanship and McAlpine's suffering labour problems.

Clearly, this is the approach to be taken by a respondent if a global claim reaches a hearing, albeit that, as in the *McAlpine* v. *McDermott* case, it will be a painstaking and substantially costly task.

### British Airways *v.* McAlpine *(1995)*

In this *British Airways* v. *McAlpine*[11] Court of Appeal case, Saville LJ said that although the plaintiff's pleading was embarrassing (in the sense that it was open to further particularization), it was not seriously prejudicial to the defendants, who were able to know the case they had to meet and were not facing an unfair hearing.

It was held that the degree of detail necessary to satisfy this requirement varies according to the knowledge of the defendant.

### Naru *v.* Matthew Hall *(1994)*

In the *Naru* v. *Matthew Hall*[12] case, Mr Justice Smith expressed the view that it may be permissible to maintain a composite delay/disruption claim where it was impossible and impractical to identify a specific nexus between each of the alleged events and the particular delay/disruption caused.

### John Holland *v.* Kvaerner *(1996)*

In the *John Holland* v. *Kvaerner*[13] case, Mr Justice Byrne found that in John Holland's pleading no causal nexus was asserted linking any breach with any item of loss or damage claimed.

However, he approached this flaw in the pleading as follows:

'The fundamental concern of the court is that the dispute between the parties should be determined expeditiously and economically and, above all, fairly . . .

In my opinion, the court should approach a total cost claim with a great deal of caution, even distrust. I would not, however, elevate this suspicion to the level of concluding that such a claim should be treated as prima facie bad . . .

[11] *British Airways Pension Trustees* v. *Sir Robert McAlpine & Sons Ltd and Others* (1995) 11 Const LJ 365, CA.
[12] *Naru Phosphate Royalties Trust* v. *Matthew Hall Mechanical & Electrical Engineers Pty Ltd* [1994] 2 vR 386.
[13] *John Holland Construction & Engineering Pty Ltd* v. *Kvaerner R.J. Brown Pty Ltd* (1996) 13 BCL 262.

Nevertheless, the point of logical weakness inherent in such claims, the causal nexus between the wrongful acts or omissions of the defendant and the loss of the plaintiff, must be addressed. I put to one side the straightforward case where each aspect of the nexus is apparent from the nature of the breach and loss as alleged . . .

But, in other cases, each aspect of the nexus must be fully set out in the pleading unless its probable existence is demonstrated by evidence or argument and further, it is demonstrated that it is impossible or impractical for it to be spelt out further in the pleading.'

In summary, Mr Justice Byrne was of the view that where it is found to be impossible or impractical to identify each aspect of the nexus, a demonstration of its probable existence is sufficient.

### Bernhard's Rugby *v.* Stockley *(1997)*

In the *Bernhard's Rugby* v. *Stockley*[14] case, Judge Lloyd, QC, held that that nexus need not always be expressed since it may be inferred. Whilst principles of natural justice require the plaintiff to set out with sufficient particularity its case, 'what is sufficient particularity is a matter of fact and degree in each case'.

Whether the claim can be sustained will depend upon the evidence in relation to it. If that evidence allows a conclusion that the plaintiff has suffered a quantifiable loss then it is open to the tribunal determining the matter to bring in a verdict for the plaintiff for the sum which it is satisfied is appropriate. It is not material that the claim is described as a global claim or given any other label.

### Pegler *v.* Wang *(2000)*

In *Pegler* v. *Wang*,[15] a broad brush assessment was made on the basis that some loss had occurred.

### Pitchmastic *v.* Birse *(2000)*

The *Pitchmastic* v. *Birse*[16] case followed this same trend, and in that case the judge made the best assessments that he could and awarded proportions of the heads of claim which he felt were appropriate.

### Amec *v.* Stork Engineering *(2002)*

In the *Amec* v. *Stork Engineering*[17] case, Mr Justice Thornton observed that:

'It is not necessary for an eyewitness to give credible evidence of the disruption that occurred and its causes, for him to have to recount the history of nearly every drawing from its first issue to final installation. The evidence of such witnesses can be compared to that of an eyewitness to a battle seeking to recount the course of a battle, its result and the causes of that result. It is not necessary for that purpose for the eyewitness to have

---

[14] *Bernhard's Rugby Landscapes Ltd* v. *Stockley Park Consortium Ltd* (1997) 82 BLR 39.
[15] *Pegler* v. *Wang* [2000] BLR 218.
[16] *Pitchmastic* v. *Birse* (2000) 97(23) LSG 41.
[17] *Amec Process and Energy Limited* v. *Stork Engineers and Contractors BV* (2002) (unreported) 15 March 2002.

account for the precise movement of every soldier for every hour of the battle. Indeed, to try and do so would run the risk of the witness's account being a meaningless confusion.'

Mr Justice Thornton found, on the facts of the Amec case, that it was unlikely in any event that had an analysis of each event been carried out it would have shown much more than had already been achieved in pleadings and evidence. The only reason which could be accredited to the additional man-hours was the many variations and instances of late information, etc.

When a global claim is commenced by a party it is commonly attacked on the basis that if one of the events that has been complained of is not the responsibility of the defendant, then the court is not able to separate the loss attributable to that individual event and deduct it from the total amount claimed. Stork Engineering therefore contended that Amec had to show detailed proof of all relevant hours expended so that the court could deduct the necessary sums from the amount claimed. To the extent that an element of the loss, however small, could be attributed to Amec's failings, the whole claim should fail. In this way, a global claim could be said to be similar to a house of cards.

Mr Justice Thornton, on the other hand, felt that, under the Civil Procedure Rules in particular, the common law and trial procedures were not so rigid as to prevent such a global claim from succeeding, provided that there was evidence to allow the court to make the necessary apportionment and to draw its own conclusions. Consequently, Amec was successful in its claim.

### John Doyle *v.* Laing *(2002)*

In the first instance, the *John Doyle* v. *Laing*[18] Scottish case, Lord MacFadyen refused an application by Laing to strike out parts of Doyle's claim on the grounds that it was global. Whilst Lord MacFadyen said that advancing a claim for loss and expense in a global form was a 'risky enterprise' he continued that:

> 'The rigour of that analysis is in my view mitigated by two considerations. The first of these is that while, in the circumstances outlined, the global claim as such will fail, it does not follow that no claim will succeed. The fact that a pursuer has been driven (or chosen) to advance a global claim because of the difficulty of relating each causative event to an individual sum of loss or expense does not mean that after evidence has been led it will remain impossible to attribute individual sums of loss or expense to individual causative events . . . The global claim may fail, but there may be in the evidence a sufficient basis to find causal connections between individual losses and individual events, or to make a rational apportionment of part of the global loss to the causative events for which the defender has been held responsible.
>
> [39] The second factor mitigating the rigour of the logic of global claims is that causation must be treated as a common sense matter (*Holland* v. *Kvaerner*, per Byrne J at 84I). That is particularly important, in my view, where averments are made attributing, for example, the same period of delay to more than one cause.
>
> [40] . . . How each of the concurrent causes ought to be viewed in determining whether the causes for which the defenders had no liability played a material part in causing the global loss is a matter that is, in my view, best left for consideration at the conclusion of

---

[18] *John Doyle Construction Ltd* v. *Laing Management (Scotland) Ltd* (2002) CILL 1870.

a proof before answer. The second consideration which leads me to allow a proof before answer on the global claim is that it would be wrong to exclude, at this stage, the possibility that the evidence properly led at proof before answer will afford a satisfactory basis for an award of some lesser sum than the full global claim.

[41] Before leaving this aspect of the debate, I would make two further observations. The first is that the risk that the pursuers' global claim will fail because a material part of the causation of the loss and expense was an event for which the defenders are not liable, if the evidence discloses no rational basis for the award of any lesser sum, remains a live one. Secondly, the allowance of a proof before answer does not afford the pursuers carte blanche to attempt to prove their loss and expense in any way they choose. Their pleadings remain the measure of what they are entitled to prove by way of computation of loss and expense. If a lesser claim is to be made out, that must be done on the basis of evidence which is properly led within the scope of the existing pleadings.'

But this matter did not end there, because Laing proceeded to appeal.

## John Doyle *v.* Laing *(2004)*

In the appeal to the above *John Doyle* v. *Laing*[19] Scottish case, in the Inner House of the Court of Session, a review of Lord MacFadyen's judgment was made.
Lord MacFadyen had held that:

'The logic of a global claim demands, however, that all the events which contribute to causing the global loss be events for which the defender is liable. If the causal events include events for which the defender bears no liability, the effect of upholding the global claim is to impose on the defender a liability which, in part, is not legally his. That is unjustified. A global claim, as such, must therefore fail if any material contribution to the causation of the global loss is made by a factor or factors for which the defender bears no legal liability. That point has been noted in Keating at paragraph 17–18, in Hudson at paragraph 8–210, more clearly in Emden at paragraph [231], in the American cases, and most clearly by Byrne J in *Holland* v. *Kvaerner* at 85H and 86D (see paragraph [25] above). The point has on occasions been expressed in terms of a requirement that the pursuer should not himself have been responsible for any factor contributing materially to the global loss, but it is in my view clearly more accurate to say that there must be no material causative factor for which the defender is not liable.'

'How each of the concurrent causes ought to be viewed in determining the causes for which the defenders had no liability played a material part in causing the global loss is a matter that is, in my view, best left for consideration at the conclusion of a proof before answer . . . [It] would be wrong to exclude, at this stage, the possibility that the evidence properly led at proof before answer will afford a satisfactory basis for an award of some lesser sum than the full global claim.'

The defendant appealed against this judgment, but the Inner House of the Court of Session held:

(1) That in the normal case, individual causal links must be demonstrated between each of the events for which the employer is responsible and particular items of loss and expense.

---

[19] *John Doyle Construction Ltd* v. *Laing Management (Scotland) Ltd* [2004] CILL 2135.

(2) That in circumstances where it is impossible to separate the specific loss and expense caused by a number of different events which are the responsibility of the employer, these can be pleaded as producing a cumulative effect. In these circumstances, it is not necessary to break down each event and isolate the loss caused by each.

(3) That where, however, a significant cause of the loss and expense is a matter for which the employer is not liable, the global claim must fail.

(4) That where it is shown that some of these events (albeit not a 'significant' amount in causal terms) are not actually the responsibility of the employer, the global claim should not necessarily fail, since it may be possible for the judge, or arbitrator, to apportion the loss as between the causes for which the employer is responsible and other causes.

(5) That it was acknowledged that this could lead to 'rough and ready' results.

(6) That when pleading the claim, the particular events and heads of loss should be set out in reasonable detail. There will, however, usually be no need in commercial cases to do more than simply to plead the proposition that the particular events caused the relevant heads of loss. Causation is largely 'a matter of inference' and is frequently based upon experts' reports. The consideration of the global claim should wait until this evidence is before the tribunal.

The above six points seem to set out the latest position in respect of global claims.

### Submitting a claim/global claim

Generally, the best practice to follow in preparing a claim involving multiple variations, delay, disruption and extra cost is to follow the guidelines laid down by the courts strictly, so far as it is possible, and to present a cause and effect claim where possible, and a global claim only in respect of any element of the claim where it is simply not possible or practical to do otherwise.

One way of presenting this type of claim that is often convenient is to have a points of claim document which deals with the legal issues, setting out the terms of the contract which are applicable, in particular the variations provisions, any terms which give rise to a claim for breach of contract and setting out in summary form in what respects the breaches occurred. This document would be relatively short and the detailed factual matters supporting the claim would be set out in a separate document.

The separate document could be in the form of a 'cause and effect' schedule which, for a claim based mainly on variations might be set out as follows:

Column 1 – event
Column 2 – clause no.
Column 3 – other term
Column 4 – effect (disruption)
Column 5 – delay (from and to)
Column 6 – claim/loss (£)

Column 1 would set out the event relied on, for example a variation order or an event which is said to be a breach of the sub-contract.

Column 2 would simply refer to an appropriate and relevant sub-contract clause number or, if the clause has subclauses, the appropriate and relevant sub-contract subclause number.

Column 3 would be used if some other express or implied terms were relied upon for events other than variations.

Column 4 would be used to demonstrate the effect. This would most likely be a narrative if the claim related to disruption, as opposed to a prolongation claim, explaining what activity was affected by the event complained of and in what way.

Column 5 would deal with prolongation, giving the precise start date of delay and the precise finish date of delay. It would also be appropriate to indicate in that column whether delay was continuous or intermittent, since it is a common failing in this type of claim to adopt a notional rather than a historical analysis in which it is assumed that where something goes wrong, a continuous delay to the contract as a whole occurs until that matter is put right, which is often not the case. It is also necessary in assessing prolongation to include an analysis of concurrent delay, so that critical delay is identified (another common failing in this type of claim being to assume that all delay is cumulative or failing to identify what is concurrent and what is cumulative).

Column 6 would include an amount which is claimable under the terms of the contract, for each individual item noted in column 1.

In attempting to produce this type of schedule it will be found in some cases that it is not possible with multiple variations, and possibly other claims, to give a precise delay period or claim figure for each one, because of the complex interaction of events.

In that event, all one can do is to break it down *as far as possible* and to present the remainder as a global claim.

Approaching the matter in this way is not only a good discipline but is also more likely to result in success, as it closely replicates the latest requirements of the courts.

If faced with a global claim, it is fairly standard to request a cause and effect analysis, as this will highlight the weaknesses of the other party's case.

# Potential heads of claim for a loss and expense claim

A claim for loss and/or expense is the means of putting a sub-contractor back into the position in which he would have been but for the delay or disruption; it is not a means of turning a loss into a profit (unless, of course, the loss is because of the non-payment of loss and/or expense).

The settlement in effect amounts to common law damages which are dealt with later in this chapter; therefore, an exact establishment of the sub-contractor's additional costs must be made. At first sight it might be thought that establishment of those *additional* costs is relatively easy but in fact it can be very difficult.

The common heads of a loss and expense claim are described below:

## On-site establishment costs (site overheads)

These are generally what are normally referred to as 'preliminary items' and are usually easily ascertainable from the cost records.

These are usually time related costs, although that may not always be the case and it may not be the overall completion date which is important. This is the case where *additional* supervision may have been required during the original sub-contract period, or the costs may have resulted from delay to non-critical items (e.g. scaffold to an area of work which was not on the critical path).

The site overheads part of the claim is commonly broken down into delay costs (due to prolongation) and 'thickening' (or extra resource) costs (due to disruption).

The prolongation part of the claim must relate to those periods when delay occurred; this will not necessarily be the over-run period.

For example, if a sub-contractor had a sub-contract period of six weeks, but was delayed in week two for one week, the site overhead costs would need to be claimed for week two (i.e. the week when the delay occurred) and not for week seven (i.e. the week when the delay effect became apparent).

There is much general debate over whether a sub-contractor is entitled to additional site overhead costs where it programmes to complete a project before the contract completion date but is prevented from doing so.

The sub-contract on-site period may be, for example, eight weeks, but the sub-contractor may decide that he wishes to complete the sub-contract works in six weeks. However, the contractor issues instructions that prevent the sub-contractor from completing the sub-contract works until the end of week eight. The sub-contractor then wishes to recover the additional site overhead costs for weeks seven and eight.

In the *Glenlion* v. *The Guinness Trust*[20] case, the position in respect of JCT contracts was made reasonably clear. In that case (which related to a contractor/employer relationship), it was found that the contractor was not entitled to an extension of time in such a situation, because under JCT contracts extensions of time are only due if completion of the works is delayed beyond the completion date (refer to clause 2.18.2 of SBCSub/C and SBCSub/D/C), and Judge Fox-Andrews QC found that there was no term to be implied into a JCT contract and that, if and so far as the programme showed a completion date before the date for completion, the employer must enable the contractor to carry out the works in accordance with the earlier date. A similar situation would equally apply to a contractor/sub-contractor relationship.

Despite the above, if a sub-contractor can show that he has been caused loss and expense by reason of the delay, he may be entitled to claim it under clause 4.19 of SBCSub/C and SBCSub/D/C, depending, of course, on the reason for the delay.

If a sub-contractor is delayed or disrupted by an instruction ordering additional work, for example, he may be entitled to loss and expense incurred (i.e. through thickening as noted above) regardless of whether the additional work delayed the sub-contractor in the completion of the works.

Under this head of claim the sub-contractor is essentially seeking to recover site overhead costs, that is those costs which are not related to one specific item of work but which are necessary for the site establishment generally.

Such costs may include the following examples:

- project manager (site based);
- site foreman;

---

[20] *Glenlion Construction Ltd v. The Guinness Trust* (1987) 39 BLR 89.

- site QS;
- Site engineer and assistants;
- site offices;
- site administration costs;
- site stores;
- welfare facilities (if provided by the sub-contractor);
- cranage/forklifts;
- general and small plant;
- scaffolding/towers;
- telephone;
- electricity;
- insurance.

What the sub-contractor is seeking to establish is that, because of events that have occurred and for which the sub-contract provides on entitlement, actual resources of the type listed above (amongst others), additional to those which otherwise would have been necessary, have had quite reasonably to be engaged on the project.

In assessing such a claim several points should be kept in mind:

- What has to be established in the first instance is that additional resources were required.

- The sub-contractor needs to establish with sufficient particularity:

  — what was included in his tender, and
  — that the tender allowance was sufficient.

- Just because resources are engaged during a period of over run it does not mean they are additional. For example, if a sub-contractor originally expected to have two foremen on site for ten weeks but actually only had one foreman on site for twenty weeks there would not (on the face of it) be any additional costs.

- The sub-contractor needs to establish that the additional resources were required due to the reasons that he is relying on.
  This may be done in several ways:

  — If the claim is for prolongation then it is necessary to show resources were retained on site during the periods of delay for reasons outside his control. This is usually one of the easier points of the claim.
    However, the sub-contractor does not need to have received an extension of time before he can pursue such prolongation costs during a period in excess of the original period.
    Conversely, not all extensions of time circumstances give rise to financial entitlement, and, even if they do, not all the additional resources may be recoverable.
  — A claim for site overhead costs may be pursued in the absence of prolongation.
    If an appropriate resource is required for longer during the original sub-contract period, or more of them are required for the same period, then they should be recovered if they can be associated with a specific event.

This may apply, for example, in the situation where a variation is issued which results in certain work activities, which were programmed to be executed consecutively now needing to be carried out concurrently, thus requiring two or more work faces to be operating and needing two (or more) sets of supervisors, etc., where previously one would have been adequate.

- The sub-contractor then has to establish the cost of the resources.
    With supervisory staff this would be ascertained using salary levels, pensions, national insurance, car costs, petrol, bonuses, etc.
    If agency staff have been used then their actual costs will be recoverable providing it was reasonable to use them.

## Inefficient or increased use of labour and plant (commonly referred to as a disruption claim)

Delay and disruption can lead to increased expenditure on plant and labour in two ways. One is that it may be necessary to employ extra resources and the other is that existing resources may be used inefficiently.

This head of claim in particular may be difficult to ascertain due to lack of records and the difficulty of establishing the nexus of cause and effect.

However, just because it is difficult to establish the above, it does not mean that the sub-contractor should not submit a claim and equally it does not mean that the contractor should not make a reasonable assessment (see comments earlier in this chapter regarding global claims).

As noted earlier in this chapter, the global claim approach should be used very much as the exception rather than the rule. It is applicable and permitted only where the events are so numerous/complex/interrelated that no amount of analysis or record keeping would establish or evidence the connection of cause and effect on individual matters. However, it is not good enough that such analysis cannot be done simply because the sub-contractor has not kept reasonable records, etc.

There are a number of methods of assessing this head of claim and indeed few people agree on the approach to use.

Ideally, what should be established is that, before a particular event occurred productive resources were achieving a certain level of productivity/financial income for the sub-contractor, but that after that particular event occurred and indeed, because of it, the same level of productivity/financial income could not be achieved.

Take, for example, the case of a bricklaying sub-contractor.

In the sub-contract the sub-contractor is told that he will be required to build wall A, followed by wall B, followed by wall C. In his tender, the sub-contractor allowed for an output of 60 bricks per hour per bricklayer.

In the first week on site things went very well; the bricklayers worked on wall A only, as planned, and by the end of that first week, the output achieved by the bricklayers was 65 bricks per hour. The sub-contractor was exceeding his tender expectations and this illustrated the sufficiency of his tender.

However, in the next week, things started to go wrong. The sequence of work changed, and the sub-contractor was instructed by the contractor to switch his bricklayers from wall A to wall B to wall C, often all in the same day.

As a consequence, the output of the bricklayers at the end of the second week reduced to 50 bricks per hour.

In such a case, the disruption claim would be based on the loss of production caused by the disruption event. In other words, the sub-contractor expected to achieve 60 bricks per hour (or the sub-contractor might argue that he actually achieved 65 bricks per hour) before the disruption event occurred; but because of the disruption event the sub-contractor's bricklayers only achieved 50 bricks per hour.

Although the above is a very simple example, the principle remains the same for much more complicated claims. Clearly, for this head of claim, more than any other, good, complete and accurate records are absolutely crucial.

As noted above, there are several types of approach in respect of a disruption claim which may be applicable, depending on the nature of the work, the circumstances and the records available. The main approaches used are:

- to show that resources were standing idle when they should otherwise have been working, thus taking longer to do the same work;
- to illustrate how the work of a particular trade has been prolonged and to apply a disruption factor in the same proportions to the cost of the trade;
- to estimate a disruption factor;
- to do an overall global claim based on costs of all trades assessed in the tender and to claim the difference between that and actual costs;
- a mini global claim using the approach as last but targeting specific trades;
- to illustrate the resources employed in producing a specific and evidenced amount of work prior to an event and to claim the extra resources for doing equivalent work after the event;
- to show that more expensive resources were required to do the same value of work because of an event, even though the level of resources was the same, that is agency resources, emergency hire of plant, expensive gangs, etc.

## Winter working

This heading amounts to a different aspect of loss of productivity resulting from the need to work in less favourable climatic conditions. This is a recoverable head of claim.[21]

What the sub-contractor has to establish in this case is that because of providing an express contractual entitlement he has had to undertake productive work in climatic conditions less favourable than would otherwise have been the case.

It is not enough to show only that work was undertaken during the winter, for example; it must also be shown that this fact did actually affect the work.

Of course, this type of claim could be problematic for sub-contractors, particularly if their actual period on site could have commenced at any time, within wide parameters.

---

[21] *Bush v. Whitehaven Port & Town* (1888) 52 JP 392.

## Head office overheads and profit

Head office overheads are sometimes called 'off-site overheads' or 'establishment charges'.

Head office overheads are those administrative and management costs of running the head office of a sub-contractor's business over and above the site costs.

Generally, head office overheads will include:

- purchase and/or rent of office and yards;
- maintenance and running costs;
- transport and mobilization costs;
- company cars;
- depreciation;
- director's emoluments and expenses;
- salary and other cost of staff;
- administration costs;
- legal and audit fees, etc.

If particular head office overhead costs are proven by the sub-contractor to have been increased as a result of a delay and disruption suffered by the sub-contractor, then these costs are recoverable.

Examples may range from the cost of extra staff recruited because the particular project was in difficulties, to the cost of extra telephone calls and postage in the period of delay.

However, substantial claims of this kind are rarely made because most sub-contractors are able to cope with delay on a particular contract using their existing resources whose cost is reasonably constant.

Despite this, claims for loss of head office overheads (and claims for loss of profit) are usually made on a different basis.

A sub-contractor's head office overheads are normally taken as being recovered out of the income from his business as a whole. Therefore, where completion of one project has been delayed a sub-contractor may claim to have suffered a loss arising from the diminution of his income from the project and hence the turnover of his business.

Despite this diminution of turnover, the sub-contractor continues to incur expenditure on head office overheads which he cannot materially reduce or, in respect of the project, can only reduce, if at all, to a limited extent.

Therefore, but for the delay and disruption, the sub-contractor's workforce would have had the opportunity of being employed on another project which would have had the effect of contributing to the head office overheads and profit during the over-run period.

There is some authority that a claim on this basis is sustainable.

In *J.F. Finnegan* v. *Sheffield City Council*,[22] Justice Sir William Stabb (talking about a contractor in that case) said:

'It is generally accepted that, on principle, a contractor who is delayed in completing a contract due to the default of his employer, may properly have a claim for head office or

---

[22] *J.F. Finnegan Limited* v. *Sheffield City Council* (1988) 43 BLR 124.

off-site overheads during the period of delay, on the basis that the workforce, but for the delay, might have had the opportunity of being employed on another contract which would have had the effect of funding the overheads during the over-run period.'

It should be noted that the entitlement to such general overheads is in itself an arguable point. However, the consensus of opinion is in favour, providing, of course, it can be established that such overheads would have been recovered were it not for the delay.

It is suggested that, in order to succeed, a sub-contractor has in principle to prove that there was other work available which, but for the delay, he would have secured but which in fact because of the delay, he did not secure.

He might do this by producing invitations to tender which he declined, with evidence that the reason for declining was that the delay in question left him insufficient capacity to undertake other work.

He might, alternatively, show from his accounts a drop in turnover and establish that this resulted from the particular delay rather than from extraneous causes. If loss of turnover resulting from delay is not established, the effect of the delay is only that receipt of the money is delayed. It is not lost.

When pursuing a claim for head office overheads and profit, a sub-contractor will frequently rely on the use of a formula.

There are three formulae that are in common usage (and, in addition, there are many other derivatives from these formulae).

These formulae are:

- the Hudson formula;
- the Emden formula; and
- the Eichleay formula.

### The Hudson formula

The Hudson formula is based upon a percentage to cover both head office overheads and profit as built into the tender and is as follows:

$$\frac{\text{HO/profit percentage}}{100} \times \frac{\text{Contract sum}}{\text{contract period (weeks)}} \times \text{period of delay (weeks)}$$

The Hudson formula is fundamentally flawed in that HO/profit percentage is applied to the full contract sum, whereas it should only be applied to the contract, sum less the head office overhead and profit value included within the contract sum.

Despite this flaw, judicial support for the Hudson formula was given in the case *J.F.Finnegan* v. *Sheffield City Council*[23] and the Canadian case of *Ellis-Don Limited* v. *The Parking Authority of Toronto.*[24]

However, its use was doubted in the *Tate & Lyle* v. *GLC*[25] case, but this was more because of the failure of the claimant to keep proper records and supporting documents.

---

[23] *J.F. Finnegan Limited* v. *Sheffield City Council* (1988) 43 BLR 124.
[24] *Ellis-Don Limited* v. *The Parking Authority of Toronto* (1978) 28 BLR 98.
[25] *Tate & Lyle Food and Distribution Limited* v. *Greater London Council* [1981] 3 All ER 717.

### The Emden formula

An alternative formula is the Emden formula, which takes the percentage for head office overheads and profit from the contractor's overall organization and is therefore somewhat more realistic in its approach than the Hudson formula. This formula is:

$$\frac{h}{100} \times \frac{c}{cp} \times pd$$

where:

> h = head office overhead and profit percentage arrived at by dividing the total head office overhead cost and profit of the contractor's organization as a whole by total turnover;
> c = the contract sum;
> cp = contract period in weeks;
> pd = period of delay in weeks.

A variation of the Emden formula was accepted in the *Whittall Builders* v. *Chester-le-Street DC*[26] case, even though no factual evidence of actual loss was provided.

In the *Alfred McAlpine* v. *Property and Land*[27] case, the use of Emden and other formulae was rejected (although in that case Property and Land carried out only one contract at a time and factual proof of 'unabsorbed overheads' was not difficult to establish).

In the interesting *Norwest Holst* v. *Co-op Wholesale Society*[28] case, Mr Justice Thornton said:

> 'Emden formula is sustainable and may be used as the basis of ascertaining a contractor's entitlement to payment for loss and/or expense in the following circumstances:
>
> (1) The loss in question must be proved to have occurred.
> (2) The delay in question must be shown to have caused the contractor to decline to take on other work which was available and which would have contributed to its overhead recovery. Alternatively, it must have caused a reduction in the overhead recovery in the relevant financial year or years which would have been earned but for that delay.
> (3) The delay must not have had associated with it a commensurate increase in turnover and recovery towards overheads.
> (4) The overheads must not have been ones which would have been incurred in any event without the contractor achieving turnover to pay for them.
> (5) There must have been no change in the market affecting the possibility of earning profit elsewhere and an alternative market must have been available.'

### The Eichleay formula

The Eichleay formula is basically an American version of the Hudson formula. This formula is:

---

[26] *Whittall Builders Limited* v. *Chester-le-Street District Council* (1987) 40 BLR 82, QBD.
[27] *Alfred McAlpine Homes North Ltd* v. *Property and Land Contractors Ltd* (1995) 76 BLR 59.
[28] *Norwest Holst Construction* v. *Co-operative Wholesale Society* (1997/1998) (unreported) 2 December 1997; 17 February 1998.

(1) $\dfrac{\text{contract turnover} \times \text{fixed overheads for contract period}}{\text{total company turnover}} = \dfrac{\text{contract}}{\text{contribution}}$

(2) $\dfrac{\text{contract contribution}}{\text{contract period}} = \text{weekly contribution from contract}$

(3) weekly contribution from contract × period of delay (in weeks)
= sum claimable

The Eichleay formula was judicially approved in the case of *Capital Electric v. US*.[29]

However, whatever formula may be used, it must be noted that the use of a formula merely provides the means for assessing quantum. The fact that there was an actual loss, and that this loss flows directly from the relevant matter relied upon must first be established.

In addition, when submitting claims for head office overheads and profit, care needs to be taken to avoid any double recovery or overlap with other claims or payments obtained by the sub-contractor, such as variations, which have been computed by using the contract prices as a basis. Normally, in such a situation the valuation of the variation will include for an element of additional head office overheads and profit recovery.

Also, generally speaking, if the matter of liability has been established, then evidence needs to be provided to show:

(1) that the profit or overhead contribution was capable of being earned elsewhere at the time of delay;
(2) that the profit and overhead percentage is a reasonable one; and
(3) that work of the same level of profitability and/or overhead recovery was available during the period of delay.

The use of formulae has been heavily criticized by many commentators, and some of the main criticisms are that:

- All formulae were evolved in the 1960s at times of high economic activity when plentiful work and a high return could be achieved. Such criteria are difficult to prove when the economic climate is not so buoyant.
- Formulae assume there has been no change in market conditions.
- Formulae can compensate a contractor (or sub-contractor) whether or not his overall business has increased or decreased for reasons that have nothing whatsoever to do with a claim on a particular project.
- Formulae are time-related only and ignore the actual level of activity on the project during the delay period.

Although there has been some debate about whether loss of profit is recoverable as part of a loss and expense claim, if the words direct loss and expense are to be interpreted as equivalent to measure of damages at common law for breach of contract (which is generally accepted to be the case), then there can really be no doubt that a sub-contractor can claim for the loss of profit that he would have earned on other sub-contracts had there been no delay and disruption to the project in question, provided that such loss of profit was foreseeable.

---

[29] *Capital Electric Company v. US 729 F (2d) 743 (1984)*.

In the *Robertson* v. *Amey-Miller*[30] case, Lord Drummond Young observed that in cases involving breach of contract, the general principle was that the innocent party would be entitled to be put in the same position as he would have been if the contract had been performed. Given that loss and expense clauses are to be interpreted as equivalent to the measure of damages at common law for breach of contract, the significance of the word 'loss' in the phrase 'loss and expense' was therefore that it could include both a loss of profit and a loss of contribution to general overheads.

Clearly, for a claim of loss of profit to be made, a sub-contractor would need to show, as with head office overheads, that at the time of the delay he could have used the lost turnover profitably. A claim for loss of profit does not, it is submitted, fail merely because the contract in question was unprofitable. The question is what the contractor would have done with the money if he had received it at the proper time. Even if, at that time, the contractor's business was making a loss, a sum analogous to loss of profit is, it is submitted, recoverable if the loss of turnover increased the loss of business.

## Increased costs

This head of claim principally comes into effect where a fixed price sub-contract has been placed (although it can equally apply where a fluctuation recovery has been agreed).

In such a situation, additional expenditure on resources due to increases in cost during an extended period is allowable. The correct measure of this head of claim is the difference between what the sub-contractor would have spent on resources and what he has actually spent.

There are a number of different methods of calculating this head of claim.

The best way is for the claim to be based on substantiated details of the level of costs that would have been expended on resources and the difference between that sum and the actual costs paid for those same resources backed up by invoices, etc. However, it is rare in practical terms that such an analysis could be economically undertaken.

Therefore an alternative method would be to use some type of formula approach. Using a formula approach, the calculation would broadly be as described below.

- Ascertain the value of resources in the original tender but using actual work values.
- Find out the level of inflation for such resources over the original sub-contract period, normally by using some standard published indices.
- Calculate how much the cost of resources would have increased during the original sub-contract period. This will then be taken as the amount deemed to be allowed in the rates.
- Undertake the same calculation but using the indexed inflation figure over the extended sub-contract period.
- The extra over cost between the above two calculations will constitute the increased cost claim.

---

[30] *Robertson Group (Construction) Ltd* v. *Amey-Miller (Edinburgh) Joint venture* (2005) Outer House Court of Session, 10 May 2005.

## Cost of claim

Where the preparation of a claim is undertaken by a sub-contractor's in-house staff, this cost will often be partly (or completely) reflected in the sub-contractor's loss and expense claim for site overheads or head office overheads.

Often, a sub-contractor may employ a consultant to assist in the preparation of a loss and expense claim, and the general rule is that the costs incurred in taking this action are generally not recoverable, except and until the matter proceeds to arbitration[31] or litigation.

The reasoning behind disallowing this head of claim is that the sub-contractor is obliged, under the sub-contract, to provide information, etc. to resolve financial issues. Therefore, the submission of a claim and the work involved must have been included in his price.

It may be the case, however, that such monies might be recovered if the contractor requested further information from the sub-contractor which is judged to be beyond the sub-contract requirements.

## Interest and finance charges

When a sub-contractor incurs loss and expense, this has to be financed by him either from his own capital resources or alternatively by increased borrowing. In either case, it is clear that the use of money costs money.

Depending on the level of inflation (and the interest rates applicable at any point in time), and the amount outstanding in respect of the loss and expense claim, this matter will either be a relatively small matter or will have major implications. In either event, it is a matter of great importance to the sub-contractor how such interest or finance charges should be recovered.

Interest in respect of monies owed is dealt with as an express term of the sub-contract (reference clause 4.10.5 and clause 4.12.4 of SBCSub/C and SBCSub/D/C which are dealt with in more detail in Chapter 8). Therefore, if it is proven that monies were outstanding in respect of the non-payment or underpayment of a loss and expense claim then simple interest would be applied, and the interest rate applicable would be 5% per annum above the official dealing rate of the Bank of England current at the date that a payment due under the sub-contract became due.

In the alternative case, a finance charge, or the cost of being stood out of one's money, is a recoverable head of a loss and expense claim in any event. This was established beyond doubt in the Court of Appeal case of *F.G. Minter v. WHTSO*.[32]

In respect of common law damages, to recover finance charges, it is necessary for the claimant to show that the finance charges fell within the second limb of *Hadley v. Baxendale* (dealt with later within this chapter) that is, that the finance charges are special damages that were in the contemplation of the parties at the time that the contract was formed.[33]

---

[31] *James Longley & Co. Ltd v. South West Thames Regional Health Authority* (1984) 25 BLR 56.
[32] *F.G. Minter Ltd v. Welsh Health Technical Services Organisation* (1980) 13 BLR 1, CA.
[33] Refer to *President of India v. Lips Maritime* [1988] AC 395, HL; *Holbeach Plant Hire v. Anglian Water Authority* (1988) 14 Con LR 101.

## Acceleration

In the case of delay, the only express action available to the contractor under SBCSub/C and SBCSub/D/C is to extend the sub-contractor's period of time on site.

However, particularly in the case of delay caused by a failing of the contractor, and mindful of the financial consequences on the contractor of finishing late (e.g. liquidated damages, and prolongation costs), the contractor is often more interested in the sub-contractor completing its works in line with the original dates than in extending the sub-contractor's period of time on site.

In such a situation it is not uncommon for a contractor to attempt to reach some form of acceleration agreement with the sub-contractor, whereby the sub-contractor will take measures (for example, weekend working, double shifts, increased resources, etc.) to complete 'on time' (or, at least, earlier than would have been achieved under the normal course of events) and, if the earlier required date is achieved, the contractor will make an agreed acceleration payment to the sub-contractor.

It is debatable whether a schedule 2 quotation provided by a sub-contractor incorporates the possibility of agreeing acceleration to the sub-contract.

On the face of it, this would be possible because clause 2.2 of schedule 2 quotation states that the quotation must comprise 'any adjustment required to the time period'. This adjustment could naturally include for both an increase and a decrease to the previously agreed time period. A decrease to the previously agreed time period could be as a result of acceleration measures, and those acceleration measures could be priced within a schedule 2 quotation.

The reason why the word 'debatable' is used above, is that clause 2.2 of the schedule 2 quotation for the main contract specifically states: 'including, where relevant, stating an earlier Completion Date than the Date for Completion given in the Contract Particulars'; but no such similar wording is included in clause 2.2 of the schedule 2 quotation document within SBCSub/C and SBCSub/D/C.

Nevertheless, this is a matter for the parties to agree, and it is therefore questionable whether the omission of the words in question has any real practical effect.

Apart from the above possibility, SBCSub/C and SBCSub/D/C do not specifically allow for acceleration agreements, and great care needs to be taken by both the contractor and the sub-contractor when entering into any such agreement.

Most acceleration agreements fall into three types:

- An agreement for the sub-contractor to take an agreed and specified course of action. In this situation, the sub-contractor is paid the additional cost of following this course of action, and the contractor obtains the benefit (if any) of the acceleration measures undertaken.

- An agreement that the sub-contractor will seek to improve on the completion date for a stated sum. This form of agreement will effectively revise both the sub-contract sum and the sub-contract completion date, with all other terms, obligations and rights remaining in place. Therefore, if a variation is issued after the acceleration agreement has been signed, the sub-contractor may be entitled to an extension of time; but if the sub-contractor does not complete by the agreed extended date he may be liable for the contractor's damages in the normal way.

• An agreement whereby the sub-contractor agrees to a completion date earlier than its contractual entitlement on the basis of a bonus payment. If the sub-contractor does not achieve the completion date it will not receive the bonus at all. If the reason that the sub-contractor did not achieve the earlier completion date was because of some act of hindrance by the contractor, then it may be expected that the completion date would be adjusted accordingly (without affecting the sub-contractor's right to the recovery of the bonus payment).

In a case[34] where this adjustment did not occur, the bonus due to a contractor (in that case) was disallowed, but a value of 50% of the bonus was awarded to the contractor in any event by way of a 'loss of chance'.

Each of the above acceleration types carry their own benefits and risks, and both parties need to carefully consider their respective positions before entering any acceleration agreement.

It is self-evident that a sub-contractor should not undertake acceleration measures (unless those measures are in mitigation of delays that the sub-contractor has caused) without reaching some form of agreement with the contractor, since if he does so, the probability is that he will not receive any payment at all for the acceleration measures undertaken.

## Common law damages claims (in respect of a breach of contract) (reservation of rights and remedies of contractor and sub-contractor)

As noted earlier in this chapter, clause 4.22 of SBCSub/C and SBCSub/D/C preserves the common law rights of the contractor and the sub-contractor.

In respect of these common law rights, a breach of contract which has not been excused gives the injured party the right to bring an action for damages (breach of contract is dealt with in detail in Chapter 12).

Damages are awarded to put the claimant as nearly as possible 'in the same position as he would have been in if he had not sustained the wrong for which he is now getting compensation or reparation'.[35]

This was put another way by Lord Pearce:

'the underlying rule of the common law ... Where a party sustains a loss by reason of a breach of contract, he is, so far as money can do it, to be placed in the same situation with regard to damages, as if the contract had been performed.'[36]

As noted by Lord Wilberforce 'The general principle for the assessment of damages is compensatory,'[37] but the courts set a limit to the loss for which damages are recoverable, and loss beyond such limit is said to be too remote.

[34] *John Barker Construction Ltd* v. *London Portman Hotel Ltd* (1996) 83 BLR 31.
[35] Lord Blackburn in *Livingstone* v. *Rawyards Coal Company* (1880) 5 App Cas 25, HL.
[36] *Robinson* v. *Harman* (1848) 1 Ex 850.
[37] *Johnson* v. *Agnew* [1980] AC 367.

## Hadley v. Baxendale

The famous rule as stated in the case of *Hadley* v. *Baxendale*[38] is:

> 'Where two parties have made a contract which one of them has broken, the damages which the other party ought to receive in respect of such breach of contract should be such as may fairly and reasonably be considered either:
>
> (1) arising naturally, i.e. according to the usual course of things from such breach of contract itself; or
> (2) such as may reasonably be supposed to have been in the contemplation of both parties at the time they made the contract, as the probable result of the breach of it.'

The rule as stated above is recognized as having two limbs, and these limbs have been indicated in the above quoted text by the insertion of the numbers (1) and (2).

### The first limb of Hadley v. Baxendale

There have been two leading cases where the *Hadley* v. *Baxendale* rule has been considered. These are the *Victoria Laundry* v. *Newman Industries*[39] case and the *Koufos* v. *Czarnikow*[40] case.

In the *Victoria Laundry* case, Lord Justice Asquith stated several propositions that emerged from the authorities as a whole. Taking these into account, together with the opinions of the House of Lords in the *Koufos* v. *Czarnikow* case, and other following cases, the first limb of the *Hadley* v. *Baxendale* rule may be elaborated into the following propositions:

(1) The aggrieved party is only entitled to recover such part of the loss actually resulting as may fairly and reasonably be considered as arising naturally, that is according to the usual course of things, from the breach of contract.

(2) The question is to be judged as at the time of the contract.

(3) In order to make the contract breaker liable it is not necessary that he should actually have asked himself what loss was liable to result from a breach of the kind which subsequently occurred. It suffices that, if he had considered the question, he would as a reasonable man have concluded that the loss of the type in question, not necessarily the specific loss, was 'liable to result'.

(4) The words 'liable to result' should be read in the sense conveyed by the expressions 'a serious possibility', 'a real danger' and 'not unlikely to occur'.

For the first limb, therefore, knowledge of certain basic facts according to the usual course of things is imputed, but not special knowledge.

---

[38] *Hadley* v. *Baxendale* (1854) 9 Ex 341.
[39] *Victoria Laundry (Windsor) Ltd* v. *Newman Industries Ltd* [1949] 2 KB 528.
[40] *Koufos* v. *Czarnikow* [1969] 1 AC 350.

### *The second limb of* Hadley v. Baxendale

The second limb of *Hadley* v. *Baxendale* depends on additional special knowledge by the defendant.

The passage from *Hadley* v. *Baxendale* quoted above is followed by:

'If the special circumstances were communicated by the plaintiffs to the defendants, and thus known to both parties, the damages resulting from the breach of such a contract, which they would reasonably contemplate, would be the amount of injury which would ordinarily follow from a breach of contract under these special circumstances so known and communicated.'

As with the first limb, the question is to be judged at the time of the contract so that damages claimed under the second limb will not be awarded unless the claimant has particular evidence to show that the defendant then knew the special circumstances relied on.

The first limb of the rule contains the necessity for knowledge of certain basic facts.

In respect of the second limb, 'additional' or 'special' knowledge, however, may extend liability to include for losses that are outside the natural course of events.

As noted by Lord Jauncey, 'It must always be a question of circumstances what one contracting party is presumed to know about the business activities of another.'[41]

Propositions 2, 3 and 4 noted above in respect of the first limb of the *Hadley* v. *Baxendale* rule apply to both limbs of the rule.

The necessary degree of likelihood of the loss occurring was considered at length in *Koufos* v. *Czarnikow*[42] case by reference to phrases such as those given at 4 above.

From case law it appears that the degree of likelihood can be less than evens, even slight,[43] but that to equate it with what is reasonably foreseeable could be misleading. This was stated in the *Koufos* v. *Czarnikow* case: 'To bring in reasonable foreseeability appears to me to be confusing measure of damages in contract with measure of damages in tort. A great many extremely unlikely results are reasonably foreseeable.'

## Date of assessment

The general rule, both in contract and in tort, is that damages should be assessed as at the date when the cause of action arises.[44] But there are many exceptions and it has been said that 'this so-called general rule . . . has been so far eroded in recent times . . . that little of practical reality remains of it'.[45]

---

[41] *Balfour Beatty* v. *Scottish Power* (1994) 71 BLR 20 HL.
[42] *Koufos* v. *Czarnikow* [1969] 1 AC 350.
[43] *Parsons* v. *Uttley Ingham* [1978] QB 791, CA.
[44] *Miliangos* v. *George Frank (Textiles) Ltd* [1976] AC 443, HL; *Dodd Properties* v. *Canterbury City Council* [1980] 1 WLR 433, CA.
[45] Ormrod LJ in *Cory & Son* v. *Wingate Investments* (1980) 17 BLR 104, CA.

Thus: 'where it is necessary in order adequately to compensate the plaintiff for the damage suffered by reason of the defendant's wrong a different date of assessment can be selected'.[46]

## Measure of damages

Sometimes the proper measure of damages is not the cost of reinstatement but the difference in value between the work actually produced and the work that should have been produced,[47] and this will particularly be so where the claimant has no prospect or intention of rebuilding, or where it would be unreasonable to award the cost of reinstatement.

The frequently quoted *Ruxley v. Forsyth*[48] House of Lords case was regarding a swimming pool that was not constructed to the required depth.

The swimming pool should have been constructed to a depth of seven feet six inches. However, when the pool was completed, its depth was only six feet nine inches.

Forsyth sought damages for the reinstatement costs (i.e. to rebuild the swimming pool so that it was seven feet six inches deep). However, the House of Lords found that where it would be unreasonable for the claimant to insist on reinstatement because the cost of the work involved would be out of all proportion to the benefit obtained, the claimant's measure of damages would simply be the difference in value (sometimes referred to as the loss of amenity value).

In the case of *Murphy v. Brentwood*,[49] it was found that a claim for diminution in value will not normally exceed the relevant costs of reinstatement. However, in one case, an employer was awarded as damages the difference between the values of houses as they should have been constructed as opposed to how they were constructed, although, in reality, the only loss the employer had suffered was the cost of voluntarily putting right defects at the request of subsequent purchasers.[50]

## Mitigation of loss

The award of damages as compensation is qualified by a principle, 'which imposes on a plaintiff the duty of taking all reasonable steps to mitigate the loss consequent on the breach, and debars him from claiming any part of the damage which is due to his neglect to take such steps'.[51] But this 'does not impose on the plaintiff an obligation to take any step which a reasonable and prudent man would not ordinarily take in the course of his business'.[52]

---

[46] Lord Browne-Wilkinson in *Smith New Court Ltd* v. *Scrimgeour Vickers* [1997] AC 254, HL.

[47] *Dodd Properties* v. *Canterbury City Council* [1980] 1 WLR 433, CA.

[48] *Ruxley Electronics and Construction Ltd* v. *Forsyth* [1995] 3 WLR 118.

[49] *Murphy* v. *Brentwood District Council* [1991] 1 AC 398, HL.

[50] *Newton Abbot Development Co. Ltd* v. *Stockman Bros* (1931) 47 TLR 616, referred to without apparent disapproval in *Linden Gardens* v. *Lenesta Sludge Disposals* [1993] 3 WLR 408.

[51] Lord Haldane in *British Westinghouse* v. *Underground Railways Co.* [1912] AC 673; *Andros Springs (Owners)* v. *World Beauty (Owners) (The World Beauty)* [1969] 3 All ER 158, CA; *Sotiros Shipping* v. *Sameiet Solholt* [1983] 1 Lloyds Rep 605, CA; *Kaines* v. *Osterreichische* [1993] 2 Lloyds Rep 1, CA.

[52] *British Westinghouse* v. *Underground Railways Co.* [1912] AC 673, referring to James LJ in *Dunkirk Colliery Co.* v. *Lever* (1878) 9 Ch D 20, CA.

## Liquidated damages

Whilst not normally specifically relevant to a contractor/sub-contractor relationship, a brief note should be made of liquidated damages.

The parties to a contract often agree that a liquidated (i.e. fixed and agreed) sum must be paid as damages for some breach of a contract.

A typical clause provides that if a contractor fails to complete by a date stipulated in the contract, or by any extended date, he must pay or allow the employer to deduct liquidated damages at the rate of a certain amount per week for the period during which the works are uncompleted.

The basic rules in respect of liquidated damages are that they cannot be a 'penalty' and they must be a genuine pre-estimate of the level of damages that would be incurred.

A contractor may be able to recover from a sub-contractor the liquidated damages that it has incurred as part of a common law damages claim (or, indeed as part of a loss and expense claim) against the sub-contractor.

# Chapter 10
# Variations

## General

### Variations under the contract

This chapter deals with variations under the sub-contract and not variations to the sub-contract.

Variations to the sub-contract are changes to the sub-contract itself (i.e. changes to the agreement, including the terms and conditions). Such changes can only be made by express agreement between the parties.

For example, an agreement to accelerate the progress of the works (where no acceleration provision is allowed for in the sub-contract) would be a variation to the sub-contract and not a variation under the sub-contract.

Variations under the sub-contract are changes to the scope of the work or conditions under which the work is carried out, only to the extent that terms in the sub-contract provide for such changes.

#### Why are variation clauses under the contract included in contracts?

The nature of construction work makes the possibility of changes to the scope of the work or the conditions under which the work is carried out susceptible to change.

This may be because of any one or more of numerous factors, including incomplete design at the commencement of the works, alterations necessary to suit the peculiarities of the site, restrictions imposed on the execution of the works after commencement of the works, or changes to suit the needs or wishes of the building owner or the planned occupier.

All of these matters are commonly encountered, and it is necessary from the building owner's point of view, and consequently from the contractor's point of view, that the sub-contract conditions must incorporate provisions for variations to the original scope of the work or the conditions under which the work is to be carried out.

If such a provision was not included within the contract then every time a change occurred this would constitute a breach of contract or would require separate sub-contracts to be raised for all additional works, etc.

For this reason all standard forms allow for variations arising under the contract (and the JCT Standard Building Sub-Contract is no exception).

## What is a variation?

A variation is, in simple terms, a change to the original scope of the sub-contract works.

The definitions section of SBCSub/C and SBCSub/D/C (clause 1.1) defines variation as 'see clause 5.1'.

Under clause 5.1 a variation is defined as being:

'5.1.1 the alteration or modification of the design, the quality or (except where the Remeasurement basis applies) the quantity of the Sub-Contract Works including:

> 5.1.1.1 the addition, omission or substitution of any work;
> 5.1.1.2 the alteration of the kind or standard of any of the materials or goods to be used in the Sub-Contract Works;
> 5.1.1.3 the removal from the site of any work executed or materials or goods brought thereon by the Sub-Contractor for the purposes of the Sub-Contract Works other than work, materials or goods which are not in accordance with this Sub-Contract.

5.1.2 the imposition in an instruction of the Architect/Contract Administrator issued under the Main Contract (or a direction of the Contractor passing on that instruction) of any obligation or restrictions in regard to the matters set out in this clause 5.1.2 or the addition to or alteration or omission of any such obligations or restrictions so imposed or imposed by the Numbered Documents and the Schedule of Information and its annexures in regard to:

> 5.1.2.1 access to the site or use of any specific parts of the site;
> 5.1.2.2 limitations of working space;
> 5.1.2.3 limitations of working hours;
> 5.1.2.4 the execution or completion of the work in any specific order.'

It should be noted that certain directions issued by the contractor under clause 6.7.3 and clause 6.8.3.1 of SBCSub/C or SBCSub/D/C (in respect of particular insurance related matters) are to be treated as variations in line with clause 6.7.4 and clause 6.8.3.2 of SBCSub/C and SBCSub/D/C.

Also, certain directions issued by the contractor under clause 6.14.1 of SBCSub/C or SBCSub/D/C (in respect of other insurance related matters) are to be treated as a variation under clause 6.14.2.

## How is a variation instructed?

All variation instructions are to be issued in writing (refer to clause 3.4 of SBCSub/C and SBCSub/D/C).

If the contractor purports to give any direction to the sub-contractor, or his authorized representative, otherwise than in writing (e.g. orally), then that direction will have no immediate effect (refer to clause 3.7 of SBCSub/C and SBCSub/D/C).

However, despite this, clause 3.7 of SBCSub/C and SBCSub/D/C does set out a procedure to be followed for the confirmation of instructions issued otherwise than in writing.

This procedure is:

(1) The sub-contractor is to confirm the direction to the contractor in writing within seven days of the direction being issued.
(2) If the contractor does not dissent in writing to the sub-contractor that the direction was issued (within seven days from receipt of the sub-contractor's confirmation), then the direction will take effect as from the expiry of that seven-day period.

If, within seven days of giving a direction other than in writing, the contractor confirms the direction in writing, then the sub-contractor will not be obliged to confirm the direction and the direction will take effect as from the date of the contractor's written confirmation (refer to clause 3.7.1 of SBCSub/C and SBCSub/D/C).

It should be noted that if action is taken on the basis of an oral instruction (more commonly referred to as a verbal instruction), without following the above procedure, then such action is taken entirely at the sub-contractor's risk.

However, clause 3.7.2 of SBCSub/C and SBCSub/D/C confirms that if a sub-contractor complies with an oral direction without following the procedure set out under clause 3.7 of SBCSub/C and SBCSub/D/C the contractor may at any time prior to the final payment under the sub-contract confirm the direction in writing with retrospective effect (refer to clause 3.7.2 of SBCSub/C and SBCSub/D/C).

The operative word in the foregoing phrase is 'may', and there is no obligation for a contractor to retrospectively confirm a purported direction at all, and in reality it is extremely unlikely that a contractor would confirm a purported direction where it disagrees that such a direction was issued in the first place.

It should be noted that any written instruction of the architect/contract administrator issued under the main contract that affects the sub-contract works which is then issued by the contractor to the sub-contractor is deemed to be a direction of the contractor (refer to clause 3.4 of SBCSub/C and SBCSub/D/C).

How this latter requirement will be implemented in practice, in terms of disclosure by the contractor, remains to be seen.

## Must a sub-contractor comply with all variation directions issued?

Clause 3.4 of SBCSub/C and SBCSub/D/C makes it clear that no variation directed by the contractor or subsequently sanctioned by him shall vitiate the sub-contract. To vitiate means to make invalid or ineffectual, and clause 3.4 is therefore noting that no variation will make the sub-contract invalid or ineffectual.

Although the above phrase is common in most standard forms, it is submitted that there must be some limit to the nature and extent of a variation which can be ordered, particularly where the variation requires the sub-contractor to carry out design works or works on site for which it professes no skill or experience whatsoever.

Clause 3.5.1 of SBCSub/C and SBCSub/D/C notes that where a contractor issues a direction which requires a variation of the type referred to in clause 5.1.2 of SBCSub/C and SBCSub/D/C (i.e. the imposition, alteration or omission of any obligation or restriction in regard to access to the site, limitation of working space, limitation of working hours, etc.), the sub-contractor need not comply to the extent that he makes reasonable objection to it in writing to the contractor.

Of course, there could be many disputes regarding what 'reasonable objections' are, and any such dispute could have a major impact on the actions that a sub-contractor can or should sensibly attempt to take.

In the case where a direction is given which pursuant to clause 5.3.1 of SBCSub/C and SBCSub/D/C requires the sub-contractor to provide a schedule 2 quotation (see further below regarding schedule 2 quotations), clause 3.5.2 of SBCSub/C and SBCSub/D/C makes clear that the variation must not be carried out until the contractor has in relation to the schedule 2 quotation either issued written acceptance, or has issued a further direction under clause 5.3.2 of SBCSub/C and SBCSub/D/C (i.e. that the work is to be valued by a valuation by the contractor in accordance with the valuation rules pursuant to clause 5.2 of SBCSub/C and SBCSub/D/C).

Under clause 3.5.3 of SBCSub/D/C if, in the sub-contractor's opinion, compliance with any direction of the contractor may injuriously affect the efficacy of the design of the sub-contractor's design portion, the sub-contractor has the right to object (providing that such objection is made within five days from receipt of the contractor's direction).

The contractor's direction will then not have any effect unless confirmed by the contractor.

Naturally, given the strict liability placed upon the sub-contractor in respect of design liability, the sub-contractor would need to make clear to the contractor that by confirming the earlier direction that was (in the opinion of the sub-contractor) injurious to the sub-contractor's design the contractor would be removing the liability for such design from the sub-contractor.

### What happens if a sub-contractor does not comply with a direction issued?

Clause 3.6 of SBCSub/C and SBCSub/D/C makes it clear that, if, within seven days after receipt of a written notice from the contractor which requires a sub-contractor to comply with a direction, the sub-contractor still does not comply with that direction, then the contractor may employ and pay other persons to execute any work which may be necessary to give effect to that direction.

In such a situation, the sub-contractor would be liable for all additional costs incurred by the contractor in connection with such employment and an appropriate deduction must either be taken into account in the calculation of the final sub-contract sum or will be recoverable by the contractor from the sub-contractor as a debt.

## *Valuation of work*

Unless otherwise agreed by the contractor and the sub-contractor, a variation must be valued either by way of an accepted schedule 2 quotation, or in accordance with clauses 5.6 to 5.12 of SBCSub/C and SBCSub/D/C (otherwise known as the valuation rules). Refer to clause 5.2 of SBCSub/C and SBCSub/D/C.

The sub-contract may be based on an adjustment basis (which means that the sub-contract sum must not be adjusted or altered in any way other than in accordance with the express provisions of SBCSub/C and SBCSub/D/C) or on a remeasurement basis (which means that the sub-contract works must be subject to complete remeasurement).

If the sub-contract is based on the adjustment basis then, in line with clause 5.2.1 of SBCSub/C and SBCSub/D/C) the valuation rules will apply to:

(1) all variations, including any sanctioned in writing by the contractor but excluding any variation to an accepted schedule 2 quotation;
(2) all work which under SBCSub/C and SBCSub/D/C conditions is to be treated as a variation;
(3) all work executed by the sub-contractor in accordance with the directions of the contractor as to the expenditure of provisional sums which are included in the sub-contract documents; and
(4) all work executed by the sub-contractor for which an approximate quantity has been included in any bills of quantities.

If the sub-contract is based on the remeasurement basis then the valuation rules will apply to all work executed by the sub-contractor in accordance with the sub-contract documents and the directions of the contractor, including any direction requiring a variation or in regard to the expenditure of a provisional sum included in the sub-contract documents (refer to clause 5.2.2 of SBCSub/C and SBCSub/D/C).

Such valuation, insofar as it relates to the sub-contractor's designed portion, must be in accordance with clause 5.10 (of SBCSub/D/C) and references in clauses 5.6 and 5.7 of SBCSub/D/C must (thus) exclude the valuation of variations in respect of the sub-contractor's designed portion analysis.

## Schedule 2 quotation

The provisions for a schedule 2 quotation are included under clause 5.3 of SBCSub/C and SBCSub/D/C.

Paragraph 178 of SBCSub/G makes it clear that the schedule 2 quotation provisions under SBCSub/C and SBCSub/D/C follow the schedule 2 quotation provisions in the JCT main contract form. However, SBCSub/C and SBCSub/D/C includes for its own schedule 2 quotation provisions under the schedule 2 appendix attached to SBCSub/C and SBCSub/D/C.

The requirements for a schedule 2 quotation are set out below.

Clause 5.3.1 of SBCSub/C and SBCSub/D/C notes that if the contractor in his direction states that the sub-contractor is to provide a schedule 2 quotation, the sub-contractor must, subject to receipt of sufficient information, provide a quotation in accordance with the procedure outlined below, unless, within four days of his receipt of that direction (or such longer period as is either stated in the direction or is agreed between the contractor and the sub-contractor) he notifies the contractor that he disagrees with the application of that procedure to that direction.

If the sub-contractor does issue such a notice of disagreement, then the sub-contractor will not be obliged to provide a schedule 2 quotation, and the work must not be carried out unless and until the contractor gives a further direction that the work is to be carried out and is to be valued using one of the valuation rules.

## Submission of quotation

Any instruction of the contractor requesting a schedule 2 quotation must provide sufficient information to enable the sub-contractor to provide that quotation.

The information provided would normally be expected to be in a similar format to that provided at tender stage.

If the sub-contractor reasonably considers that the information provided is not sufficient, then, not later than four days from the date of the instruction, he must notify the contractor who must supply that information.

The sub-contractor must submit his schedule 2 quotation to the contractor in compliance with the instruction not later than 14 days from the later of:

(1) the date of receipt of the instruction; or
(2) the date of receipt by the sub-contractor of sufficient information as referred to above.

The schedule 2 quotation must remain open for acceptance by the contractor for 14 days from its receipt by the contractor.

It should be noted that the contractor and the sub-contractor may agree to increase or reduce the number of days stated; and any agreement must be confirmed by the contractor to the sub-contractor in writing.

The variation, for which the sub-contractor has submitted his schedule 2 quotation, must not be carried out by the sub-contractor until receipt by the sub-contractor of the confirmed acceptance issued by the contractor; or until the contractor gives a further direction that the work is to be carried out and is to be valued by one of the valuation rules.

In summary, therefore, and assuming that the contractor and sub-contractor do not agree upon extended time periods:

- the contractor in his direction states that the sub-contractor is to provide a schedule 2 quotation;
- if the sub-contractor disagrees with the application of that procedure to that direction, he is to notify the contractor of that fact within four days of his receipt of the direction;
- if the sub-contractor does issue such a notice of disagreement, then the sub-contractor will not be obliged to provide a schedule 2 quotation, and the work will not be carried out unless and until the contractor gives a further direction that the work is to be carried out and is to be valued using one of the valuation rules;
- if the sub-contractor does not issue such a notice of disagreement, but considers that the information provided is not sufficient, then, not later than four days from the date of the instruction, he must notify the contractor who must supply the necessary information;
- the sub-contractor must submit his schedule 2 quotation to the contractor not later than 14 days from the later of:

  — the date of receipt of the instruction; or
  — the date of receipt by the sub-contractor of sufficient information as referred to above.

- the schedule 2 quotation must remain open for acceptance by the contractor for days from its receipt by the contractor.

## Content of the quotation

It is important to note that the schedule 2 quotation is to separately comprise:

(1) the amount of the adjustment to the final sub-contract sum (excluding any amount to be paid in lieu of any ascertainment for direct loss and/or expense, but including allowance, where appropriate, for preliminary items) supported by all necessary calculations, which must be made by reference, where relevant, to the rates and prices in the sub-contract sum or the sub-contract tender sum;

(2) any adjustment to the time required for completion (see commentary regarding acceleration in Chapter 9) of the sub-contract works and/or of any works in any section by reference to the period or periods stated in the sub-contract particulars (item 5) to the extent that such adjustment is not included in any revision to the completion date previously issued by the contractor or in the contractor's confirmed acceptance of any other schedule 2 quotation;

(3) the amount to be paid in lieu of any ascertainment, under clause 4.19 of SBCSub/C or SBCSub/D/C, of direct loss and expense not included in any other accepted schedule 2 quotation or in any previous ascertainment under clause 4.19;

(4) a fair and reasonable amount in respect of the cost of preparing the schedule 2 quotation; and

(5) where specifically required by the instruction, must provide indicative information in statements on:

(a) the additional resources (if any) required to carry out the variation; and
(b) the method of carrying out the variation.

Each part of the schedule 2 quotation must contain reasonably sufficient supporting information to enable that part to be evaluated by or on behalf of the contractor.

## Acceptance of the quotation

If the contractor wishes to accept a schedule 2 quotation then he must notify the sub-contractor in writing not later than the last day of the period of acceptance.

If the contractor accepts a schedule 2 quotation, the contractor must, immediately upon that acceptance, confirm such acceptance in writing to the sub-contractor:

(1) that the sub-contractor is to carry out the variation;
(2) the adjustment to the final sub-contract sum;
(3) any adjustment to the time required by the sub-contractor for completion of the sub-contract works or (where applicable) such works in any relevant section.

## Quotation not accepted

If the contractor does not accept the schedule 2 quotation by the expiry of the period of acceptance, he must, on the expiry of that period either:

(1) instruct that the variation is to be carried out and is to be valued under the valuation rules; or
(2) instruct that the variation is not to be carried out.

## Cost of quotation

If a schedule 2 quotation is not accepted, a fair and reasonable amount must be added to the final sub-contract sum in respect of the preparation of the schedule 2 quotation, provided that the schedule 2 quotation has been prepared on a fair and reasonable basis.

Although there is no definition of when a schedule 2 quotation has been prepared on a fair and reasonable basis, it is clear that the non-acceptance by the contractor will not, of itself, be evidence that the quotation was not prepared on a fair and reasonable basis.

## Restriction on use of quotation

Unless the contractor accepts a schedule 2 quotation, neither the contractor nor the sub-contractor may use that quotation for any purpose whatsoever.

## What happens if work covered by a schedule 2 quotation that has been accepted by the contractor is itself varied?

Where a schedule 2 quotation has been accepted by the contractor, then, if the contractor subsequently issues a direction varying the work that was the subject matter of that schedule 2 quotation, the contractor must make a valuation of that variation on a fair and reasonable basis having regard to the content of the schedule 2 quotation and must include in that valuation the direct loss and/or expense, if any, incurred by the sub-contractor because the regular progress of the sub-contract works or any section of it is materially affected by compliance with the direction (refer to clause 5.3.3 of SBCSub/C and SBCSub/D/C).

## Giving effect of valuations, agreements, etc.

Clause 5.5 of SBCSub/C and SBCSub/D/C notes that effect must be given, in the calculation of the final sub-contract sum, to each agreement by the contractor and the sub-contractor under clause 5.2.1 of SBCSub/C and SBCSub/D/C, to each valuation and each schedule 2 quotation accepted by the contractor.

## *The valuation rules*

Where a schedule 2 quotation does not apply, the valuation rules are applicable to the extent that a valuation relates to the execution of additional or substituted work which can properly be valued by measurement.

It should be noted that in respect of all the valuation rules, the contractor is to give the sub-contractor an opportunity to be present and to allow him to take such notes and measurements as he may require at the time when it is necessary to measure work for the purpose of valuation (refer to clause 5.4 of SBCSub/C and SBCSub/D/C).

Although it is for the contractor to value the works, it is obviously sensible for the sub-contractor to issue its own valuation assessment to minimize the risk of under recovery.

There are certain general rules that apply (refer to clauses 5.8.1 to 5.8.4 of SBCSub/C and SBCSub/D/C), namely:

(1) Where there are bills of quantities, the measurement of variations must be in accordance with the same principles as those governing the preparation of those bills of quantities.

The default position in respect of the preparation of bills of quantities is that SSM7, produced by the Royal Institution of Chartered Surveyors and the Construction Confederation, will have been used.

Of course, if the sub-contract documents specifically state that an alternative method of measurement has been used then the default position will not apply.

In such a situation, it is good practice for the sub-contractor to identify the effect of the changed method of measurement, and to make due allowance for such effect in its tender.

(2) When valuing variations, allowances must be made for any percentage or lump sum adjustments in any bills of quantities and/or other sub-contract documents.

Therefore, if within the bills of quantities and/or any other sub-contract documents a percentage uplift (or a lump sum uplift) was shown and included within the sub-contract sum (or the sub-contract tender sum, as appropriate) for overheads and profit, for example, then that same percentage uplift (or the relevant proportion of the lump sum uplift) should be used when valuing variations.

(3) Where the adjustment basis applies, an allowance (where appropriate) must be made for any addition to or reduction from any preliminary items of the type referred to in SMM7.

SMM7 lists out many preliminary items, including site accommodation, site storage, site plant and scaffolding, etc.

Therefore if a variation meant that, for example, further scaffolding was required, then the value of that additional scaffolding should be included with the value of the variation.

The above principle does not apply where a variation relates to the contractor's direction for the expenditure of a provisional sum for defined work, for the reasons outlined below.

The SMM7 general rule 10.3 defines a provisional sum for defined work as being a sum for work which is not completely designed but for which the following information is provided:

(a) the nature of the construction of the work;
(b) a statement of how and where the work is fixed to the building and what other work is to be fixed to it;
(c) a quantity or quantities which indicate the scope and extent of the works;
(d) any specific limitations on the method, sequence or timing of the works.

If the information specified in rule 10.3 is not available then the contract bills should describe the provisional sum as undefined.

General rule 10.4 of SMM7 states that 'where Provisional Sums are given for defined work the Contractor [and the sub-contract or] will be deemed to have made due allowance in programming, planning and pricing preliminaries', whilst general rule 10.6 of SMM7 states that the contractor is deemed to have made *no such allowance* for undefined work.

Therefore the categorization of defined or undefined provisional sums has significant implications, and the sub-contractor should ensure that the sub-contract has not been amended to alter this important provision.

It is often found in bills of quantities and other sub-contract documents that provisional sums are stated as being defined but, because the information required by general rule 10.3 of SMM7 has not been provided (or has not been sufficiently provided) the provisional sum in question cannot be construed as being defined. In other words a provisional sum is either defined or undefined based upon its content and not upon its title.

If such a situation were to arise, where a provisional sum was incorrectly described as defined, then clause 2.9.1 of SBCSub/C and SBCSub/D/C makes it clear that if there is any error in, or omission of, information in any item which is the subject of a provisional sum for defined work, the description must be corrected so that it does provide that information.

Clause 2.9.2 of SBCSub/C and SBCSub/D/C then adds that any such correction, alteration or modification to the provisional sum must be treated as a variation.

(4) Where the remeasurement basis applies, any amounts priced in the preliminaries section of any bills of quantities must be adjusted (where appropriate) to take into account any variations of any contractor's directions for the expenditure of provisional sums for undefined work included in the sub-contract documents.

## The valuation rule dealt with under clauses 5.6.1.1 and 5.7.1 of SBCSub/C and SBCSub/D/C: variation works of similar character, conditions and quantity

Under this valuation rule, where variation work is carried out that is of similar character to, is executed under similar conditions to, and does not significantly change the quantity of work set out in any bills of quantities and/or other sub-contract documents (including the contractor's requirements in the case of SBCSub/D/C), the rates and prices for the work in those documents must determine the valuation.

This is subject only to clause 5.10 of SBCSub/D/C in the case of sub-contractor design portion works (see below).

Similar conditions are those conditions which are to be derived from the express provisions of the sub-contract. Extrinsic evidence of, for instance, the parties' subjective expectations is not admissible.[1]

### The valuation rule dealt with under clause 5.6.1.2 and 5.7.2 of SBCSub/C and SBCSub/D/C: variation works of similar character, but dissimilar conditions and/or dissimilar quantity

Under this valuation rule, where variation work is carried out that is of similar character to work set out in any bills of quantities and/or other sub-contract documents, but is not executed under similar conditions and/or significantly changes its quantity, the rates and prices for the work set out in the documents above must be the basis for determining the valuation and the valuation value must include a fair allowance for such difference in conditions and/or quantity (refer to clauses 5.6.1.2 and 5.7.2 of SBCSub/C and SBCSub/D/C).

Dissimilar conditions might include physical site conditions such as high compared with low, wet compared with dry, confined space compared with ample working space and winter working compared with summer working, but only where the sub-contract documents show that the measured rate prices were based on such conditions.

This matter was considered in the first instance decision of *Wates* v. *Bredero Fleet*,[2] which was an appeal on a point of law against an arbitrator's interpretation of the valuation rules under the 1980 edition of the JCT Standard Form of Contract (where similar variation valuation rules apply).

The arbitrator had decided the question of whether there was a change of conditions under clause 13.5 by reference to:

(1) the conditions the contractor had in mind when pricing his tender;
(2) the knowledge gained from negotiations;
(3) the express terms of the contract bills; and
(4) the physical conditions relating to time, the contract period, the nature of the works and other matters impinging upon the working environment.

However, Judge Forbes QC held that any information as to conditions not stated in the contract was inadmissible extrinsic evidence and that therefore the arbitrator had erred in law by considering such evidence.

He stated that the conditions referred to in clause 13.5.5 were the conditions under which the relevant contract works were to be executed, as could be determined from the express provisions of the contract bills, drawings and other contract documents.

Therefore, for example, if a sub-contractor's price for rendering had been based upon the fact that the rendering would all be carried out from the ground (in line with the contract details) but the actual rendering needed to be carried out from

---

[1] *Wates Construction v. Bredero Fleet* (1993) 63 BLR 128.
[2] *Wates Construction v. Bredero Fleet* (1993) 63 BLR 128.

access towers, then a fair allowance would need to be made for the extra time in loading out materials to height, for the extra time in operatives ascending and descending the access towers, and for the cost of providing the access towers (if these are provided by the sub-contractor).

In addition, the sub-contract includes as an explicit consideration of whether a variation significantly changes the quantity of the work.

Therefore, for example, if a sub-contractor originally priced 5 metres of skirting, but through a variation this increased to 500 metres of skirting, then a fair allowance may need to be allowed for the increased quantity. When reviewing this matter, improved bulk purchase discounts, lower average material production costs and changed tradesmen/labour ratios may need to be considered.

The normal way to determine a fair allowance is to analyse the original rate and pro rata elements of the rate and/or apply pricing constants (either from price books or from other data held by the sub-contractor) where allowances for changed conditions need to be made.

### How does the reference to a fair allowance apply where pricing errors have been made in the tender?

If the pricing error is where a sub-contractor has correctly stated his price per unit of measurement but has incorrectly extended the price and this error has been incorporated into the lump sum (i.e. the sub-contract sum) or, in the case of the remeasurement basis into his estimated total of prices (i.e. the sub-contract tender sum), and neither party noticed the error before the sub-contract was made and there is no ground for rectification, then, if the contract is for a lump sum, there is, it is submitted, no implied right to have the contract price adjusted to take account of the error. However, if the contract is formed on a remeasurement basis where the sub-contractor is to be paid 'x' pounds per unit of measurement, the pricing error will disappear on valuation of the measured work at completion and the estimated total of prices will have no significance.

In the case where a sub-contractor has simply submitted a rate in error, it has been found that such a mistake would not prevent the use of those rates to value a subsequent variation.[3] Therefore, if a sub-contractor inadvertently inserted a rate of £5 per metre for an element of the work, whereas he intended to insert a rate of £50 per metre, the sub-contractor would not only be required to stand by his tendered rate for the original quantity but would also be required to stand by that rate for any increase in quantity because of variation instructions issued.

It should be noted that if, in the normal course of events, a change in conditions has the effect of reducing or increasing a tendered rate (for example a large increase in quantity which may reduce the manufacturing costs per unit) then any such reduction or increase should be applied to the tendered rate irrespective of whether or not the tendered rate contains an error.

Of course, conversely to the above, if the sub-contractor had written £50 per metre in the tender when he intended to write £5 per metre he obtains the benefit of his error when variations are issued that impact upon that rate.

---

[3] *Henry Boot Construction Limited v. Alstom* [1999] BLR 123.

## The valuation rule dealt with under clause 5.6.1.3 and 5.7.3 of SBCSub/C and SBCSub/D/C: variation works not of a similar character

Under this valuation rule, where variation work is carried out that is not of similar character to work set out in any bills of quantities and/or other sub-contract documents, the work must be valued at fair rates and prices.

Fair rates and prices were considered in the *Crittall Windows* v. *T.J. Evers*[4] case.

In that case, Judge Humphrey Lloyd QC stated that: 'a fair valuation generally means a valuation which will not give the contractor more than his actual costs reasonably and necessarily incurred plus similar allowance for overheads and profit'.

Judge Humphrey Lloyd QC has since repeated this approach in two other cases.[5]

However, a more common view is that fair rates and prices must have regard to the contractor's general pricing level, and therefore a valuation below actual cost would be fair where the contract price is below actual costs or market pricing.

## The valuation rule dealt with under clause 5.6.1.4 of SBCSub/C and SBCSub/D/C: reasonably accurate approximate quantity

This valuation rule is only applicable when the sub-contract is let on the adjustment basis (i.e. where a sub-contract sum is used as the basis of the sub-contract rather than where a sub-contract tender sum is used). The rule is that where an approximate quantity is included in the bills of quantities and/or other sub-contract documents, and that approximate quantity is a reasonably accurate forecast of the quantity of work actually required, then the rate or price for that approximate quantity must determine the valuation of the varied works (refer to clause 5.6.1.4 of SBCSub/C and SBCSub/D/C).

Unfortunately, there is no definition of what a 'reasonably accurate forecast' is, but a margin of plus or minus 10% of the approximate quantity allowance would normally be accepted as being within the parameters of a 'reasonably accurate forecast'.

This rule only applies where the work required has not been altered or modified in any way from that specified other than in terms of quantity (refer to clause 5.6.1. of SBCSub/C and SBCSub/D/C).

## The valuation rule dealt with under clause 5.6.1.5 of SBCSub/C and SBCSub/D/C: not reasonably accurate approximate quantity

This valuation rule is again also only applicable when the sub-contract is let on the adjustment basis (i.e. where a sub-contract sum is used as the basis of the sub-contract rather than where a sub-contract tender sum is used). The rule is that

---

[4] *Crittall Windows* v. *T.J. Evers Ltd* (1996) 54 Con LR 66.
[5] *Floods Queensferry Limited* v. *Shand Construction Limited* [1999] BLR 315; *Weldon Plant Limited* v. *The Commissioner for the New Towns* [2000] BLR 496.

where an approximate quantity is included in the bills of quantities and/or other sub-contract documents and that approximate quantity is not a reasonably accurate forecast of the quantity of work actually required, then the rate or price for that approximate quantity must be used as the basis of determining the valuation of the varied works, with a fair allowance being made for such difference in quantity.

It should be noted that this fair allowance may be nil, and a positive or negative value for this fair allowance would only be appropriate if the change in quantity had an actual effect on the measured rate that should apply. This may be, for example, because of increased or decreased bulk discounts, or increased or decreased material production costs, etc.

Again, this rule only applies where the work required has not been altered or modified in any way from that specified other than in terms of quantity (refer to clause 5.6.1. of SBCSub/C and SBCSub/D/C).

## The valuation rule dealt with under clause 5.6.2 of SBCSub/C: omission of work

This valuation rule is only applicable when the sub-contract is let on the adjustment basis (i.e. where a sub-contract sum is used as the basis of the sub-contract rather than a sub-contract tender sum).

The rule states that to the extent that a valuation relates to the omission of work set out in any bills of quantities and/or other sub-contract documents, the rates and the prices for such work set out must determine the valuation of the work omitted, subject only to clause 5.10 of SBCSub/D/C, in the case of sub-contractor design portion works.

It is commonly argued that where a rate in a bill of quantities and/or other sub-contract document appears to be too high (perhaps because a sub-contractor may allege that this rate includes for other activities, or that a larger proportion of the sub-contractor's overheads and profit has been allocated to the rate in question) or appears to be too low (perhaps where a contractor alleges that a sub-contractor has made a mistake in its tender) some adjustment should be made to the bill of quantity and/or other sub-contract document rate when omissions are made.

However, clause 5.6.2 of SBCSub/C and SBCSub/D/C makes it quite clear that only the rate in the bill of quantities and/or in other sub-contract documents (and no adjustment to that rate) may be used where omissions of work set out in the bills of quantities and/or other sub-contract documents is made.

## Change of conditions for other work

This valuation rule notes that if there is a substantial change in the conditions under which any other work (including, in the case of SBCSub/D/C, sub-contractor design portion works) is executed, as a result of any of the following then such other work shall be treated as if it had been the subject of a direction of the contractor requiring a variation and shall be valued on the basis of the rates and prices for the work set out in any bills of quantities and/or other sub-contract documents together with a fair allowance for the difference in conditions:

(1) compliance with any direction requiring a variation (except a variation for which a schedule 2 quotation has been accepted or where the variation is to works covered by a schedule 2 quotation; and except for any direction as a result of which work due to be executed in the sub-contract on the remeasurement basis is not executed);

(2) compliance with any direction as to the expenditure of a provisional sum for undefined work;

(3) compliance with any direction as to the expenditure of a provisional sum for defined work, to the extent that the direction for that work differs from the description given for such work in any bills of quantities; or

(4) where the adjustment basis applies, the execution of the work for which an approximate quantity is included in any bills of quantities, to the extent that the quantity is more or less than the quantity ascribed to that work in those bills.

## Sub-contractor's designed portion – valuation

Valuations relating to the sub-contractor's designed portion are to be made under clause 5.10 of SBCSub/D/C.
In such valuation:

- Allowance must be made for work involved in the preparation of the relevant design work (refer to clause 5.10.1 of SBCSub/D/C).

- The valuation must be consistent with the values of work of a similar character set out in the sub-contractor's designed portion analysis, making due allowance for any change in the conditions under which work is carried out and/or any significant change in the quantity of work so set out.
  Where there is no work of a similar character set out in the sub-contractor's designed portion analysis, a fair valuation must be made (refer to clause 5.10.2 of SBCSub/D/C).

- The valuation of the omission of the work set out in the sub-contractor's designed portion analysis must be in accordance with the values for such work (refer to clause 5.10.3 of SBCSub/D/C).

- Clauses 5.8.2, 5.8.3, 5.8.4, 5.9 and 5.11 of SBCSub/D/C (as dealt with elsewhere in this chapter) will apply so far as is relevant (refer to clause 5.10.4 of SBCSub/D/C).

## *Daywork*

Daywork is often an emotive issue in the relationship between contractors and sub-contractors.
Daywork is a means of valuing variations which is based upon recorded time, and recorded material and plant usage.
Contractors often consider that sub-contractors use daywork for the valuation of variations that should be measured and for which measured rates should apply,

because daywork is seen as a means of recovering monies on effectively a cost plus basis.

Sub-contractors often consider that despite the fact that daywork is the only way of valuing certain variations, contractors will attempt to ignore daywork sheets submitted in favour of a measurement approach that is not appropriate to the variation in question.

Despite these (sometimes legitimate) concerns SBCSub/C and SBCSub/D/C provides for the use of daywork for the valuation of certain variations.

Clause 5.9 of SBCSub/C and SBCSub/D/C makes it clear that these circumstances are only when a variation cannot properly be valued by measurement.

Of course, there may well be disputes about what can and what cannot properly be valued by measurement, and SBCSub/C and SBCSub/D/C does not give any further guidance on this matter.

Clearly, certain events cannot be valued by measurement, for example providing labourers that are required simply to attend upon other trades. But other events which could be measured may not properly be able to be valued by measurement, perhaps, for example, breaking out concrete around live services.

The time and resources spent on a variation that it is proposed is valued on a daywork basis should be recorded by the sub-contractor on a daywork voucher (or sheet), and that daywork voucher should have sections to record the operatives' names and trades, the time spent by the operatives on the variation, and also the plant and materials used.

The completed daywork voucher must be delivered for verification to the contractor not later than the Wednesday following the week in which the work was executed (refer to clause 5.9 of SBCSub/C and SBCSub/D/C). The reason for this relatively short timetable is clearly to enable the contractor to contemporaneously check the details on the daywork voucher.

The various elements of the daywork voucher are valued on the basis of the schedule of daywork rates included in the numbered documents. If no such schedule is included in the numbered documents then the various elements of the daywork voucher are valued on the following basis.

## Labour

This will be valued either at the all-in rate stated for each grade of operative as noted under item 11 of the sub-contract particulars (refer to clause 5.9.1 of SBCSub/C and SBCSub/D/C), or alternatively, if an all-in rate is not stated, then the prime cost for each grade of operative calculated in accordance with the *Definition of Prime Cost of Daywork carried out under a Building Contract* together with percentage additions to the prime cost at the rates set out in the sub-contract particulars (item 11) (refer to clause 5.9.1 of SBCSub/C and SBCSub/D/C).

The *Definition of Prime Cost of Daywork carried out under a Building Contract* is a document issued by the Royal Institution of Chartered Surveyors and the Construction Confederation, with the edition to be used being that current at the sub-contract base date (as stated in the sub-contract particulars).

If the operatives in question are within the province of any specialist trade where the Royal Institution of Chartered Surveyors and the appropriate body representing the employers in that trade have agreed and issued a definition of prime cost of daywork, then the prime cost for each grade of operative calculated in accordance with that definition is to be used together with the percentage addition to the prime cost at the rates set out in the sub-contract particulars (refer to clause 5.9.2 of SBCSub/C and SBCSub/D/C).

There are currently three definitions to which clause 5.9.2 refers. These are those agreed between the Royal Institution of Chartered Surveyors and the Electrical Contractors' Association, the Electrical Contractors' Association of Scotland, and the Heating and Ventilating Association, respectively.

When tendering for work, it is important to note that the prime cost for operatives is based upon the union wage rates and the percentage addition (which is to be provided by the sub-contractor) and is deemed to include, amongst other things:

- head office charges;
- site staff including site supervision;
- time lost due to inclement weather;
- the additional cost of bonuses and all other incentive payments in excess of any guaranteed minimum union wage rates;
- fares and travel allowance;
- sick pay or insurance in respect thereof;
- third party and employer's liability insurance;
- liability in respect of redundancy payments to employees;
- employer's National Insurance contributions.

## Plant

Plant is valued in accordance with one or more of the three definitions noted above to obtain the plant daywork base rate at the time that the daywork was carried out, and then the appropriate percentage addition is applied as inserted against the appropriate definition (refer to clause 5.9.1 of SBCSub/C and SBCSub/D/C).

When tendering for work, it is important to note that the prime cost for plant on a daywork basis may, on occasion, be based on a basic plant schedule which may pre-date the date when the dayworks are being carried out by several years, and the percentage addition allowed by the sub-contractor needs to reflect this fact.

## Materials

Materials are priced at the prime cost for materials, together with the percentage addition to the prime cost at the rates set out in the sub-contract particulars (item 11) (refer to clause 5.9.1 of SBCSub/C and SBCSub/D/C).

When tendering for work, it is important to note that the prime cost for materials is usually taken to be the cost of materials after all trade and cash discounts have been deducted.

# Additional provisions

## Fair valuation

If a valuation does not relate to the execution of work or the omission of work, or if the valuation of any work or liabilities directly associated with a variation cannot reasonably be effected in the valuation by any of the valuation rules noted above, then clause 5.12.1 of the SBCSub/C notes that a fair valuation must be made.

It must be noted that this additional provision does not relate to the execution of additional work or the omission of work, and only relates to any work or liabilities directly associated with a variation.

## Non-recovery of loss and/or expense

Clause 5.12.2 of SBCSub/C notes that no allowance for any effect upon the regular progress of the works (or any part of them) or for any other direct loss and/or expense for which the sub-contractor would be reimbursed by any other provision in SBCSub/C or SBCSub/D/C is to be made under the valuation rules.

# Chapter 11
# Injury, Damage and Insurance

SBCSUB/C AND SBCSUB/D/C: SECTION 6

## Definitions

The definitions detailed under clause 6.1 of SBCSub/C and SBCSub/D/C are listed under Definitions at the beginning of this book.

## Injury to persons and property

Clauses 6.2 to 6.5 inclusive of SBCSub/C and SBCSub/D/C deal with injury to persons and property.

### Liability of sub-contractor – personal injury or death

The sub-contractor is liable for, and must indemnify the contractor against, any expense, liability, loss, claim or proceedings whatsoever in respect of personal injury to or the death of any person arising out of, or in the course of, or caused by the carrying out of the sub-contract works (as clause 6.2 of SBCSub/C and SBCSub/D/C). Clause 6.5.2 notes that this indemnity does not include for personal injury to, or the death of, any person due to the effect of an excepted risk (refer to definitions at the beginning of this book).

The above is known as an indemnity clause, where one party agrees to make good a loss suffered by another.

Although this indemnity may exist, from the contractor's perspective it should be noted that it is no defence to a contractor sued by a third person that he is indemnified by a sub-contractor,[1] although he can join the sub-contractor as a Civil Procedure Rules part 20 defendant to court proceedings between himself and the third person.

Under clause 6.2 of SBCSub/C and SBCSub/D/C, the sub-contractor is not liable where the personal injury to, or the death of, any person arising out of, or in the course of, or caused by, the carrying out of the sub-contract works is due to any act of neglect, breach of statutory duty, omission or default of:

(1) the contractor;
(2) any of the contractor's persons;

---

[1] *Dalton v. Angus* (1881) 6 App Cas 740, HL.

(3) the employer;
(4) any of the employers persons; or
(5) any statutory undertaker.

In addition to the sub-contractor's obligation to indemnify the contractor as detailed above, the sub-contractor is also to take out and maintain insurance in respect of claims arising under this liability (as clause 6.5.1 of SBCSub/C and SBCSub/D/C). This is an entirely sensible requirement because the indemnity of a sub-contractor that is insubstantial may effectively be worthless. The insurance cover required by clause 6.5.1 of SBCSub/C and SBCSub/D/C supports and protects both the sub-contractor and the contractor, although it must be noted that clause 6.5.1 makes it clear that such insurance is without prejudice to the sub-contractor's obligations to indemnify the contractor under clause 6.2 of SBCSub/C and SBCSub/D/C.

In other words, the fact that clause 6.5.1 requires insurance to be taken out does not affect or lessen the sub-contractor's obligation to indemnify the contractor under clause 6.2 of SBCSub/C and SBCSub/D/C.

The level of the insurance cover required under clause 6.5 is to be entered under item 12 of the sub-contract particulars in SBCSub/A and SBCSub/D/A. The insurance cover to be provided is to be not less than that entered under item 12 of the sub-contract particulars, but footnote 10 to clause 6.5.1.2 makes it clear that the sub-contractor may, if he wishes, insure for a sum greater than that stated under item 12 of the sub-contract particulars.

The insurance cover is to be for any one occurrence or series of occurrences arising out of one event. Clause 6.5.1.1 of SBCSub/C and SBCSub/D/C makes it clear that, where the insurance is in respect of claims for personal injury to, or the death of, any employee of the sub-contractor arising out of and in the course of such person's employment, the insurance must comply with all relevant legislation.

The sub-contractor needs to note that the cover granted under typical public liability policies taken out pursuant to clause 6.5.1 may not entirely cover the level of indemnity required under clause 6.2 of SBCSub/C and SBCSub/D/C. The reason for this may be, for example, because each claim may be subject to the excess (i.e. that part of the insurance claim that is to be paid by the insured) in the policy.

Under clause 6.5.3 of SBCSub/C and SBCSub/D/C, the contractor has the right to require the sub-contractor to produce evidence that he has taken out the insurances required by clause 6.5.1 and that they are being maintained. The contractor may require the sub-contractor to produce the relevant policy and policies and related premium receipts, but such a request is not to be made unreasonably or vexatiously.

Clause 6.5.4 of SBCSub/C and SBCSub/D/C adds that if the sub-contractor does not take out or maintain the insurance as required by clause 6.5.1, then the contractor may take out an insurance to cover the sub-contractor's default and the cost of doing so must either be deducted from any monies due, or to become due, to the sub-contractor under the sub-contract, or shall be recoverable from the sub-contractor as a debt. Of course, before deducting monies a withholding notice in line with clause 4.10.3 or clause 4.12.3 of SBCSub/C and SBCSub/D/C or SBCSub/D/C would need to be given by the contractor to the sub-contractor.

## Liability of sub-contractor – injury or damage to property

The sub-contractor is liable for, and must indemnify the contractor against, any expense, liability, loss, claim or proceedings in respect of any loss, injury or damage whatsoever to any property real or personal in so far as such loss, injury or damage arises out of, or in the course of, or by reason of, the carrying out of the sub-contract works (as clause 6.3 of SBCSub/C and SBCSub/D/C).

Clause 6.5.2 of SBCSub/C and SBCSub/D/C notes that this indemnity does not include for loss, injury or damage due to the effect of an excepted risk (see Definitions at the front of this book).

In line with clause 6.3 of SBCSub/C and SBCSub/D/C, the sub-contractor is only liable where the said loss, injury or damage is due to any negligence, breach of statutory duty, omission or default of the sub-contractor or any of the sub-contractor's persons.

Clause 6.4 of SBCSub/C and SBCSub/D/C makes it clear that the above liability and indemnity must not include any liability or indemnity in respect of loss or damage to the main contract works (including the sub-contract works) and/or to any site materials (including the sub-contract site materials) by any of the specified perils (see Definitions at the front of this book), even if this loss or damage is caused by the negligence, breach of statutory duty, omission or default of the sub-contractor or any of the sub-contractor's persons for the period up to and including the relevant terminal date. The terminal date is defined under clause 6.1 of SBCSub/C and SBCSub/D/C as being:

'(a)  the date of practical completion of the Sub-Contract Works or, in respect of a Section, of such works in the Section, as determined in accordance with clause 2.20; or

(b)  the date of termination of the Sub-Contractor's employment under this Sub-Contract, however arising:

whichever occurs first.'

In addition to the sub-contractor's obligation to indemnify the contractor as detailed above, the sub-contractor is also to take out and maintain insurance in respect of claims arising under this liability (as clause 6.5.1 of SBCSub/C and SBCSub/D/C).

As noted above, this is an entirely sensible requirement because the indemnity of a sub-contractor that is insubstantial may effectively be worthless. The insurance cover required by clause 6.5.1 of SBCSub/C and SBCSub/D/C supports and protects both the sub-contractor and the contractor, although it must be noted that clause 6.5.1 makes it clear that such insurance is without prejudice to the sub-contractor's obligations to indemnify the contractor under clause 6.3 of SBCSub/C and SBCSub/D/C. In other words, the fact that clause 6.5.1 requires insurance to be taken out does not affect or lessen the sub-contractor's obligation to indemnify the contractor under clause 6.3 of SBCSub/C and SBCSub/D/C.

In respect of the loss or damage to the sub-contract works and/or to any of the sub-contract site materials by any of the specified perils, clause 6.5.1 of SBCSub/C and SBCSub/D/C makes it clear that the sub-contractor is not obliged to take out and maintain insurance for this eventuality. The reason for this is that the sub-contract works (when they form part of the main contract works) and the sub-contract site materials (when they form part of the main contract site materials) are covered by the main contract joint names policy for loss or damage by the specified perils (as noted in clause 6.6.1 of SBCSub/C and SBCSub/D/C, which is dealt with later in this chapter) up to and including the terminal date (as defined above).

However, the sub-contractor needs to particularly note that the main contract joint names policy referred to above is only in respect of the specified perils. Other risks, such as subsidence, impact, theft or vandalism, are not covered (although these other risks do not include for excepted risks (see definitions at the front of this book) which clause 6.5.2 of SBCSub/C and SBCSub/D/C specifically excludes from the sub-contractor's liability).

Also, the main contract joint names policy referred to above does not cover the sub-contract works when they are still in the custody and control of the sub-contractor (i.e. when they do not form part of the main contract works), or the sub-contract site materials when they are still in the custody and control of the sub-contractor (i.e. when they do not form part of the main contract site materials).

It should be noted that the parties are to list under sub-contract particulars item 13, those elements of the sub-contract works that the contractor is prepared to regard as fully, finally and properly incorporated into the main contract works prior to practical completion of the sub-contract works or section as applicable; and also to detail the extent to which each of the listed elements needs to be carried out to achieve the status of being fully, finally and properly incorporated into the main contract works.

The provision of this information reduces the probability of a dispute arising as to whether the sub-contract works and/or the sub-contract site materials are still in the custody and control of the sub-contractor or not.

Because of the shortfall on cover noted above, the sub-contractor should seriously consider taking out his own works insurance policy to provide the cover which he does not get under the main contract joint names policy referred to above.

In addition, unlike the other joint names policies taken out under the provisions of the main contract conditions, the joint names policy referred to in paragraph C.1 of insurance option C in schedule 3 of the main contract conditions does not provide any cover to sub-contractors.

This particular joint names policy relates to the insurance by the employer of existing structures and works in extensions to them. The sub-contractor will be liable for any loss or damage caused to existing structures and/or their contents by the negligence, omission or default of the sub-contractor or any person for whom he is responsible.

Such liability is a third party liability for which the sub-contractor is expressly liable under clause 6.3 of SBCSub/C or SBCSub/D/C and against which he is required to insure under clause 6.5.1 of SBCSub/C or SBCSub/D/C.

The level of the insurance cover required under clause 6.5 is to be entered under item 12 of the sub-contract particulars in SBCSub/A or SBCSub/D/A. The insurance cover to be provided is to be not less than that entered under item 12 of the sub-contract particulars, but footnote 10 to clause 6.5.1.2 makes it clear that the sub-contractor may, if he wishes, insure for a sum greater than that stated under item 12 of the sub-contract particulars. The insurance cover is to be for any one occurrence or series of occurrences arising out of one event.

The sub-contractor needs to note that the cover granted under typical public liability policies taken out pursuant to clause 6.5.1 may not entirely cover the level of indemnity required under clause 6.2 of SBCSub/C and SBCSub/D/C. The reason for this may be, for example, because each claim may be subject to the excess (i.e. that part of the insurance claim that is to be paid by the insured) in the policy, and/or because cover may not be available in respect of loss or damage due to gradual pollution.

Under clause 6.5.3 of SBCSub/C and SBCSub/D/C, the contractor has the right to require the sub-contractor to produce evidence that he has taken out the insurances required by clause 6.5.1 and that they are being maintained. The contractor may require the sub-contractor to produce the relevant policy and policies and related premium receipts, but such request is not to be made unreasonably or vexatiously.

Clause 6.5.4 of SBCSub/C and SBCSub/D/C adds that if the sub-contractor does not take out or maintain the insurance as required by clause 6.5.1, then the contractor may take out an insurance to cover the sub-contractor's default and the cost of doing so must either be deducted from any monies due or to become due to the sub-contractor under the sub-contract or shall be recoverable from the sub-contractor as a debt.

Of course, before deducting monies a withholding notice in line with clause 4.10.3 or clause 4.12.3 of SBCSub/C and SBCSub/D/C or SBCSub/D/C would need to be given by the contractor to the sub-contractor.

The entire area of insurance is extremely complicated, and it is normally safest to seek professional advice before taking out policies and when pursuing claims.

For example, a Court of Appeal case in 2004[2] considered the issue of consequential loss in respect of insurances covering loss or damage to property. In that case, the sub-contractor installed suspended ceilings to sixteen cinemas within a Multiplex Cinema project. Five of those suspended ceilings were negligently installed incorrectly, and one of those incorrectly installed ceilings collapsed, causing the entire Multiplex Cinema to be closed for one month for safety reasons.

The sub-contractor made a claim on its insurance, and that claim included for the consequential loss associated with the closure of the entire Multiplex Cinema.

The insurers rejected the claim on the basis that the policy extended only to damage to property, and that as no physical damage had occurred to fifteen of the cinemas, the policy did not cover them at all. The Court of Appeal agreed that, as there was no actual damage to property in fifteen of the cinemas, the consequential losses being claimed in respect of the closure of the entire Multiplex Cinema could not have been caused by damage to property.

## Insurance – loss or damage to work and site materials

### Specified perils cover under joint names all risks policies

Clause 6.6.1 of SBCSub/C and SBCSub/D/C sets out the cover which the sub-contractor obtains from the all risks policy taken out under the main contract. This cover is limited to loss or damage to the sub-contract works and the sub-contract site materials by the specified perils (see definitions at the front of this book).

[2] *Horbury Building Systems Ltd v. Hampden Insurance NV* [2004] EWCA Civ 418.

The benefit of this cover is only available to sub-contractors to whom the contractor sub-lets works in accordance with clause 3.7.1 of the main contract conditions.

Paragraph 185 of SBCSub/G notes that the all risks policy cover is not available to any company, firm or person to whom the sub-contractor sub-lets any of the sub-contract works (i.e. the cover is not available to a sub-subcontractor).

The cover which the sub-contractor obtains from the all risks policy is provided for by clause 6.9 of the main contract conditions which states:

'6.9.1 The Contractor, where Insurance Option A applies, and the Employer, where Insurance Option B or C applies, shall ensure that the Joint Names Policy referred to in paragraph A.1, A.3, B.1 or C.2 of Schedule 3 shall either:

6.9.1.1 provide for recognition of each sub-contractor as an insured under the relevant Joint Names Policy; or

6.9.1.2 include a waiver by the relevant insurers of any right of subrogation which they may have against any such sub-contractor;

in respect of loss or damage by the Specified Perils to the Works or relevant Section, work executed and Site Materials and that this recognition or waiver shall continue up to and including the date of issue of any certificate or other document which states that in relation to the Works, the sub-contractor's works are practically complete or, if earlier, the date of termination of the sub-contractor's employment. Where there are Sections and the sub-contractor's works relate to more than one Section, the recognition or waiver for such sub-contractor shall nevertheless cease in relation to a Section upon the issue of such certificate or other document for his work in that Section.

6.9.2 The provisions of clause 6.9.1 shall apply also in respect of any Joint Names Policy taken out by the Employer under paragraph A.2, or by the Contractor under paragraph B.2.1.2 or C.3.1.2 of Schedule 3.'

The sub-contractor has the right (under clause 6.6.2 of SBCSub/C and SBCSub/D/C) to require the contractor to produce evidence that he has taken out the insurances required by clause 6.6.1 and that they are being maintained.

The sub-contractor may require the contractor to produce documentary evidence that the insurance has been taken out and is being maintained, but such request is not to be made unreasonably or vexatiously.

In the case where the employer is a local authority and insurance options B or C apply (under the main contract) the contractor's obligation is limited to the production of a certificate certifying that terrorism cover is being provided. The reason for this exclusion is because under the main contract conditions both clause B.2.1 and C.3.1 make it clear that if the employer is a local authority, then evidence of insurance does not need to be provided to the contractor.

Clause 6.6.3 of SBCSub/C and SBCSub/D/C adds that if the contractor does not take out or maintain the insurance as required by clause 6.6.1, then the sub-contractor may take out an insurance to cover the contractor's default and the cost of doing so shall either be taken into account in the calculation of the final sub-contract sum, or shall be recoverable from the contractor as a debt.

Of course, many sub-contractors will in any event have their own annually renewable all risks insurance. This insurance normally covers for any physical loss or damage to work executed and site materials. This is a more comprehensive cover than specified perils insurance, although there are normally certain exclusions including the cost necessary to repair, replace or rectify property which is defective due to wear and tear, obsolescence, deterioration, rust or mildew; defective design or workmanship; loss or damage arising out of civil and other war; and also an excepted risk.

The sub-contractor is normally required under the provisions of that insurance cover to add the project details and the contractor's name and/or the employers' name to the policy mandate to make sure that cover is in place.

## Sub-contractor's responsibility – damage to the sub-contract works, etc.

The sub-contractor's responsibility in the event that there is damage to the sub-contract works, etc. is set out under clause 6.7 of SBCSub/C and SBCSub/D/C.

The sub-contractor is responsible for the cost of restoration of the sub-contract works and replacement or repair of any sub-contract site materials that are lost or damaged before the terminal date and of the removal and disposal of any resultant debris except to the extent that the loss or damage to the sub-contract works or sub-contract site materials is due to:

- Any of the specified perils (whether or not caused by the negligence, breach of statutory duty, omission or default of the sub-contractor or any of the sub-contractor's persons) or any excepted risk. The exception in respect of the specified perils applies irrespective of any excess under, and whether or not the employer or the contractor as joint insured under the joint names policy covering the main contract works and the site materials under insurance options A, B, or C of the main contract conditions make a claim under that policy.

- Any negligence, breach of statutory duty, omission or default of the contractor or any of the contractor's persons or of the employer or of any of the employer's persons or of any statutory undertaker executing work solely in pursuance of his statutory rights or obligations.

Clause 6.7.2 of SBCSub/C and SBCSub/D/C makes it clear that, if, during the progress of the sub-contract works, sub-contract materials or goods have been fully and finally incorporated into the main contract works before the terminal date, then the sub-contractor is only responsible for the cost of restoration of such work lost or damaged and the removal and disposal of any debris to the extent that such loss or damage is caused by the negligence, breach of statutory duty, omission or default of the sub-contractor or any of the sub-contractor's persons.

Clause 6.7.8 of SBCSub/C and SBCSub/D/C notes that, for the purposes of clause 6.7.2 only, materials and goods forming part of the sub-contract works will be deemed to have been fully, finally and properly incorporated into the main contract works when in each case they have been completed by the sub-contractor to the extent indicated or referred to in item 13 of the sub-contract particulars (contained in SBCSub/A and SBCSub/D/A). The efficacy (or otherwise) of the entries made under item 13 of the sub-contract particulars is dealt with in Chapter 2.

In the event that any loss or damage is occasioned to the sub-contract works or the sub-contract site materials, by whatever means, then, upon its occurrence or later discovery, the sub-contractor is to immediately give notice in writing to the contractor of its extent, nature and location.

The contractor is then to issue his directions (which, for the avoidance of future doubt, should be in writing) and if those directions are such, the sub-contractor is, with due diligence (i.e. with careful and persistent application or effort) to restore the lost or damaged sub-contract works, and/or replace or repair any lost or damaged sub-contract site materials, remove and dispose of any debris, and proceed with the carrying out of the sub-contract works.

Where, in line with the above provisions, the sub-contractor is not responsible for the cost of compliance, such compliance must be treated as a variation (variations, generally, are dealt with in Chapter 10).

It must be noted, however, that the occurrence of loss or damage affecting the sub-contract works occasioned by any of the specified perils must be disregarded in computing any amounts payable to the sub-contractor under the sub-contract, although it may be payable as part of an insurance claim under the joint names all risks policy noted above.

Where monies are to be paid out by the insurer under that joint names all risks policy, the sub-contractor must not object to the payment of the monies to the employer (as clause 6.7.5 of SBCSub/C and SBCSub/D/C).

Clause 6.7.6 of SBCSub/C and SBCSub/D/C makes it clear that, except in the case of loss or damage caused by the negligence, breach of statutory duty, omission or default of the sub-contractor or any of the sub-contractor's persons, the sub-contractor will not be responsible for loss or damage to the sub-contract works occurring after the terminal date.

Finally, clause 6.7.7 of SBCSub/C and SBCSub/D/C states that nothing in clause 6.7 (of SBCSub/C and SBCSub/D/C) will in any way modify the sub-contractor's obligations in regard to defects in the sub-contract works as set out in clause 2.22. (The issue of defects is dealt with in detail in Chapter 6.)

## Terrorism cover – non-availability – employers' options

This issue is dealt with under clause 6.8 of SBCSub/C and SBCSub/D/C.

Under this condition, if the insurers named in the joint names policy referred to in clause 6.6 of SBCSub/C or SBC/Sub/D/C notify the contractor or the employer that terrorism cover will cease on a particular date (and will no longer be available) the contractor is to immediately inform the sub-contractor.

When the employer has notified the contractor of his decision (in the light of this lack of cover) either to continue with the main contract works or to terminate the contractor's employment under the main contract, the contractor is to notify the sub-contractor accordingly.

If the employer gives notice terminating the contractor's employment under the main contract, then upon and from the date stated by the employer in his notice to the contractor the sub-contractor's employment under the sub-contract will terminate. The provisions of clause 7.11 (except clause 7.11.3.5) of SBCSub/C or SBCSub/D/C must apply (for commentary on clause 7.11 refer to Chapter 12).

If the employer does not gives notice terminating the contractor's employment under the main contract, then upon and from the date that terrorism cover will cease:

- if the sub-contract work that has been executed and/or the sub-contract site materials suffer physical loss or damage caused by terrorism, the sub-contractor must with due diligence restore lost or damaged work, repair or replace any lost or damaged site materials, remove and dispose of any debris, and proceed with the carrying out of the sub-contract works;

- the restoration, replacement or repair of such loss or damage and any removal and the disposal of debris must be treated as a variation with no reduction in any amount payable to the sub-contractor by reason of any act or neglect of the sub-contractor or of any of his own sub-contractors (i.e. any sub-subcontractors) which may have contributed to the physical loss or damage; and

- where insurance option C applies, the requirement that the sub-contract works continue to be carried out shall not be affected by any loss or damage to the existing structures and/or their contents caused by terrorism.

## Damage to sub-contractor's plant, etc.

Clause 2.9 of SBCSub/C and SBCSub/D/C makes clear that the contractor is only responsible for the loss of or damage to temporary works, plant tools, equipment or other property belonging to or provided by the sub-contractor or the sub-contractor's persons or to any materials or goods of the sub-contractor which are not sub-contract site materials, when such loss or damage is due to any negligence, breach of statutory duty, omission or default of the contractor or any of the contractor's persons.

In all other instances, the sub-contractor is responsible, and the sub-contractor needs to ensure that its insurance arrangements with its own sub-contractors (i.e. the sub-subcontractors) are compatible with this liability.

## *Sub-contractor's design portion professional indemnity insurance*

Professional indemnity insurance (usually referred to as PI insurance) covers a designer's liability for damages and costs arising from breach of professional duty caused by neglect, error or omission.

PI insurance is normally an annual policy and a renewal proposal form must be completed every year, giving details of the previous year's activities together with those anticipated for the future. The insured is obliged to disclose any material information, which may influence the underwriter in his acceptance of the risk, and the premium he might charge.

There is no doubt that onerous conditions of appointment and collateral warranties fall within the realm of material information and should therefore be disclosed annually.

Most PI insurances are based on a claims made basis, which means that the PI insurance policy that applies is that operative when a claim is made, rather than that operative when the alleged defect occurred.

Unfortunately, the fundamental inadequacy of PI insurance cover in its protection of building owners remains. Claims can only succeed if the designer is found to be legally liable. In the same way that there is no obligation on the designer to actually claim upon his PI insurance policy, the building owner who effects the cover and pays the premium does not have the direct right of claim on the policy (although he has the comfort that the cover is in force).

However, there is some limited protection afforded by section 1 of the Third Parties (Rights Against Insurers) Act 1930, which provides that if an insured becomes bankrupt or goes into liquidation either before or after incurring an insured liability to a third party, the insured's right to indemnity is transferred to the third party.

Under SBCSub/D/C, Clause 6.10 places an obligation on the sub-contractor to insure in respect of PI insurance.

The sub-contractor is required to take out this insurance of the type and with the level of cover of at least that stated in item 14 of the sub-contract particulars (SBCSub/D/A).

If no level of cover is stated there, then no PI insurance is required. This is an important point to note.

Clause 6.10.2 of SBCSub/D/C notes that, provided that it remains available at commercially reasonable rates, the sub-contractor is to maintain the PI insurance until the expiry of the period stated in the sub-contract particulars (item 14 of SBCSub/D/A) from the date of practical completion of the sub-contract works.

Normally, where the sub-contract is executed as a deed, the period will be twelve years, but if the sub-contract is executed under hand, the period should be six years.

The sub-contractor is required to provide documentary evidence that the PI insurance is being maintained, as and when the contractor reasonably requires (as clause 6.10.3 of SBCSub/D/C).

If the PI insurance is not available at commercially reasonable rates, clause 6.11 of SBCSub/D/C requires the sub-contractor to notify the contractor immediately so that the contractor and sub-contractor can discuss the means of best protecting their respective positions in the absence of such PI insurance.

SBCSub/G makes the very good point that not all sub-contractors will have, or can obtain, PI insurance and that, therefore, contractors need to recognize this when seeking to operate the terms of clause 6.10. SBCSub/G also adds that some sub-contractors may have product guarantee insurance and this may be acceptable as an alternative to PI insurance.

However, care needs to be taken when relying on a product guarantee insurance as an alternative to PI insurance, particularly as a product guarantee insurance often only guarantees the product and not the design of the product, and often the product guarantee insurance will have a plethora of maintenance and other requirements that must be met in respect of the installed product to prevent the insurance from being invalidated.

# Joint fire code

## Joint fire code – compliance

Clause 6.12 of SBCSub/C and SBCSub/D/C notes that the joint fire code only applies where the main contract particulars state that the joint fire code applies.

The provisions of the joint fire code are dealt with in more detail under item 6.2 of the sub-contract particulars (which is dealt with in Chapter 2).

Clause 6.13 of SBCSub/C and SBCSub/D/C notes that where the joint fire code applies, the parties must comply with the joint fire code; the contractor is to ensure compliance by all contractor's persons and the sub-contractor must ensure such compliance by all sub-contractor's persons.

## Breach of joint fire code

Clause 6.14 of SBCSub/C and SBCSub/D/C sets out the procedure to be followed in the event of a breach of the joint fire code:

- if there is a breach of the joint fire code, the insurers may require the contractor to carry out remedial measures;
- if the insurers do require the contractor to carry out remedial measures, the contractor is to send a copy of the notice requiring the remedial measures to the sub-contractor, immediately;
- upon receipt of the notice, the sub-contractor is to comply with any direction of the contractor that is reasonably necessary in respect of the sub-contract works to correct the breach of the joint fire code;
- except where the breach of the joint fire code was as a result of an act, omission or default of the sub-contractor or of any of the sub-contractor's persons, then the compliance by the sub-contractor with any direction of the contractor, as noted above, will be treated as a variation;
- if the breach of the joint fire code was as a result of (or was partly as a result of) an act, omission or default of the sub-contractor or of any of the sub-contractor's persons, then the sub-contractor must be liable for the appropriate additional costs incurred by the contractor for the remedial measures and either an appropriate deduction must be made in calculating the final sub-contract sum or it will be recoverable by the contractor from the sub-contractor as a debt;
- if the sub-contractor, within four days of receipt of a direction as outlined above, does not begin to comply with it or thereafter fails without reasonable cause regularly and diligently to comply with it, then the provisions of clause 3.6 of SBCSub/C or SBCSub/D/C will apply.

Clause 3.6 of SBCSub/C or SBCSub/D/C makes it clear that if, within seven days after receipt of a written notice from the contractor which requires a sub-contractor to comply with a direction, the sub-contractor still does not comply with that direction, then the contractor may employ and pay other persons to execute any work whatsoever which may be necessary to give effect to that direction.

In such a situation, the sub-contractor would be liable for all additional costs incurred by the contractor in connection with such employment and an appropriate deduction would either be taken into account in the calculation of the final sub-contract sum or would be recoverable by the contractor from the sub-contractor as a debt.

## Joint fire code – amendments/revisions

Clause 6.15 of SBCSub/C and SBCSub/D/C makes clear that, if a joint fire code that is amended after the sub-contract base date is to be applied in respect of the main contract works, the cost, if any, of compliance by the sub-contractor with any amendment or revision to the joint fire code must be added into the calculations of the final sub-contract sum.

# Chapter 12
# Termination

## Summary of general law

There are essentially four ways in which a contract can come to an end:

- performance;
- agreement;
- frustration;
- breach.

### Performance

The usual rule is that a contract is discharged when it is performed exactly to the terms of the contract.[1]

Exceptions to this rule are:

- where the other party agrees to accept partial performance;
- divisible or severable contracts (which are payable by instalments);
- where one party is prevented from completing the contract by some wrongful action of the other party.

### Agreement

The parties can mutually agree to abandon or discharge the contract.

The only restriction in this regard is that where one party has fully fulfilled his performance under the contract, any subsequent agreement to abandon or discharge the contract would need to be executed in the form of a deed, or would need to be supported by some fresh consideration, to avoid the rule in *Pinnel's*[2] case that 'Payment of a lesser sum on the day in satisfaction of a greater sum cannot be any satisfaction of the whole.'

Where one of the parties has not fully discharged its performance, then by giving up their rights to compel each other to perform, each party is giving something to

---

[1] *Cutter v. Powell* (1795) 2 Sm LC.
[2] *Pinnel's case* (1602) 5 Co Rep 117a.

the new agreement and this would be taken to be the consideration needed to bind the agreement.

## Frustration

Frequently, sub-contractors will claim that they are no longer bound by the obligations under a sub-contract because the contract has been frustrated.

In their view the sub-contract has been frustrated because the contract has taken much longer to complete than they had originally envisaged, or because they are being denied access to a particular area of the site, or because the original agreed sequence has been changed and the project appears (to them) to be an uncontrollable mess.

Although, in such situations, sub-contractors may themselves be frustrated, these circumstances would not mean that the sub-contract is frustrated.

In reality, a sub-contract can only be frustrated where, after the sub-contract was concluded, events occur which make performance of the sub-contract impossible, illegal, or something radically different from that which was in the contemplation of both parties when they entered the sub-contract.

By the operation of a rule of law, a contract which is discharged on the ground of frustration is brought to an end automatically, irrespective of the wishes of the parties.

In the *Davis Contractors* v. *Fareham UDC*[3] case the position was summarized as follows:

> 'Frustration occurs whenever the law recognizes that without default of either party a contractual obligation has become incapable of being performed because the circumstances in which performance is called for would render it a thing radically different from that which was undertaken by the contract.'

If a contract is frustrated then the Law Reform (Frustrated Contracts) Act 1943 states that any money paid is recoverable and any money not yet paid ceases to be payable.

## Breach

A breach of contract is committed when a party without lawful excuse fails or refuses to perform what is due from him under the contract, or performs defectively or incapacitates himself from performing.

When a term of a contract is breached various remedies are available:

- Damages (dealt with in Chapter 9).

- Injunction
  An injunction is an order of the court whereby a party is prohibited from acting in a certain way or is required to act in a prescribed manner. Where this relates to a contract it is usually to stop a party doing something that breaches the contract terms.

---

[3] *Davis Contractors Ltd* v. *Fareham UDC* [1956] 2 All ER 148.

- Specific performance

  An order of specific performance is an order of the court which requires the party in breach to perform its obligations under the contract. Although, traditionally specific performance has been considered as being a supplementary remedy, only to be granted when damages were considered to be inadequate, there has been some movement in the courts to use the remedy when it is the more appropriate remedy.[4]

- Termination

  There are several equivalent terms for the word 'termination' in common usage, for example rescission, determination, renunciation, etc. In this book the term 'termination' is used as SBCSub/C and SBCSub/D/C uses this particular terminology.

  The terms of a sub-contract are usually split into two (or sometimes three) categories:

  — conditions, which are essential terms of the sub-contract which go to the root of the contract;
  — warranties, which are lesser subsidiary terms of the sub-contract;
  — innominate terms, which are terms somewhere between conditions and warranties.[5]

If there is a breach of a condition of the sub-contract then a party has a right to terminate the sub-contract.

This right may also exist in respect of the breach of an innominate term, but this would largely be dependent upon the seriousness of the breach and its effects.

## Termination

Termination is effected by taking proceedings to have the sub-contract set aside by the court (in the case of misrepresentation) or by giving notice to the other party of one's intention to treat the sub-contract as at an end.

In the latter case it is normal to say that a party's 'employment under the sub-contract' has been terminated, for fear that, without adding that caveat, the termination of the sub-contract *per se* would mean that there would be no right for the innocent party to recover damages.

However, it is now reasonably well established that termination in the latter situation means that first, the innocent party is released from all further performance of his obligations and second, he is entitled to recover full damages. Despite this there is no reason why the above noted caveat should not be used.

Termination may come into effect either through:

- common law termination; or
- contractual termination.

---

[4] *Beswick* v *Beswick* [1968] AC 58.
[5] *Hong Kong Fir Shipping Co. Ltd* v. *Kawaski Kisen Kaisha Ltd* [1962] 1 All ER 474.

## Common law termination

If there is a breach of a condition (or, in some cases an innominate term) of the sub-contract then a party has a right to terminate the sub-contract.

This breach may be either:

- a repudiatory breach, which is a breach whereby one party clearly indicates (by words or by conduct) that it no longer intends to be bound by the terms of the sub-contract; or perhaps where the default of a party has rendered itself unable to perform its outstanding contractual obligations (for example where a contractor needs to be NHBC Registered but he is removed from the NHBC Register); or

- an anticipatory breach, where one party states that he will not carry out his obligations under the sub-contract before the time for carrying out those obligations has arrived.

Once either a repudiatory breach or an anticipatory breach has occurred, then the innocent party may either affirm the sub-contract or may elect to accept the repudiatory breach or the anticipatory breach as appropriate, and then terminate the sub-contract.

In the case of an anticipatory breach, in particular, the innocent party may postpone the right to rescind while retaining the right to damages.

It is necessary for the repudiatory breach or the anticipatory breach to be accepted, because as noted in the *White & Carter* v. *McGregor*[6] case: 'repudiation by one party standing alone does not terminate the contract. It takes two to end it, by repudiation on the one side, and acceptance of the repudiation on the other.'

Although acceptance of the repudiatory breach or anticipatory breach may be by word or by deed, it is always safer to use words since deeds may be misinterpreted.

As noted by Lord Simon in *Heyman* v. *Darwins*[7], the innocent party must make it plain that 'in view of the wrongful act of the party who has repudiated he claims to treat the contract as at an end'.

In the case of common law termination, a high degree of freedom and flexibility of remedy is available, but the innocent party has the burden of establishing that the breach of sub-contract was sufficiently fundamental to justify the termination in law.

A fundamental point to note is that it appears clear that the right to common law termination coexists with any contractual termination provisions.

In respect of SBCSub/C and SBCSub/D/C, clause 7.3.1 makes it clear that 'the provisions of clause 7.4 to 7.8, 7.10 and 7.11 are without prejudice to any other rights and remedies which the Contractor or the Sub-Contractor may possess'.

This clause therefore effectively confirms that the parties' common law termination rights are preserved.

---

[6] *White & Carter (Councils) Ltd* v. *McGregor* [1962] AC 413, HL.
[7] *Heyman* v. *Darwins* [1942] AC 356, HL.

*Contractual termination*

In respect of contractual termination, a party may lawfully terminate the sub-contract by exercising powers provided for in a contractual termination clause.

In a contractual termination it will not be necessary to show that the (alleged) breach was of a sufficiently fundamental nature to warrant termination; all that will need to be shown is that the breach relied upon has been defined within the contract as being a breach that allows contractual termination to take place.

However, for a contractual termination to be successful it will be necessary for the contract procedures for notices, etc. to be followed meticulously, otherwise the party pursuing the contractual termination route may find itself accused of a reputiatory breach.

In addition, it should be noted that the rights and remedies under a common law termination may be different from those under a contractual termination. For example, under a common law termination (unlike most contractual termination clauses) the innocent party may decide not to complete the project but may still claim damages. On the other hand, under contractual termination, the innocent party is frequently entitled to seize materials to complete the works which would not be the case under a common law termination.

Also, under a contractual termination, the consequences arising from the termination are clearly set out, and in respect of SBCSub/C and SBCSub/D/C, this includes (in cases where the sub-contractor is not at fault) the payment to the sub-contractor of all sums due, within a relatively short timescale (as clause 7.11.5 of SBCSub/C and SBCSub/D/C).

# Termination under SBCSub/C and SBCSub/D/C

Section 7 of SBCSub/C and SBCSub/D/C deals with termination.

# Termination by contractor

The contractor may terminate the sub-contractor's employment under SBCSub/C and SBCSub/D/C for three principal reasons:

- default by the sub-contractor;
- insolvency of the sub-contractor;
- corruption.

## Default by the sub-contractor (refer to clause 7.4 of SBCSub/C and SBCSub/D/C)

The basic process that is followed is:

- The sub-contractor defaults in a way stipulated in the sub-contract (see further comments on specified defaults below).

- The contractor gives a notice to the sub-contractor specifying the default.

- If the sub-contractor continues a specified default for ten calendar days (excluding any public holiday days) after receipt of the contractor's notice, then the contractor may on, or within ten calendar days (excluding any public holiday days) from the expiry of the initial notice period, issue a further notice terminating the sub-contractor's employment under the sub-contract.

- If the contractor does not give the further notice referred to above (for any reason), but the sub-contractor repeats a specified default (whether previously repeated or not), then, upon or within a reasonable time after such repetition, the contractor may by notice to the sub-contractor terminate the sub-contractor's employment under the sub-contract.

- Clause 7.3.2 of SBCSub/C and SBCSub/D/C makes it clear that irrespective of the grounds of termination, the sub-contractor's employment may at any time be reinstated if and on such terms as the parties may agree. In other words, there is nothing to prevent the parties reinstating their contractual relationship on such terms as they may agree, after the sub-contractor's termination of employment under the sub-contract has taken place. It should be noted, however, that section 178 of the Insolvency Act 1986 gives a liquidator or a receiver the statutory right to disclaim unprofitable contracts, and this may close this avenue in certain circumstances.

Generally:

- The notices given above must be in writing, and must be given by actual, special or recorded delivery.

- Where the notice is given by special or recorded delivery it will, subject to proof to the contrary, be deemed to have been received on the second business day after the date of posting (clause 7.2.3 of SBCSub/C and SBCSub/D/C).

   The words 'subject to proof to the contrary' are important, because this means that (subject to proof) a notice could have been actually received prior to the second business day after the date of posting.[8]

- Termination will take effect on receipt of the relevant notice (clause 7.2.2 of SBCSub/C and SBCSub/D/C).

- Notice of termination of the sub-contractor's employment must not be given unreasonably or vexatiously (clause 7.2.1 of SBCSub/C and SBCSub/D/C).

- In the case where a contractor does not (for any reason) give a notice of termination following the further ten-day notice period referred to above, and the sub-contractor repeats a specified default which allows the contractor to terminate the sub-contractor's employment, it is submitted that this repeated default should be a specified default that the sub-contractor has already committed and has been given notice of rather than another specified default from the list below. However, this proposition is untested.

---

[8] *Lafarge (aggregates) Ltd* v. *Newham Borough Council* [2005] (unreported) QBD, 24 June 2005.

*Specified defaults*

The specified defaults (which to be defaults must be committed before practical completion of the sub-contract works) are dealt with under clause 7.4.1 of SBCSub/C and SBCSub/D/C.

These specified defaults are outlined below.

### Clause 7.4.1.1 of SBCSub/C and SBCSub/D/C – 'the Sub-Contractor without reasonable cause wholly or substantially suspends the carrying out of the Sub-Contract Works'

This ground for termination does not apply where the sub-contractor has reasonable cause for suspending the carrying out of the sub-contract works.

Therefore, a suspension because of non-payment in line with clause 4.11 of SBCSub/C or SBCSub/D/C would be a reasonable cause for suspending the carrying out of the sub-contract works.

However, if the suspension was made on the false premise that further monies were due; or if the suspension continued because of extended remobilization periods, it is likely that these would not be found to be reasonable causes for suspension.

### Clause 7.4.1.2 of SBCSub/C and SBCSub/D/C – 'the Sub-Contractor fails to proceed regularly and diligently with the Sub-Contract Works'

'Regularly and diligently' should be construed together and mean that the sub-contractor must carry out the work in a way in which it will achieve its contractual obligations. To do this it will be necessary to plan the work, lead and manage the workforce, provide sufficient materials of the proper sort and employ competent tradespeople so the works are carried out to an acceptable standard and to time.

In the *West Faulkner* v. *London Borough of Newham*[9] case, Lord Justice Brown said the following in respect of the phrase 'regularly and diligently':

> 'My approach to the proper construction and application of the clause would be this. Although the contractor must proceed both regularly and diligently with the works, and although each word imports into that obligation certain discrete concepts which would not otherwise inform it, there is a measure of overlap between them and it is thus unhelpful to seek to define two quite separate and distinct obligations.'

The word 'regularly' suggests a requirement to attend for work on a regular daily basis with sufficient men, materials and plant to have the physical capacity to progress the works substantially in accordance with the sub-contract obligations.

'Diligently' adds the concept of the need to apply the above physical capacity industriously and efficiently towards that same end.

Taken together, the obligation upon the sub-contractor is essentially to proceed continuously, industriously and efficiently with appropriate physical resources so as to progress the works steadily towards completion substantially in accordance with the contractual requirements as to time, sequence and quality of work.

---

[9] *West Faulkner Associates v. London Borough of Newham* (1994) 71 BLR 1.

It is submitted that circumstances would need to be quite extreme before a contractor could be sure that a sub-contractor was not proceeding regularly and diligently with the sub-contract works.

### Clause 7.4.1.3 of SBCSub/C and SBCSub/D/C – 'the Sub-Contractor refuses or neglects to comply with a written direction from the Contractor requiring him to remove any work, materials or goods not in accordance with this Sub-Contract and by such refusal or neglect the Main Contract Works are materially affected'

The contractor has an express power under clauses 3.11.1 and 3.11.2 of SBCSub/C and SBCSub/D/C to issue instructions in regard to the removal of any work, materials or goods which are not in accordance with the sub-contract.

If the sub-contractor refuses or neglects to comply with the contactor's written directions in this regard this is a specified default.

### Clause 7.4.1.4 of SBCSub/C and SBCSub/D/C – 'the Sub-Contractor fails to comply with clause 3.1 or 3.2'

Clause 3.1 relates to non-assignment and clause 3.2 relates to the requirement for the sub-contractor to obtain the contractor's consent before sub-letting any part of the sub-contract works.

In respect of clause 3.2 particularly (where the contractor's consent for the sub-contractor's request to sub-let part of the sub-contract works is not to be unreasonably delayed or withheld) it is submitted that, in view of the fact that a notice of termination of the sub-contractor's employment is not to be given unreasonably or vexatiously, it would require a substantial breach of this term before the contractor could operate the termination provisions because of this default.

### Clause 7.4.1.5 of SBCSub/C and SBCSub/D/C – 'the Sub-Contractor fails to comply with clause 3.20'

Clause 3.20 relates to the requirement of the sub-contractor to comply with the CDM Regulations.

It is submitted that incidental failures (e.g. providing information for the health and safety file) may not be considered to be a default given that a notice of termination of the sub-contractor's employment is not to be given unreasonably or vexatiously.

## Insolvency of the sub-contractor (refer to clause 7.5 of SBCSub/C and SBCSub/D/C)

Clause 7.5.1 of SBCSub/C and SBCSub/D/C states that if the sub-contractor is insolvent, the contractor may at any time by notice to the sub-contractor terminate the sub-contractor's employment under the sub-contract.

It is interesting to note that upon a sub-contractor becoming insolvent, its employment under the sub-contract is not automatically terminated and a notice from the contractor (which may be made at any time) is required to bring the termination of employment into effect.

However, the sub-contractor's obligations to carry out and complete the sub-contract works (in the case of SBCSub/C and SBCSub/D/C) and the design of the sub-contractor's designed portion (only in the case of SBCSub/D/C) are immediately suspended upon the sub-contractor becoming insolvent (as clause 7.5.3.1 of SBCSub/C and SBCSub/D/C).

The reason why there is not automatic termination, but there is a period of suspension, is to allow a period following the happening of a specified insolvency event, during which the sub-contract arrangements will remain in force while the parties attempt to agree a way forward.

In addition, clause 7.5.3.2 of SBCSub/C and SBCSub/D/C states that as from when the sub-contractor becomes insolvent (which may be at a date earlier than when the sub-contractor's employment under the sub-contract has been terminated) the contractor may take reasonable measures to ensure that the sub-contract works and the sub-contract site materials are retained on site. The sub-contractor must allow and must not hinder or delay the taking of those measures.

Under clause 7.1 of SBCSub/C and SBCSub/D/C (and for the purposes of the sub-contract conditions only) a sub-contractor is insolvent if:

7.1.1   he enters into an arrangement, compromise or composition in satisfaction of his debts (excluding a scheme of arrangement as a solvent company for the purposes of amalgamation or reconstruction); or

7.1.2   without a declaration of insolvency, he passes a resolution or makes a determination that he be wound up; or

7.1.3   he has a winding up order or bankruptcy order made against him; or

7.1.4   he has appointed to him an administrator or administrative receiver; or

7.1.5   he is the subject of an analogous arrangement, event or proceedings in any other jurisdiction; or

7.1.6   (additionally in the case of partnerships) each partner is the subject of an individual arrangement or any other event or proceedings referred to in clause 7.1.1 to 7.1.5.

Clause 7.5.2 of SBCSub/C and SBCSub/D/C requires the sub-contractor to immediately inform the contractor in writing if he makes any proposal, gives notice of any meeting or becomes the subject of any proceedings or appointment to any of the matters referred to above.

## Corruption (refer to clause 7.6 of SBCSub/C and SBCSub/D/C)

Clause 7.6 of SBCSub/C and SBCSub/D/C gives the contractor the right to terminate the sub-contractor's employment under the sub-contract if the sub-contractor or any persons employed by him or acting on his behalf engages in 'corruption', as defined.

This clause is restrictive in terms of what comprises corruption but is broad in terms of its application.

For example, 'corruption' as defined only exists when the sub-contractor or any person employed by him or acting on his behalf has committed an offence under the Prevention of Corruption Acts 1889 to 1916, or (where the employer is a local authority) has given any fee or reward, the receipt of which is an offence under subsection 2 of section 117 of the Local Government Act 1972.

However, the effect of this corruption will be felt *even if it relates to an entirely separate sub-contract between the contractor and the sub-contractor.*

In other words, the contractor's right to termination may arise from some form of defined corruption that the sub-contractor took part in on a separate sub-contract between the parties.

## Consequences of termination under clauses 7.4 to 7.6 of SBCSub/C and SBCSub/D/C

If the sub-contractor's employment is terminated by the contractor for any of the reasons outlined above, the following consequences arise:

- The contractor may employ and pay other persons to carry out and complete the sub-contract works (and/or the design for the sub-contractor's designed portion, as applicable) and to make good any defects of the kind referred to in clause 2.22 of SBCSub/C and SBCSub/D/C (refer to clause 7.7.1 of SBCSub/C and SBCSub/D/C).

- The contractor (and/or the persons employed by the contractor) may enter upon and take possession of the sub-contract works and (subject to obtaining any necessary third party consents) may use all the sub-contractor's temporary buildings, plant, tools, equipment and sub-contract site materials for those purposes.

It should be noted that sub-subcontractors or the sub-contractor's suppliers could originally have provided some of the sub-contractor's temporary buildings, plant, tools, equipment and site materials. In the event that a sub-contractor becomes insolvent, it may be that a sub-subcontractor or a sub-contractor's supplier may attempt to reclaim those items (normally in the case where it has not been paid for those items). The sub-subcontractor's likelihood of success in this attempt may depend upon the efficacy of any retention of title clause that may be in place. In respect of retention of titles, etc. in respect of unfixed materials on site, reference should be made to the comments in Chapter 8.

There are certain restrictions imposed in respect of the above by the Insolvency Act 1986. Once a petition has been presented to the court for the appointment of an administrator to a company, section 10(1)(b) of the Insolvency Act 1986 states that no steps may be taken to repossess goods in the company's possession under any retention of title agreement, other than with the leave of court.

Also, there may be restrictions placed on the contractor's right to use site materials, etc. of a sub-contractor that has become insolvent, because this may prejudice other creditors of the sub-contractor and may not be compliant with the mandatory equal discharge of liabilities by a liquidator or receiver, something that cannot be excluded by contract.[10]

- When required by the contractor in writing (but not beforehand) the sub-contractor must remove or procure the removal from the site of all temporary buildings, plant, tools, equipment, goods and materials belonging to the

---

[10] *British Eagle International Air Lines* v. *Compagnie Nationale Air France* [1975] 1 WLR 758.

sub-contractor or to the sub-contractor's persons (as clause 7.7.2.1 of SBCSub/C and SBCSub/D/C).

• Where applicable, and without charge, the sub-contractor is to provide the contractor with three copies of all the sub-contractor's design documents then prepared, whether or not previously provided. This consequence is required irrespective of any request by the contractor (as clause 7.7.2.2 of SBCSub/D/C).

• If required by the contractor, the sub-contractor must within 14 days of the date of termination, assign to the contractor, without charge, the benefit of any agreement for the supply of materials or goods and/or for the execution of any work for the purposes of the sub-contract (as clause 7.7.2.2 of SBCSub/C and 7.7.2.3 of SBCSub/D/C.
  Of course, the above will only be possible to the extent that:

  — the sub-contractor can legally carry out or is legally required to carry out such assignments (noting that assignment may not be legally possible in the case of the sub-contractor's insolvency);
  — the sub-contractor's own sub-contractor or supplier is prepared for the agreement to be assigned.

• The contractor is not required to make any further payment or release any outstanding retention, other than as provided for under clause 7.7.4 of SBCSub/C and SBC/D/C (as clause 7.7.3 of SBCSub/C and SBCSub/D/C).

• Clause 7.7.4 of SBCSub/C and SBCSub/D/C sets out the sub-contractor's rights in respect of future payment as follows:

  — Upon the completion of the sub-contract works and the making good of defects (i.e. by the contractor or by a person appointed by the contractor), or upon the termination of the contractor's employment under the main contract, whichever occurs first; the sub-contractor may apply to the contractor for payment.
  The contractor needs to take no action until an application for payment is received from the sub-contractor that is compliant with the above time restriction.
  — When such an application for payment is received by the contractor, the contractor is to pay the sub-contractor the value of any work executed or goods and materials supplied by the sub-contractor to the extent not included in previous payments. Without prejudice to any other rights that the contractor may have, the contractor may deduct from the sum determined

    — the amount of any direct loss and/or damage caused to the contractor as a result of the termination (e.g. the cost of completing the works, and other delay costs, etc.);
    — any other amount payable to the contractor under the sub-contractor.

In that the amount due to the contractor exceeds the amount due to the sub-contractor, the balance is recoverable from the sub-contractor as a debt.
  Given that clause 7.7.3 of SBCSub/C and SBCSub/D/C states that the 'other provisions of the this Sub-Contract which require any further payment or any release of Retention to the Sub-Contractor shall cease to apply', it is submitted that when a deduction as outlined above is made from the sub-contractor's

payment, a withholding notice in line with clause 4.12.3 of SBCSub/C and SBCSub/D/C is unnecessary.

## *Termination by sub-contractor, etc.*

The contractor may terminate the sub-contractor's employment under SBCSub/C and SBCSub/D/C for three principal reasons:

- default by the contractor;
- termination under the main contract;
- insolvency of the contractor.

### Default by the contractor (refer to clause 7.8 of SBCSub/C and SBCSub/D/C)

The basic process that is followed is:

- The contractor defaults in a way stipulated in the sub-contract (see further comments on specified defaults below).

- The sub-contractor gives a notice to the contractor specifying the default.

- If the contractor continues a specified default for ten calendar days (excluding any public holidays) after receipt of the sub-contractor's notice, then the sub-contractor may on, or within ten calendar days (excluding any public holidays) from the expiry of the initial notice period, issue a further notice terminating the sub-contractor's employment under the sub-contract.

- If the sub-contractor does not give the further notice referred to above but the contractor repeats a specified default (whether previously repeated or not), then, upon or within a reasonable time after such repetition, the sub-contractor may by notice to the contractor terminate the sub-contractor's employment under the sub-contract.

- Clause 7.3.2 of SBCSub/C and SBCSub/D/C makes it clear that irrespective of the grounds of termination, the sub-contractor's employment may at any time be reinstated if and on such terms as the parties may agree. In other words, there is nothing to prevent the parties reinstating their contractual relationship on such terms as they may agree, after the sub-contractor's termination of employment under the sub-contract has taken place. It should be noted, however, that section 178 of the Insolvency Act 1986 gives a liquidator or a receiver the statutory right to disclaim unprofitable contracts, and this may close this avenue in certain circumstances.

Generally:

- The notices given above must be in writing, and must be given by actual, special or recorded delivery.

- Where the notice is given by special or recorded delivery it must, subject to proof to the contrary, be deemed to have been received on the second business day after the date of posting (as clause 7.2.3 of SBCSub/C and SBCSub/D/C).

The words 'subject to proof to the contrary' are important, because this means that (subject to proof) a notice could have been actually received prior to the second business day after the date of posting.[11]

- Termination must take effect on receipt of the relevant notice (as clause 7.2.2 of SBCSub/C and SBCSub/D/C).

- Notice of termination of the sub-contractor's employment must not be given unreasonably or vexatiously (as clause 7.2.1 of SBCSub/C and SBCSub/D/C).

- In the case where a sub-contractor does not (for any reason) give a notice of termination following the further ten-day notice period referred to above, and the contractor repeats a specified default which allows the sub-contractor to terminate the sub-contractor's employment, it is submitted that this repeated default must be a specified default that the contractor has already committed and has been given notice of rather than another specified default from the list below However, this proposition is untested.

### Specified defaults

The specified defaults are dealt with under clause 7.8.1 of SBCSub/C and SBCSub/D/C.

These specified defaults are described below:

**Clause 7.8.1.1 of SBCSub/C and SBCSub/D/C – 'the Contractor without reasonable cause wholly or substantially suspends the carrying out of the Main Contract Works'**

This ground for termination does not apply where the contractor has reasonable cause for suspending the carrying out of the main contract works.

Therefore, a suspension for non-payment in line with clause 4.14 of the main contract would be a reasonable cause for suspending the carrying out of the main contract works.

However, if the suspension was made on the false premise that further monies were due, or if the suspension continued because of extended remobilization periods, it may be found that these were not reasonable causes for suspension.

**Clause 7.8.1.2 of SBCSub/C and SBCSub/D/C – 'the Contractor without reasonable cause fails to proceed regularly and diligently with the Main Contract Works so that the reasonable progress of the Sub-Contract Works is seriously affected'**

As noted previously, 'regularly and diligently' should be construed together and mean that the contractor must carry out the work in a way in which it will achieve its contractual obligations. To do this it will be necessary to plan the work, lead and manage the workforce, provide sufficient materials of the proper sort and employ competent tradespeople so the works are carried out to an acceptable standard and to time:

---

[11] *Lafarge (aggregates) Ltd v. Newham Borough Council* [2005] (unreported) QBD, 24 June 2005.

Reference should be made to the *West Faulkner* v. *London Borough of Newham*[12] case referred to above.

Interestingly, it should be noted that a specified default for a sub-contractor is that it 'fails to proceed regularly and diligently with the Sub-Contract Works, etc.', but in respect of the contractor, the specified default is amended to read that 'without reasonable cause it fails to proceed regularly and diligently with the Main Contract Works, etc.'. What actual effect the additional words 'without reasonable cause' have, is not fully clear.

However, irrespective of that point, it is submitted that circumstances would need to be quite extreme before a sub-contractor could be sure that a contractor was not proceeding regularly and diligently with the main contract works, particularly given that the sub-contractor would normally be operating from a base of minimal knowledge.

### Clause 7.8.1.3 of SBCSub/C and SBCSub/D/C – 'the Contractor fails to make payment in accordance with the Sub-Contract'

This is an interesting clause, particularly given that the sub-contractor has the express right to interest payments (as clause 4.10.5 and clause 4.12.4 of SBCSub/C and SBCSub/D/C), and the right to suspend performance (as clause 4.11 of SBCSub/C and SBCSub/D/C) when payments are not made on time.

Of course, before using this sanction, which would almost certainly end any goodwill between the parties, a sub-contractor would need to carefully consider its past, and perhaps more relevantly, its expected future trading relationship with the contractor.

It may be worth considering how the procedure may work in practice:

Scenario 1

| | |
|---|---|
| 30 June | Final date for payment. |
| 1 July | No payment received. |
| | Sub-contractor issues suspension notice in line with clause 4.11 of SBCSub/C and SBCSub/D/C. |
| | Sub-contractor issues default notice in line with clause 7.8.1 of SBCSub/C and SBCSub/D/C. |
| 2 July | Contractor confirms receipt of the sub-contractor's suspension notice and 'default' notice. |
| 10 July | No payment received from contractor. |
| | Sub-contractor notifies the contractor that it is suspending its performance with immediate effect (in compliance with the clause 4.11 suspension notice). |
| 13 July | No payment received. |
| | Sub-contractor notifies that it is terminating its employment under the sub-contract (in compliance with the clause 7.8.1 default notice). |

Scenario 2

| | |
|---|---|
| 30 June | Final date for payment. |
| 1 July | No payment received. |

---

[12] *West Faulkner Associates* v. *London Borough of Newham* (1994) 71 BLR 1.

> Sub-contractor issues suspension notice in line with clause 4.11 of SBCSub/C and SBCSub/D/C.
>
> Sub-contractor issues default notice in line with clause 7.8.1 of SBCSub/C and SBCSub/D/C.
>
> 2 July    Contractor confirms receipt of the sub-contractor's suspension notice and 'default' notice.
>
> 10 July    No payment received from contractor.
>
> Sub-contractor notifies that it is suspending its performance with immediate effect (in compliance with the clause 4.11 suspension notice).
>
> 12 July    Payment received.
>
> 13 July    Sub-contractor recommences work on site.
>
> 30 July    Final date for payment.
>
> 1 August    No payment received.
>
> Sub-contractor notifies that it is terminating its employment under the sub-contract (in compliance with clause 7.8.3 of SBCSub/C and SBCSub/D/C). By terminating its employment this quickly, the sub-contractor will be complying with the SBCSub/C and SBCSub/D/C. The use of the words 'reasonable time after such repetition' within clause 7.8.3 of SBCSub/C and SBCSub/D/C in respect of a repeat of a specified default does not limit how quickly a termination may be given effect, but rather restricts how long after certain events occur that a termination may be given effect. It is submitted that in the case of non-payment these words would have a limited effect in any case.

As can be seen from the above, there is a very short timescale in respect of the above termination process, and the sub-contractor needs to understand the full implications of following the above path before embarking upon it.

Further, the contractor is only in default if it has failed 'to make payment in accordance with the Sub-Contract'. Therefore, the sub-contractor needs to be certain that the amount claimed is actually due before taking the step of terminating its employment.

The amount actually due may be affected by things such as:

(1) abatement (e.g. a reduction in the value of the works because of defective or incomplete works); and
(2) set-off or a withholding amount.

A sub-contractor may obviously be at risk in operating this sanction where the contractor relies in any future action upon an abatement or a set-off.

A wrongful termination by the sub-contractor would almost certainly amount to a repudiatory breach, which, if accepted by the contractor, may lead to the sub-contractor suffering all of the contractor's damages arising from that repudiatory breach.

### Clause 7.8.1.4 of SBCSub/C and SBCSub/D/C – 'the Contractor fails to comply with the CDM Regulations'

It is submitted that minor or trifling failures would not be considered to be a default, given that a notice of default and/or termination of the sub-contractor's employment is not to be given unreasonably or vexatiously.

## Termination under the main contract (refer to clause 7.9 of SBCSub/C and SBCSub/D/C)

In the event that the contractor's employment under the main contract is terminated, the sub-contractor's employment under the sub-contract shall terminate at the same time, and the contractor is to immediately notify the sub-contractor of that fact.

## Insolvency of the contractor (refer to clause 7.10 of SBCSub/C and SBCSub/D/C)

Clause 7.10.1 of SBCSub/C and SBCSub/D/C states that if the contractor is insolvent or makes any proposal, gives notice of any meeting or becomes the subject of any proceedings relating to any of the matters referred to in clause 7.1, the contractor must immediately inform the sub-contractor.

Under clause 7.1 of SBCSub/C and SBCSub/D/C (and for the purposes of the sub-contract conditions only) a contractor is insolvent if any of the conditions listed earlier in this chapter on page 290 apply.

Clause 7.10.2 states that if the contractor is insolvent (as defined), then:

- The sub-contractor is entitled by notice to the contractor to terminate the sub-contractor's employment under the sub-contract.

- If, before the contractor became insolvent, the sub-contractor had already issued a notice of 'default' then the sub-contractor may issue the 'termination' notice immediately that the contractor becomes insolvent.

- If, however, before the contractor became insolvent the sub-contractor had *not* issued a notice of 'default' then, when the contractor became insolvent, the sub-contractor may either:

  — issue a notice of termination of the sub-contractor's employment (in line with clause 7.10.2.1 of SBCSub/C or SBCSub/D/C); or
  — issue a notice of default pursuant to clause 7.8.1 of SBCSub/C and SBCSub/D/C.

Whichever of these notices the sub-contractor decides to issue, clause 7.10.2.1 of SBCSub/C and SBCSub/D/C makes it clear that the sub-contractor is not to exercise its rights to actually terminate its employment under the sub-contract prior to the expiry of a period of three weeks (or such further periods as the parties may agree) from the date upon which the contractor became insolvent.

The reason for this three-week period is to allow a period following the happening of a specified insolvency event, during which the sub-contract arrangements will remain in force while the parties attempt to agree a way forward.

In line with clause 7.10.2.2 of SBCSub/C and SBCSub/D/C, the sub-contractor's obligations to carry out and complete the sub-contract works (in the case of SBCSub/C and SBCSub/D/C) and the design of the sub-contractor's designed portion (only in the case of SBCSub/D/C) are immediately suspended pending termination of the sub-contractor's employment; and such suspension will for the purposes of clause 2.19 (Relevant sub-contract events – for extension of time) and for clause 4.20 (Relevant sub-contract matters – for loss and expense) be deemed to be a default of the contractor (as clause 7.10.2.3 of SBCSub/c and SBCSub/D/C).

## Consequences of termination under clauses 7.8 to 7.10 of SBCSub/C and SBCSub/D/C

If the sub-contractor's employment is terminated by the sub-contractor for any of the reasons outlined above, the following consequences arise:

- The contractor is not required to make any further payment or release any outstanding retention, other than as provided for under clause 7.11 of SBCSub/C and SBC/D/C (as clause 7.11.1 of SBCSub/C and SBCSub/D/C).

- In line with clause 7.11.2 of SBCSub/C (clause 7.11.2.1 of SBCSub/D/C) the sub-contractor must with all reasonable dispatch remove or procure the removal from the site of any temporary buildings, plant, tools and equipment belonging to the sub-contractor and the sub-contractor's persons together with all goods and materials including site materials (except only for those site materials that will be valued in the sub-contractor's concluding account (refer to clause 7.11.3.4 of SBCSub/C and SBCSub/D/C) and which will be paid for by the contractor (refer to clause 7.11.5 of SBCSub/C and SBCSub/D/C).

- Where applicable, and without charge, the sub-contractor is to provide the contractor with three copies of all the sub-contractor's design documents then prepared. This consequence is required irrespective of any request by the contractor (refer to clause 7.11.2.2 of SBCSub/D/C).

- In line with clause 7.11.3 of SBCSub/C and SBCSub/D/C, the sub-contractor must with reasonable dispatch prepare and submit to the contractor an account setting out:
  - The total value of the work properly executed at the date of termination ascertained and valued in line with the sub-contract conditions.
  - The total value of any other amounts due to the sub-contractor at the date of termination ascertained and valued in line with the sub-contract conditions, for example any applicable fluctuation payments.
    Also, given that the next item on the list only relates to any ascertained direct loss and/or expense, these 'other amounts' would presumably also include for other loss and/or expense items claimed by the sub-contractor, but not ascertained by the contractor.
  - Any sums ascertained in respect of direct loss and/or expense under clause 4.19 of SBCSub/C and SBCSub/D/C.
  - The reasonable cost of removal from the site of any temporary buildings, plant, tools and equipment belonging to the sub-contractor and the sub-contractor's persons together with all goods and materials including site materials, as appropriate.
  - The cost of materials and goods (including sub-contract site materials) properly ordered for the sub-contract works for which the sub-contractor has

already made payment, or for which the sub-contractor is legally liable to make payment.
— Any direct loss and/or damage caused to the sub-contractor by the termination of his employment. This could, of course, include, if applicable, the loss of future overheads contribution and the loss of future profit.

In respect of the sub-contractor's recovery of direct loss and/or damage caused to the sub-contractor by the termination of his employment, this will *not* apply where:

— the termination of the sub-contractor's employment is due to the termination of the contractor's employment (by either the employer or the contractor) by reason of those events (qualified as indicated in parentheses below) covered by clause 8.11.1 of the main contract conditions (refer to clause 7.11.4.1 of SBCSub/C and SBCSub/D/C), namely:

8.11.1.1 *force majeure*;
8.11.1.2 architect/contract administrator's instructions under clause 2.15, 3.14 or 3.15 issued as a result of the negligence or default of any statutory undertaker;
8.11.1.3 loss or damage to the works occasioned by one of the specified perils (except where such loss or damage was caused by the negligence or default of the employer or any of the employer's persons);
8.11.1.4 civil commotion or the use or threat of terrorism and/or the activities of the relevant authorities in dealing with such an event or threat;

— the exercise by the United Kingdom Government of any statutory power which directly affects the execution of the works.

• The termination of the sub-contractor's employment follows the employer's option in terminating the contractor's employment in the situation where there was a non-availability of insurance cover for terrorism (refer to clause 6.10.2.2 of the main contract conditions) (as clause 7.11.4.2 of SBCSub/C and SBCSub/D/C).

• The termination of the sub-contractor's employment is due to the termination of the contractor's employment (by either the employer or the contractor) because it is considered just and equitable to do so following loss or damage to the works occasioned by any of the risks covered by the joint names all risks policy referred to in paragraph C.2 of schedule 3 of the main contract conditions (refer to paragraph C.4.4 of schedule 3 to the main contract conditions) (as clause 7.11.4.2 of SBCSub/C and SBCSub/D/C).

• The contractor must pay to the sub-contractor the amount properly due in respect of the account submitted by the sub-contractor (as detailed above), without any deduction for retention, but after taking into account amounts already paid, within *28 days* of its submission by the sub-contractor to the contractor.
    Payment by the contractor for any materials and goods within the account must be made on the basis that the materials and goods become the property of the contractor upon payment being made (refer to clause 7.11.5 of SBCSub/C and SBCSub/D/C).

# Chapter 13
# Settlement of Disputes

## Introduction

It is an unfortunate reality of life in the construction industry that disputes will arise. Certain aspects of the construction process and of the industry's culture make it more likely that disputes will arise than perhaps in other industries.

Some of the more obvious factors that increase the probability of disputes arising are:

- new (or unfamiliar) materials;
- new (or unfamiliar) construction techniques;
- multi-party involvement in projects;
- uncertainty regarding the physical environment within which the works will be performed;
- uncertainty regarding the commercial environment within which the works will be performed;
- variable performance against untested methods;
- low margins;
- tender errors;
- incorrect allocation of risk;
- misunderstanding of allocation of risk;
- excessive risk taking;
- problems in contract documentation, including errors, contradictions and ambiguities;
- incomplete design before construction, causing information and documentation to be issued late.

The vast majority of disputes within the construction industry are resolved (often on a daily basis) by negotiation. In those relatively few occasions where disputes cannot be resolved by negotiation, the dispute must be referred to a more formal dispute resolution process.

The reality that disputes may need to be resolved by some formal dispute resolution process is recognized by the SBCSub/C and SBCSub/D/C by way of section 8 of SBCSub/C and SBCSub/D/C.

The four methods of dispute resolution allowed for within SBCSub/C and SBCSub/D/C are:

- mediation;
- adjudication;

- arbitration;
- legal proceedings.

Before considering each of the above dispute resolution methods, the effect of the final payment notice must be dealt with.

The final payment notice is covered by clause 1.9 of SBCSub/C and SBCSub/D/C (which is dealt with in Chapter 3).

The effect of the final payment notice is far-reaching for both parties. If it remains unchallenged, the final payment notice cannot be opened up in respect of the quality of any materials or goods or of the standard of an item of workmanship which had expressly been for the approval of the architect/contract administrator; the valuation of the sub-contract works; the sub-contractor's entitlement to extensions of time; the sub-contractor's entitlement to direct loss and expense; or the contractor's reimbursement entitlement.

If a sub-contractor disagrees with the final payment notice he should commence adjudication, arbitration or other proceedings within ten days of receipt of the notice if he wishes to avoid the notice being deemed conclusive evidence as detailed above.

It should be noted that if the proceedings only relate to one issue (for example, the sub-contractor's entitlement to loss and expense) then the final payment notice will still be taken as being conclusive evidence of all other matters (refer to clause 1.9.3 of SBCSub/C and SBCSub/D/C).

In the situation where either party has already commenced proceedings before the final payment notice is issued, unless the notice is to be taken as being conclusive evidence of all matters outlined above, either or both parties will have to take a further step in such proceedings within twelve months from or after the date the final payment notice was given (as clause 1.9.2.2 of SBCSub/C or SBCSub/D/C).

At the conclusion of such proceedings, the final payment notice will be taken as being conclusive evidence of the matters outlined above, subject only to the terms of any decision, award or judgment in, or settlement of, those proceedings (refer to clause 1.9.2.1 of SBCSub/C or SBCSub/D/C).

If an adjudicator gives his decision on a dispute or difference after the date of the final payment notice and either party then wishes to have the dispute determined by arbitration or legal proceedings, such proceedings must be commenced within 28 days of the date on which the adjudicator gives his decision if the decision is not to be final in respect of the matters referred to in clause 1.9.1 of SBCSub/C and SBCSub/D/C as outlined above.

## Mediation

### What is mediation?

Mediation is a way of settling disputes in which a third party, a neutral person called a mediator, helps both parties to come to an agreement which each party considers is acceptable.

In the UK, mediators are trained to be facilitative (that is, are trained to facilitate between the parties to enable them to reach a settlement) rather than being evaluative (where the mediator evaluates the merits and/or the quantum of the dispute and then effectively imposes his decision on the parties).

There is some debate regarding whether facilitative or evaluative mediation is more successful; and there is also informal evidence that parties often expect an evaluative mediation rather than a facilitative mediation. However, these matters cannot be considered further within the scope of this book.

Mediation differs from litigation, arbitration or adjudication as each of these processes leads to a third party imposing a binding (in the case of adjudication, a temporarily binding) decision on the parties. Any negotiated settlement between the parties in mediation is not binding until the parties draw up a binding agreement.

## How is mediation incorporated into the JCT Standard Building Sub-Contract?

Mediation is incorporated into the sub-contract by way of clause 8.1 of SBCSub/C or SBCSub/D/C, which says that: 'the Parties may by agreement seek to resolve any dispute or difference arising under this Sub-Contract through mediation.'

## Are the parties obliged to go to mediation to resolve their disputes?

Mediation is voluntary, and clause 8.1 of SBCSub/C or SBCSub/D/C simply states that the parties 'may' seek to resolve disputes through mediation.

Therefore, for example, the parties can go to adjudication at any time (irrespective of whether or not they have already gone to mediation).

Also, there is no obligation to go to mediation before commencing an arbitration action.

## Refusing to mediate and legal costs

The standard position for litigation is that the winner recovers most of its costs; this follows the principle that costs follow the event. In simple terms, this means that the party that loses is responsible for the costs of the party that is successful in the court action.

However, in litigation, the overriding objective of the Civil Procedure Rules (CPR) is to enable the courts to deal with cases justly, and the CPR encourages the use of alternative dispute resolution (ADR) (for example mediation) to help in resolving disputes both quickly and cheaply.

If a party does not use ADR before proceeding to court, then the consequence could be that a cost sanction could be imposed. This is because, in deciding what order (if any) to make about costs, the court must have regard to all of the circumstances, including the conduct of the parties, as CPR 44.3(4)(a).

In a series of recent cases, commencing with *Dunnett* v. *Railtrack*[1] and continuing to the Court of Appeal case of *Burchell* v. *Bullard*[2] the importance the court

---

[1] *Dunnett* v. *Railtrack plc* [2002] 1 WLR 2434, CA.
[2] *Burchell* v. *Bullard* [2005] EWCA Civ 358.

places on the use of ADR to resolve disputes has been emphasized. The conclusion was that, ordinarily (but with each case being decided on its own facts), the court should be prepared to penalize a party, in terms of costs, who unreasonably refuses to use ADR.

Both legal advisors and their clients must consider the possible uses of ADR whenever a dispute arises and, of course, follow the relevant approved pre-action protocols (dealt with later within this chapter). Even if the dispute is not covered by a pre-action protocol the court will require the parties to comply with the spirit of the overriding objective of enabling the court to deal with cases justly, fairly and proportionately in terms of costs.

This has been reinforced by virtue of a CPR practice direction issued in April 2006, which now includes a mandatory provision that the court must in future take into account, when considering the issue of costs, whether the parties have given proper consideration to the use of ADR.

To emphasize the court's attitude to this issue, the practice direction states that 'The Courts take the view that litigation should be a last resort.'

In respect of construction disputes, there is a specific pre-action protocol that applies (as dealt with later within this chapter), and it is considered that if this pre-action protocol is followed then the use of mediation may be unnecessary. However, this matter is not entirely clear, and the position continues to evolve.

The CPR practice direction issued in April 2006 specifically states that, despite the possible future sanction on costs, it is recognized that no party can or should be forced to mediate or enter into any form of ADR.

The practice direction summarizes some of the options for resolving disputes without litigation, as follows:

- discussion and negotiation;
- early neutral evaluation by an independent third party;
- mediation.

## The mediation process

Although mediation is a flexible process which can be adapted to suit particular circumstances or needs, typically the stages might include: brief written summaries of the dispute submitted to the mediator *in advance of the mediation*

*Then at the mediation* there would be signing of the mediation agreement in which the parties confirm their participation in the process and their understanding that information given in the mediation will be on a without prejudice basis and is to be kept confidential unless otherwise agreed.

There should be an initial joint meeting between the parties and the mediator with presentations setting out the parties' respective positions.

Private sessions (sometimes referred to as caucus sessions) in which the mediator meets privately with each party in turn. The mediator goes between the parties, gaining a better understanding of the issues, and endeavours to identify potential settlement options.

Further joint meetings are held as appropriate during which face-to-face negotiations can take place and agreements can be reached.

Where an agreement is reached it can be written up as a legally enforceable contract if this is what the parties require.

Mediation is a confidential process to the extent that the parties wish it to be. This includes confidentiality inside the mediation and also in respect of any agreement reached. The mediator will keep any information given by a party in private session confidential unless the party agrees that the information can be shared.

The advantages of mediation are its speed, flexibility and relative low cost. It also allows for non-legal interests to be taken into account in any agreement reached between the parties.

For example, a compromise might include a promise of future work. This non-adversarial approach often allows relationships to be maintained or rebuilt in a way which would be unlikely to happen with more adversarial methods of dispute resolution.

Mediators encourage parties to consider their interests, and not merely to concentrate on their legal rights.

The use of mediation is growing, partly as a result of parties seeking simpler and cheaper means of resolving disputes and partly as a result of encouragement from the courts that now require parties to make every effort to resolve their dispute before it is referred to the court.

As noted above, mediation is conducted on a 'without prejudice' basis and therefore the parties are encouraged to be proactive in putting forward ideas and seeking solutions which are not binding unless they are incorporated into the final agreement.

Despite the above, mediation should not be seen as a soft option. The parties need to prepare thoroughly and should be prepared for a long and demanding day (typically mediation is scheduled for one day but can be longer).

## Adjudication

### What is adjudication?

Adjudication is a statutory procedure by which any party to a construction contract has a right to have a dispute decided by an adjudicator. Adjudication is not arbitration. The adjudication process is a means by which an independent third party (i.e. the adjudicator) makes a (relatively) quick decision when the parties to a contract are in disagreement.

### Why have adjudication provisions in the JCT Standard Building Sub-Contract?

The Housing Grants, Construction and Regeneration Act 1996 (the Construction Act), Part II Construction Contracts provides under s.108 (1) that: 'A party to a construction contract has the right to refer a dispute arising under the contract for adjudication under a procedure complying with this section. For this purpose "dispute" includes any difference.'

The construction contract must be in writing.

Therefore, except for certain circumstances (e.g. the contract is with a residential occupier that has not received professional advice before entering into the contract), all parties to a written construction contract, as defined under s.104 of the Construction Act, have the right to refer a dispute arising under the contract for adjudication under a procedure complying with s.108 (1) of the Construction Act.

If there are not express terms in the written construction contract regarding adjudication then a term to that effect will be implied into the contract.

Obviously, parties prefer to have some control over the adjudication provisions that will apply to their contract, and rather than rely on an implied term they prefer to incorporate their own conditions, which is why there are adjudication provisions in the JCT Standard Building Sub-Contract.

## How are the adjudication provisions incorporated into the JCT Standard Building Sub-Contract?

Article 4 of SBCSub/A and SBCSub/D/A states that if any dispute or difference arises under the sub-contract either party may refer it to adjudication in accordance with clause 8.2 of SBCSub/C or SBCSub/D/C.

Clause 8.2 of SBCSub/C and SBCSub/D/C states that the Scheme rules must apply, subject to the following conditions:

(1) Clause 8.2.1. For the purposes of the Scheme the adjudicator must be the person (if any) stated in the sub-contract particulars (either item 14 of SBCSub/A or item 15 of SBCSub/D/A), and the nominating body of the adjudicator must be that stated in the same sub-contract particulars items.

(2) Clause 8.2.2. Where the dispute or difference is or includes a dispute or difference relating to clause 3.11.3 (i.e. instructions to open up for inspection or to test any further similar non-compliant works) and whether a direction issued under that clause is reasonable in all of the circumstances, then the adjudicator to decide such a dispute must (where practicable) be an individual with appropriate expertise and experience in the specialist area or discipline relevant to the direction or issue in dispute. Where the adjudicator does not have the appropriate expertise and experience, the adjudicator must appoint an independent expert with such expertise and experience to advise and report in writing on whether or not the direction under clause 3.11.3 is reasonable in all of the circumstances.

## What is the Scheme?

The Scheme is defined under clause 1.1 of SBCSub/C and SBCSub/D/C as being, Part 1 of the Schedule to the Scheme for Construction Contracts (England and Wales) Regulations 1998 (i.e. that part of the schedule that relates to adjudication).

Those regulations were made by the Secretary of State in exercise of the powers conferred on him by sections 108(6), 114 and 146(1) and (2) of the Construction Act.

The Scheme, as defined above, has 26 paragraphs split under the following headings:

- notice of intention to seek adjudication;
- powers of the adjudicator;

- adjudicator's decision;
- effects of the decision.

The Scheme sets out the adjudication provisions that apply.

## What can be referred to adjudication?

Article 4 of SBCSub/A and SBCSub/D/A states that *any dispute or difference* arising *under* the sub-contract may be referred by either party to adjudication.

Paragraph 8(1) of the Scheme notes that the adjudicator may, with the consent of all the parties to those disputes, adjudicate at the same time on more than one dispute under the same contract.

Although paragraph 1(1) of the Scheme refers to the singular word 'dispute', it is generally accepted that this can mean a range of matters within a single dispute. This position is strengthened by the fact that under paragraph 20 of the Scheme, the adjudicator is to decide 'the matters in dispute', and this position is also supported by case law.[3]

In the *Fastrack* v. *Morrison*[4] case, Judge Thornton QC held that the dispute encompassed 'whatever claims, heads of claim, issues, contentions or causes of action that are then in dispute which the referring party has chosen to crystallize into an adjudication reference'.

## What cannot be referred to adjudication?

Article 4 of SBCSub/A and SBCSub/D/A makes it clear that the dispute or difference must have arisen 'under the Sub-Contract', and therefore disputes not arising under the sub-contract and tortuous liability matters are excluded by the wording of the Construction Act.[5]

Any dispute that has already been decided by an adjudicator following an earlier adjudication action cannot be referred to adjudication (i.e. there is a bar on re-adjudication of the same dispute[6]). In line with paragraph 9(2) of the Scheme, an adjudicator must resign where the dispute is the same or substantially the same as one which has previously been referred to adjudication and a decision has been taken in that adjudication.

However, provided that an adjudicator is dealing with a different dispute, he is not bound by an earlier adjudication decision.

Therefore, for example, a claim for variations which are required to be remeasured and revalued on the final account is not the same as a similar claim under a previous interim application.

However, if part of the dispute is a claim which is the same as decided in a previous application, then:

---

[3] *Fastrack Contractors Limited* v. *Morrison Construction Ltd and Impregilo UK Ltd* [2000] BLR 168; *Balfour Kilpatrick Limited* v. *Glauser International SA* (2000) (unreported) QBD (TCC).
[4] *Fastrack Contractors Limited* v. *Morrison Construction Ltd and Impregilo UK Ltd* [2000] BLR 168.
[5] *Shepherd Construction* v. *Mecright Limited* [2000] BLR 489.
[6] *Sherwood & Casson Limited* v. *Mackenzie Engineering Limited* [2000] CILL 1577; *VHE Construction plc* v. *RBSTB Trust Co. Ltd.* [2000] CILL 1592.

- The second adjudicator has no jurisdiction to set aside, revise or vary the first adjudicator's decision.
- The contractual machinery is superseded by the first adjudicator's decision and there is an obligation to comply with his decision.
- The second adjudicator must resign from the dispute in its entirety or confirm his appointment in its entirety. The adjudicator cannot resign from part of that which has been referred, whilst retaining jurisdiction to determine the balance of the reference.

Something that has not yet crystallized into becoming a dispute cannot be referred to adjudication.

## How can you tell that a dispute has crystallized?

Simply making a claim and submitting it may not be enough to indicate that a dispute has crystallized.

For a dispute to exist a claim has to be notified and rejected; however, a rejection may stem from the other party's refusal to consider/answer a claim.[7]

## When can a dispute be referred?

A dispute may be referred at any time.

Even if a dispute is the subject of existing proceedings in court it can still be referred to an adjudicator and enforced pending a decision by the court. There is no provision in the Construction Act for a stay of execution of the adjudication action pending a decision by the court.[8]

Similarly, if there is an arbitration clause in the contract a stay of the adjudication will not be granted pending the outcome of the arbitration.[9]

## How is the adjudication process commenced?

Paragraph 1(1) of the Scheme states that 'any party to a construction contract may give written notice of his intention to refer any dispute arising under the contract, to adjudication'.

Paragraph 1(2) of the Scheme notes that the written notice must be given to every other party to the contract.

The party that commences an adjudication action is normally referred to as the referring party, and the written notice is normally referred to as the notice of adjudication.

---

[7] *Amec Civil Engineering Ltd* v. *Secretary of State for Transport* [2005] CILL 2189; *Fastrack Contractors Limited* v. *Morrison Construction Ltd and Impregilo UK Ltd* [2000] BLR 168.
[8] *A. & D. Maintenance and Construction Ltd* v. *Pagehurst Construction Services Ltd* (1999) CILL 1518.
[9] *Absolute Rentals Limited* v. *Gencor Enterprises Limited* [2000] CILL 1637.

## What should the notice of adjudication contain?

Paragraph 1(3) of the Scheme states that the notice of adjudication should set out briefly:

(1) the nature and a brief description of the dispute and the parties involved;
(2) details of where and when the dispute has arisen;
(3) the nature of the redress sought; and
(4) the names and addresses of the parties to the contract (including, where appropriate, the addresses which the parties have specified for the giving of notices).

It is very important to note that the notice of adjudication, which must be in writing,[10] sets out the jurisdiction of the adjudicator[11] (i.e. sets out the boundaries of the adjudicator's authority).

There must be a dispute between the parties for the adjudicator to have jurisdiction and, as noted above, a dispute means that whatever claims, heads of claim, issues, contentions or causes of action were in dispute at the moment when the referring party first intimated an adjudication reference and chose to crystallize it into a notice of adjudication.[12]

The notice of adjudication must comply with the contract requirements and describe precisely the dispute being referred and only those so described will be enforced.[13]

However, when the notice of adjudication is not precise about the dispute being referred a broad, rather than a narrow, construction should be put on it,[14] the court taking the view that these documents are drafted by practitioners rather than lawyers, but that the same latitude ought not to be given to one drafted with the benefit of outside advice.[15]

When asking for a decision from the adjudicator for a sum of money it is normal to add after the request 'or such other sum as the adjudicator may decide'. Similarly, if a time period is requested it is normal to add the words 'or such other time period as the adjudicator may decide'. The purpose of these caveats is to extend the adjudicator's jurisdiction and to prevent the adjudicator from deciding on an all-or-nothing basis.

## How is an adjudicator appointed, and who will be the appointed adjudicator?

Paragraph 2(1) of the Scheme makes it clear that an adjudicator can only be appointed after a notice of adjudication has been issued.

Section 108(2)(b) of Part II of the Construction Act requires the procedure being followed to provide a timetable with the object of securing the appointment of an adjudicator and referral of the dispute to him within seven days of the date of the notice of adjudication.

---

[10] *Strathmore Building Services Ltd* v. *Colin Scott Greig* (2001) 17 Const LJ 72.
[11] *Fastrack Contractors Limited* v. *Morrison Construction Ltd and Impregilo UK Ltd* [2000] BLR 168.
[12] *Fastrack Contractors Limited* v. *Morrison Construction Ltd and Impregilo UK Ltd* [2000] BLR 168.
[13] *Ken Griffin & John Tomlinson t/a K. & D. Contractors* v. *Midas Homes Ltd* (2002) 18 Const LJ 67.
[14] *Karl Construction (Scotland) Ltd* v. *Sweeney Civil Engineering (Scotland) Ltd* (2000) SCLR 766.
[15] *Ken Griffin & John Tomlinson t/a K. & D. Contractors* v. *Midas Homes Ltd* (2002) 18 Const LJ 67.

After the notice of adjudication has been issued, the parties may agree between themselves who shall act as adjudicator (ref: paragraph 2(1) of the Scheme). However, bearing in mind that, at the stage when a dispute has been referred to adjudication the relationship between the parties has often deteriorated somewhat, it is unlikely that the parties will be able to agree upon a person to act as adjudicator. There is no obligation on either party to attempt to agree upon the name of an adjudicator after the notice of adjudication has been issued, but the opportunity to do so exists.

However, the more normal way of appointing an adjudicator is for the referring party to request the person named (if one is so named) under sub-contract particulars item 14 of SBCSub/A or sub-contract particulars item 15 of SBCSub/D/A to act as the adjudicator.

When agreeing upon the name of an adjudicator to be included under sub-contract particulars item 14 of SBCSub/A or sub-contract particulars item 15 of SBCSub/D/A, the parties should note that clause 8.2.2.1 of SBCSub/C and SBCSub/D/C states that the adjudicator must decide a dispute or difference relating to clause 3.11.3, and whether a direction issued is reasonable in all the circumstances; he must also (where practicable), be an individual with appropriate expertise and experience in the specialist area or discipline relevant to the direction or issue in dispute.

Therefore, the possibility exists that, even in the case where an adjudicator is named in sub-contract particulars item 14 of SBCSub/A or sub-contract particulars item 15 of SBCSub/D/A, and that adjudicator has confirmed that he or she is willing and able to act in respect of the dispute, a different adjudicator may need to be appointed where a dispute or difference exists relating to clause 3.11.3, and as to whether a direction issued is reasonable in all the circumstances, when it is being referred to adjudication.

It is noted in SBCSub/G (under paragraph 67) that an adjudicator should not normally be named in the sub-contract without his prior agreement and the same person should not normally be named in both a sub-contract and the main contract.

The first reason is obvious, the second less so.

Presumably, the latter reason is because of concerns regarding bias (actual or perceived) by the adjudicator when dealing with a sub-contractor/contractor dispute if he or she has already dealt with, or will deal with in future, a similar dispute between the contractor and the employer.

The fact that the same person will not normally be named as adjudicator in both a sub-contract and the main contract obviously makes the joinder of adjudication actions unlikely (unless all parties agree, as paragraph 8(2) of the Scheme); although, in reality, it is unlikely that, unless a discrete dispute needed to be decided based upon common terms, a joinder of adjudication actions would be possible in any event.

Paragraph 2(1)(b) of the Scheme makes it clear that if a person is not named in SBCSub/A or SBCSub/D/A, or if the person named has already indicated that he is unwilling or unable to act, and if the contract provides under sub-contract particulars item 14 of SBCSub/A or sub-contract particulars item 15 of SBCSub/D/A for a specified nominating body to select a person, the referring party must request the specified nominating body to select a person to act as adjudicator.

Under paragraph 6(2) of the Scheme, the person named in SBCSub/A and SBCSub/D/A as the adjudicator must indicate whether or not he is willing to act within two days of receiving the request. Therefore, if a request is made to the adjudicator named in SBCSub/A or SBCSub/D/A but no response is received within two days then it would be in order for the parties to the dispute to assume that he is not willing to act.

Because of the above, paragraph 67 of SBCSub/G advises that even if an adjudicator is named, a nominating body should also be selected to cover for the eventuality that the named adjudicator is unwilling or unable to act at the time that a dispute arises.

If the dispute or difference relates to clause 3.11.3 of SBCSub/C and SBCSub/D/C and as to whether a direction issued is reasonable in all the circumstances, the nominating body needs to be advised that the selected adjudicator needs to be an individual (where practicable) with appropriate expertise and experience in the specialist area or discipline relevant to the direction or issue in dispute.

The choice of nominator of adjudicator under sub-contract particulars item 14 of SBCSub/A or sub-contract particulars Item 15 of SBCSub/D/A, are the President or a Vice-President or Chairman or a Vice-Chairman of:

• the Royal Institute of British Architects;
• the Royal Institution of Chartered Surveyors;
• the Construction Confederation;
• the National Specialist Contractors Council;
• the Chartered Institute of Arbitrators.

When completing sub-contract particulars item 14 of SBCSub/A (or sub-contract particulars item 15 of SBCSub/D/A) the parties should delete all but one of the five nominating bodies listed above.

In the event that an adjudicator has not been named and a nominating body has not been selected (or where an adjudicator has been named but has said that he or she is unwilling or unable to act, and a nominating body has not been selected), then the referring party may select any one of the five listed nominating bodies.

Again, noting that if the dispute or difference relates to clause 3.11.3 of SBCSub/C or SBCSub/D/C, and as to whether a direction issued is reasonable in all the circumstances, the nominating body needs to be advised that the selected adjudicator needs to be an individual (where practicable) with appropriate expertise and experience in the specialist area or discipline relevant to the direction or issue in dispute.

An application for the nomination of an adjudicator to a nominating body is usually made by way of a standard form produced by each nominating body. The form requires details such as the names of the parties, the nature of the dispute, any required qualifications of the proposed adjudicator and the adjudication provisions that apply.

Paragraph 3 of the Scheme also requires that a copy of the notice of adjudication is submitted with the application. The application must be accompanied by a fee required by the nominating body to nominate an adjudicator.

In line with paragraph 2(2) of the Scheme, any person requested to act as an adjudicator must indicate whether or not he is willing to act within two days of receiving the request.

Also, any person requested or selected to act as adjudicator (however that request or selection is made) must be a natural person acting in his personal capacity, and that person must not be an employee of any of the parties to the dispute and must declare any interest, financial or otherwise, in any matter relating to the dispute (refer to paragraph 4 of the Scheme).

The nominating body must communicate the selection of an adjudicator to the referring party within five days of receiving a request to do so (as paragraph 5(1) of the Scheme).

If the nominating body fails to comply with this requirement, paragraph 5(1) of the Scheme states that the referring party may agree with the other party to the dispute to request a specified person to act as adjudicator, or may ask any other nominating body to select a person to act as adjudicator (in the context of the JCT Standard Building Sub-Contract, the phrase 'any other nominating body' may be construed as meaning any of the five listed adjudication nominating bodies listed under sub-contract particulars item 14 of SBCSub/A or sub-contract particulars item 15 of SBCSub/D/A, particularly if none of those listed adjudication nominating bodies have been deleted by the parties).

## What happens if either party objects to the appointment of a particular person as adjudicator?

Paragraph 10 of the Scheme makes it clear that if any party to a dispute objects to the appointment of a particular person as adjudicator, the objection will not invalidate the adjudicator's appointment nor any decision he may eventually reach.

However, if an objection was made which the adjudicator considered was a justified complaint then the adjudicator may resign (refer to paragraph 9(1) of the Scheme), in which case the adjudication appointment procedure would need to re-commence.

## What happens after the adjudicator is appointed?

The referring party must, not later than seven days from the date of the notice of adjudication, refer the dispute in writing (the referral notice) to the adjudicator (refer to paragraph 7(1) of the Scheme).

At the same time, the referring party must send the referral notice (and all accompanying documents) to every other party to the dispute.

## What happens if the referral notice is issued to the adjudicator more than seven days after the date of the notice of adjudication?

There has been no reported case law on the use of the mandatory language 'shall' in paragraph 7(1) of the Scheme, but this matter was considered in respect of the JCT Standard Form of Contract Private with Quantities, 1998 edition.[16]

Although the mandatory word 'shall' is stronger than 'with the object of' in section 108(2)(b) of the Construction Act, no sanction for non-compliance is

---

[16] *William Verry Ltd v. North West London Communal Mikvah* [2004] BLR 308.

included within the Scheme, and it is considered that (depending, of course, upon the actual period in excess of the seven days) the courts are unlikely to strictly apply this technicality when faced with an action to enforce an adjudicator's decision.

In practice, most adjudicators appear to give a small degree of leeway (for example two or three days) before they would consider disallowing the referral notice.

## What is the referral notice?

The referral notice is effectively the referring party's statement of case, and normally details the particulars of the dispute, details the contentions being relied upon, and provides a statement of the relief or remedy sought.

Paragraph 7(2) of the Scheme makes it clear that the referral notice must be accompanied by copies of, or relevant extracts from, the construction contract and such other documents as the referring party intends to rely upon.

## What are the terms and the liability of the adjudicator?

The adjudicator's terms will normally be as the adjudication agreement (Adj) (a document that has been prepared by the JCT for use when appointing an adjudicator) or the adjudication agreement (named adjudicator) (Adj/N) (a document that has been prepared by the JCT for use when appointing a named adjudicator), as appropriate.

The reason why the word 'normally' is used is because, unlike clause 9A.2.1 of the JCT Standard Form of Domestic Sub-Contract 2002 edition, there is no clause in the JCT Standard Building Sub-Contract that stipulates that no adjudicator will be agreed or nominated who will not execute the JCT adjudication agreement.

The main terms of the standard JCT adjudication agreements are:

- The adjudicator must accept the appointment in respect of the dispute referred to in the notice of adjudication.
- The adjudicator must observe the adjudication provisions agreed between the parties.
- In the event of the adjudicator becoming unavailable to act on account of illness or other cause, he must notify the parties to that effect immediately.
- The parties will be jointly and severally liable to the adjudicator for his fee and expenses. The adjudicator is to provide his fee rate schedule within the agreement document.
- The adjudicator's appointment may be terminated jointly by the parties at any time by written notice. In such an event, except where he failed to give (which presumably means deliver rather than reach) his decision within the relevant timescale, the adjudicator must be entitled to any part of his fee still outstanding plus expenses reasonably incurred before the termination.

Paragraph 26 of the Scheme makes it clear that neither the adjudicator, nor any employee or agent of the adjudicator, will be liable for anything done or omitted in the discharge or purported discharge of his functions as adjudicator unless the act or omission is in bad faith. The definition of 'bad faith' could be argued, but probably requires a definite 'bad action' rather than just a nebulous alleged lack of 'good faith'.

## What is the jurisdiction of the adjudicator?

The jurisdiction of an adjudicator may be considered as being the boundaries within which an adjudicator may act.

These boundaries are partly established by the terms of the sub-contract and partly by the wording of the notice of adjudication.

In respect of the boundaries established by the terms of the sub-contract, article 4 of SBCSub/A and SBCSub/D/A makes it clear that the dispute or difference must have arisen 'under the sub-contract', and therefore matters arising outside the sub-contract and tortuous liability matters are excluded.

Therefore, disputes regarding settlement agreements are often outside the jurisdiction of an adjudicator because they are not construction contracts.[17]

Also, disputes regarding matters which preceded the making of the sub-contract, such as questions as to whether the sub-contract was entered into as a result of misrepresentation, etc., are not disputes under the sub-contract.[18]

In respect of the jurisdictional boundaries established by the wording of the notice of adjudication, an adjudicator only has jurisdiction to deal with matters specifically raised in the notice of adjudication, although it has been found, in certain circumstances, that the notice of adjudication and the referral notice can be read in conjunction with one another.

Therefore, if an adjudicator was asked within a notice of adjudication to reach a decision that the referring party was entitled to a further payment of £500,000 it is considered that the adjudicator could not decide that the referring party was entitled to a further payment of £450,000 (although, if an adjudicator was asked within a notice of adjudication to reach a decision that the referring party was entitled to a further payment of £500,000, *or such other sum as the adjudicator may decide*, then the adjudicator could decide that the referring party was entitled to a further payment of £450,000).

## What happens if there is a query regarding the jurisdiction of the adjudicator?

An adjudicator has no power to rule on his own jurisdiction[19] although the parties can agree to give him that power.[20]

Nevertheless, if an adjudicator's jurisdiction is challenged (which would most usually be by the non-referring party (normally referred to as the responding party)), the adjudicator has to consider the challenge.

---

[17] *Lathom Construction Limited* v. *Brian Cross and Ann Cross* (1999) CILL 1565.
[18] *Tim Butler Contracts Limited* v. *Merewood Homes Limited* (2002) 18 Const LJ 84.
[19] *The Project Consultancy Group* v. *The Trustees of the Gray Trust* [1999] BLR 377.
[20] *Nolan Davis Limited* v. *Steven Catton* (2000) (unreported), QBD (TCC) No. 590.

In considering the challenge, the adjudicator appears to have three options:

- he can ignore the challenge and proceed, and leave it to a party to challenge his jurisdiction at enforcement proceedings; or
- he can investigate the issue and decide that he has jurisdiction (but not rule that he has jurisdiction) in which case he will proceed; or
- he can decide he has no jurisdiction and decline to continue.[21]

In considering a challenge, an adjudicator would normally ask both parties for their views, and would also have in mind that where a decision is subsequently challenged and the lack of jurisdiction is clearly arguable then the court may decline to enforce the decision pending its judgment on the jurisdiction issue.[22]

Any points which are to be made as to whether the adjudicator has jurisdiction must be raised in a timely manner during the adjudication itself.[23] If it is not made in a timely manner the objection may be ignored by a court in any subsequent enforcement action.

When an adjudication has started and a court holds, before a decision is reached, that the adjudicator lacks jurisdiction, an injunction may be granted to restrain the referring party from taking any substantial step in the adjudication action or seeking to enforce or implement any award which may be made.[24]

## What are the powers of the adjudicator?

The adjudicator's powers relate solely to the contract under which he is appointed. The parties themselves give him such powers as he has.

The powers granted to the adjudicator are contractual not statutory.

In terms of the JCT Standard Building Sub-Contract, the powers of the adjudicator are set out under paragraphs 12 to 19 inclusive of the Scheme.

Under Paragraph 12 of the Scheme the adjudicator is to:

'(a) act impartially in carrying out his duties and shall do so in accordance with any relevant terms of the contract and shall reach his decision in accordance with the applicable law in relation to the contract; and
(b) avoid incurring unnecessary expense.'

Paragraph 13 of the Scheme makes it clear that the adjudicator has fairly wide powers in respect of the procedure to be followed in the adjudication action, and paragraph 14 of the Scheme states that the parties must comply with any request or direction of the adjudicator in relation to the adjudication.

Some of the particular powers that the adjudicator has at his disposal, under paragraph 13 of the Scheme, are to:

---

[21] *Christiani & Nielsen Limited* v. *The Lowry Centre Development Limited* (2000) (unreported) TCC, 16 June 2000.
[22] *The Project Consultancy Group* v. *The Trustees of the Gray Trust* [1999] BLR 377.
[23] *Maymac Environmental Services Ltd* v. *Faraday Building Services Limited* [2000] CILL 1686.
[24] *John Mowlem & Co. plc* v. *Hydra-Tight Ltd* (2000) CILL 1649.

'(a) Request any party to the contract to supply him with such documents as he may reasonably require including, if he so directs, any written statement from any party to the contract supporting or supplementing the referral notice or any other document given under paragraph 7(2) [of the Scheme].

(b) Decide the language or languages to be used in the adjudication action and whether a translation of any document is to be provided and if so by whom.

(c) Meet and question any of the parties to the contract and their representatives.

(d) Subject to obtaining any necessary consent from a third party or parties, make such site visits and inspections as he considers appropriate, whether accompanied by the parties or not.

(e) Subject to obtaining any necessary consent from a third party or parties, carry out any tests or experiments.

(f) Obtain and consider such representations and submissions as he requires, and, provided he has notified the parties of his intention, appoint experts, assessors or legal advisers.

(g) Give directions as to the timetable for the adjudication, any deadlines, or limits as to the length of written documents or oral representations to be complied with.

(h) Issue other directions relating to the conduct of the adjudication.'

Also, under clause 8.2.2.2 of SBCSub/C and SBCSub/D/C, if the dispute or difference is a dispute or difference relating to clause 3.11.3 (of SBCSub/C and SBCSub/D/C) and whether a direction issued is reasonable in all the circumstances, and if the adjudicator does not have the appropriate expertise and experience in the specialist area or discipline relevant to the direction or issue in dispute, the adjudicator must (not *may*) appoint an independent expert with such expertise and experience to advise and report in writing on whether or not the direction under clause 3.11.3 is reasonable in all the circumstances.

Of course, often the adjudicator does not use all of the above listed powers, and the powers that he does use depend more upon the circumstances of each case.

## What happens if a party does not comply with the adjudicator's direction?

Paragraph 14 of the Scheme states that the parties must comply with any request or direction of the adjudicator in relation to the adjudication.

Paragraph 15 of the Scheme makes it clear that if, without showing sufficient cause, a party fails to comply with any request, direction or timetable of the adjudicator made in accordance with his powers, fails to produce any document or written statement requested by the adjudicator, or in any other way fails to comply with a requirement under these provisions relating to the adjudication, the adjudicator may:

(1) continue the adjudication in the absence of that party or of the document or written statement requested;

(2) draw such inferences from that failure to comply as circumstances may, in the adjudicator's opinion, be justified, and make a decision on the basis of the information before him attaching such weight (i.e. the degree of reliance that an adjudicator places upon evidence provided to him, the heavier the weight of evidence, the more reliance; and the lighter the weight of evidence, the less reliance) as he thinks fit to any evidence submitted to him outside any period he may have requested or directed.

## When does the responding party submit its response to the referral notice?

Under the Scheme, the responding party has no right to issue a response.

However, the adjudicator has the power to request a response, and to satisfy the requirements of natural justice (i.e. in this case that both parties must be given a fair opportunity to present their case) an adjudicator will almost invariably ask for a response to be issued.

As the Scheme does not specifically refer to a response from the responding party, no timetable is set for the issue of same. However, it is commonly the case that an adjudicator will direct that a response should be issued by no later than seven days after the issue of the referral notice.

## Who can assist or represent the parties?

Unless agreed otherwise by the parties, paragraph 16(1) of the Scheme makes it clear that the parties can be assisted by or represented by such advisers or representatives (whether legally qualified or not) as they consider appropriate.

However, unless the adjudicator gives directions to the contrary, paragraph 16(2) of the Scheme makes clear that a party to a dispute may not be represented by more than one person where the adjudicator is considering oral evidence or representations.

## What is the timetable for the adjudication action?

Under the Scheme there is no set timetable for the adjudication action other than that the adjudicator must reach his decision not later than either 28 days after the date of the referral notice (as paragraph 19(1)(a) of the Scheme), or 42 days after the date of the referral notice if the referring party consents (as paragraph 19(1)(b) of the Scheme).

Any period in excess of 42 days after the date of the referral notice can only be agreed if both parties to the dispute consent (as paragraph 19(1)(c) of the Scheme).

Apart from the above long-stop dates, the Scheme does not set out any timetable for the submission of documents, the holding of meetings, etc., and any such timetable dates are set by the adjudicator to suit the circumstances of each individual case.

Any documents, etc. that the adjudicator directs should be issued to him by one party must also be simultaneously issued by that party to the other party.

## What can go wrong in an adjudication action?

First, an adjudicator may resign at any time on giving notice in writing to the parties to the dispute (as paragraph 9(1) of the Scheme).

If an adjudicator does simply resign then the referring party may serve a fresh notice of adjudication and may start the adjudication process anew (as paragraphs 9(3)(a) and 9(3)(b) of the Scheme).

Also, if an adjudicator dies or is otherwise unavailable, the parties may agree to replace the adjudicator or may apply for the nomination of an alternative adjudicator.

An adjudicator must resign because the dispute being referred to him is the same or substantially the same as one which has previously been referred to adjudication. Alternatively, an adjudicator may resign because the dispute varies significantly from the dispute referred to him in the notice of adjudication, and for that reason he is not competent to decide it. In such cases, the adjudicator will be entitled to payment of his reasonable fees and expenses, and will be entitled to apportion his fees as he sees fit (although the parties will remain jointly and severally liable for the adjudicator's fees and expenses) (refer to paragraph 9(4) of the Scheme).

Under paragraph 11(1) of the Scheme, the parties to a dispute (but not a single party unilaterally) can agree to revoke the appointment of the adjudicator – but if they do so they must pay the reasonable fees and expenses of the adjudicator other than if the revocation is because of some default or misconduct of the adjudicator (as paragraph 11(2) of the Scheme).

## How does an adjudicator reach his decision?

An adjudicator is required to ascertain the facts and decide upon the law. He then applies the decided law to the ascertained facts, and from this he produces his decision.

It should be noted that a basic principle is that 'He who asserts must prove'. In other words, a party putting forward a case must prove that case. This is what is known as the 'burden of proof'.

As to evidence, an adjudicator must decide on which party's evidence is preferred on the 'balance of probabilities' principle.

If the evidence of the party with the burden of proof is no more convincing than the evidence of the other party, then the party with the burden fails.

Paragraph 17 of the Scheme requires the adjudicator to consider any *relevant* information submitted to him by any of the parties to the dispute and must make available to them any information he has taken into account in reaching his decision.

The adjudicator is not obliged to give reasons for his decision, unless one of the parties requests that he does (refer to paragraph 22 of the Scheme).

It is normally sensible to ask the adjudicator to give reasons because:

(1) if reasons are given, it is more likely that the parties will understand the adjudicator's decision, and therefore more likely that they will accept the adjudicator's decision as being the final resolution of the dispute; and
(2) a silent decision is more susceptible to attack in enforcement proceedings than a reasoned one.[25]

---

[25] *Joinery Plus Ltd (in administration)* v. *Laing Ltd* [2003] BLR 184.

## What should the adjudicator's decision contain?

The adjudicator must decide the matters in dispute, and he may take into account those matters which the parties to the dispute agree should be within the scope of the adjudication or are matters under the sub-contract which the adjudicator decides are necessarily connected with the dispute (refer to paragraph 20 of the Scheme). It should be noted that it has been found that repudiation arises under a contract[26] (and would therefore, similarly arise under a sub-contract).

In particular the adjudicator may:

(1) open up, revise and review any decision taken or any other certificate given by any person referred to in the sub-contract unless the sub-contract states that the decision or certificate is final and conclusive (refer to paragraph 20(a) of the Scheme;

(2) decide that any of the parties to the dispute is liable to make a payment under the sub-contract and, subject to section 111(4) of the Construction Act, when that payment is due and the final date for payment (as paragraph 20(b) of the Scheme);

(3) having regard to any term of the sub-contract relating to the payment of interest, decide the circumstances in which, and the rates at which, and the periods for which, simple or compound rates of interest must be paid (as paragraph 20(c) of the Scheme).

The adjudicator should give directions as to the time for performance of his decision (e.g. make payment within seven days), but if the adjudicator does not make such directions, then the parties are required to comply with any decision of the adjudicator immediately on delivery of the decision to the parties (as paragraph 21 of the Scheme).

Finally, in his decision, the adjudicator may, if he thinks fit, order any of the parties to comply peremptorily with his decision or any part of it (as paragraph 23(1) of the Scheme).

The word 'peremptorily' simply means to preclude debate, discussion or opposition, and, as will become clear when the enforcement process is considered below, the inclusion of a direction by the adjudicator that the parties should comply peremptorily with his decision does little more than strengthen paragraph 20 of the Scheme (i.e. that the parties are required to comply with any decision of the adjudicator immediately on delivery of the decision to the parties).

## What is the effect of an adjudicator's decision?

The decision of the adjudicator is binding on the parties, and the parties must comply with it until the dispute is finally determined by legal proceedings, by arbitration (if the contract provides for arbitration or the parties otherwise agree to arbitration) or by agreement between the parties (as paragraph 23(2) of the Scheme).

---

[26] *Northern Developments (Cumbria) Ltd v. J. & J. Nichol* [2000] BLR 158.

It should be noted that when the dispute is finally determined by legal proceedings or by arbitration (as applicable) the adjudicator's decision is not open for review or appeal; the dispute is simply considered again with a replacement process as though the adjudicator's decision had not taken place.

Because of the above, an adjudicator's decision is often referred to as being temporarily binding.[27]

The papers used in an adjudication action are not generally privileged and they are open for all to see in a later arbitration or litigation action.

However, paragraph 18 of the Scheme states that the adjudicator and any party to the dispute must not disclose to any other person any information or document provided to him in connection with the adjudication which the party supplying it has indicated is to be treated as confidential, except to the extent that it is necessary for the purposes of, or in connection with, the adjudication.

Obviously, if at some later date the parties finally decide to settle their dispute by arbitration or by litigation, all documents other than those that are privileged would be discoverable in any event.

## Who pays the adjudicator's fees and expenses, and what level of fees and expenses is the adjudicator entitled to?

The adjudicator will normally inform the parties of his fees and expenses schedule at the time of his appointment.

The adjudicator is entitled to the payment of such reasonable amount as he may determine by way of fees and expenses reasonably incurred by him, and it is for the adjudicator to decide and direct how his fees and expenses should be apportioned between the parties (as paragraph 25 of the Scheme). If he does not apportion his fees and expenses then his fees and expenses fall equally between the parties.

Although a challenge may be made to the level of the fees and expenses raised by the adjudicator, in the case of *Stubbs Rich Architects* v. *T.H. Tolley,*[28] where the then current JCT Adjudication Agreement had been used, it was found that a challenge could only effectively be made if the adjudicator had acted in bad faith.

The parties to the adjudication are jointly and severally liable (which means that a party can be sued jointly with another party, or can be sued individually for the whole sum, leaving it to the party sued to recover the monies from the other party with whom he is jointly liable) for the adjudicator's fees and expenses (as paragraph 25 of the Scheme); therefore if one party defaults on any payment due to the adjudicator based upon the adjudicator's directions, then the other party must make payment, and must then sue the defaulting party for the amount paid on their behalf.

Frequently, an adjudicator holds a lien against his decision, in that he will not release his decision until the payment of his fees and expenses.

---

[27] *Macob Civil Engineering Ltd* v. *Morrison Construction Ltd* (1999) 1 BLR 93; *Herschel Engineering Ltd* v. *Bream Property Ltd* [2000] BLR 272.
[28] *Stubbs Rich Architects* v. *W.H. Tolley & Son Ltd* (unreported) (8 August 2001).

Although this is frowned upon by certain professional bodies,[29] and can therefore have serious consequences for the adjudicator in that respect, it is still not an uncommon occurrence.

When such a situation arises, it is obviously normally the referring party that makes payment of the adjudicator's fees and expenses to obtain release of the adjudicator's decision, even though the referring party may not be the party that the adjudicator apportions all or any of his fees to.

Obviously, the adjudicator makes provision for such a situation in his decision, but it is still incumbent upon the referring party to recover from the responding party any fees and expenses that the referring party paid (effectively) on behalf of the responding party.

## Who pays the party's own costs in connection with the adjudication action?

The general position is that each party pays for its own legal and other costs in connection with the adjudication action, irrespective of whether the party wins or loses in the adjudication action. Any application fee paid to an adjudicator nominating body is normally considered to be part of the costs of the referring party in going to adjudication.

If both parties expressly give the adjudicator the jurisdiction to apportion the parties' own legal and other costs, then the adjudicator may (rather than must) do so.

## How does the adjudicator issue his decision, and what happens if the adjudicator is late in issuing his decision?

The adjudicator must issue his decision to both parties (or their representatives) simultaneously, as soon as possible after he has reached his decision (as paragraph 19(3) of the Scheme).

Paragraph 19(2)(a) of the Scheme notes that where the adjudicator fails, for any reason, to reach his decision in accordance with the stipulated time, any of the parties to the dispute may serve a fresh notice of adjudication and may request an alternative (properly appointed) adjudicator to act.

Although the position is not entirely clear, it would appear from case law that if the adjudicator issued his decision a day or two late, this would not automatically invalidate his decision.[30] However, if the period of delay was more than this, and if the parties terminated the adjudicator's appointment, or if one of the parties to the dispute served a fresh notice of adjudication in line with paragraph 19(2)(a) of the Scheme, the initial adjudicator's decision might well be considered as being null and void.[31]

---

[29] The RICS, for example.

[30] *St Andrews Bay Development* v. *HBG Management Ltd and Mrs Janey Milligan* [2003] ScotCS 103; *Barnes & Elliott* v. *Taylor Woodrow* [2003] EWHC 3100 (TCC).

[31] *Simons Construction Ltd* v. *Aardvark Developments Ltd* (2003) CILL 2053; *Ritchie Brothers (PWC) Ltd* v. *David Philp (Commercials) Ltd* (2005) SLT 341.

# What happens if the adjudicator makes a mistake in his decision?

In the adjudication action between *Bouygues* v. *Dahl-Jensen*,[32] when making calculations to answer the question of whether the payments made under the sub-contract represented an overpayment or an underpayment, the adjudicator overlooked the fact that that assessment should be based on the contract sum less retention, rather than on the gross contract sum. This was an error, but an error made within the jurisdiction of the adjudication. The question was could this error be rectified.

When this matter was reviewed by the Court of Appeal, it was held that, provided that an adjudicator acts within his jurisdiction, his decision will stand and be enforceable, even if a mistake is made.

In doing so the Court of Appeal followed the principle stated by Mr Justice Knox in *Nikko* v. *MEPC*[33] who said, of an arbitrator: 'If he has answered the right question in the wrong way, his decision will be binding. If he has answered the wrong question, his decision will be a nullity.'

It did not matter that this appeared to be unjust; this was simply a by-product of the adjudication system. It provided for a quick summary process, which meant that mistakes might well be made.

In *C. & B. Scene* v. *Isobars*[34] the Court of Appeal found that even if an error on a matter of law was made, provided that the error was within the scope of the dispute between the parties, then the adjudicator's decision would be enforceable. In other words, the adjudicator had therefore answered the right question in the wrong way, and the referring party was entitled to enforce the decision.

However, these somewhat harsh effects may not apply if the adjudicator has made a simple slip or mathematical error.

In *Bloor* v. *Bowmer & Kirkland*,[35] Judge Toulmin QC, ruled that a term could be implied into adjudication agreements, giving an adjudicator the power to correct decisions containing accidental errors or omissions, or to clarify any ambiguity.

This therefore introduced into adjudication a slip rule without which an adjudicator would not be able to correct decisions at all, since, without this slip rule, once he has made his decision an adjudicator would be *functus officio*, which means that having discharged his duty he had no further power to rescind his decision or retry the case.

The slip rule was confirmed in a second case, *Edmund Nuttall* v. *Sevenoaks*,[36] provided that the power was exercised within a reasonable time and did not cause prejudice (i.e. injury) to either party.

Following *Bloor* v. *Bowmer & Kirkland*,[37] the following types of error are covered by this slip rule:

(1) accidental error;
(2) omission; and
(3) clarification and/or removal of ambiguity.

---

[32] *Bouygues (UK) Ltd* v. *Dahl-Jensen (UK) Ltd* [2000] BLR 522, CA.
[33] *Nikko Hotels (UK) Ltd* v. *MEPC Ltd* (1991) 28 EG 86.
[34] *C. & B. Scene Concept Design Ltd* v. *Isobars Ltd* (2002) CILL 1829.
[35] *Bloor Construction* v. *Bowmer & Kirkland* (2000) CILL 1626.
[36] *Edmund Nuttall Limited* v. *Sevenoaks District Council* (2000) (unreported), QBD (TCC).
[37] *Bloor Construction* v. *Bowmer & Kirkland* (2000) CILL 1626.

Therefore, provided that the adjudicator has made a simple mistake, that mistake can be corrected, provided that the adjudicator is prepared to acknowledge the slip and make the appropriate amendment.

It must be noted that just because the slip rule now exists, adjudicators are not obliged to apply the slip rule (although in practice they almost always do).

## How can you appeal against an adjudicator's decision?

You cannot appeal against an adjudicator's decision.

If either party is not prepared to accept the adjudicator's decision as being a final decision on the dispute or difference, it can refer the dispute to either legal proceedings or to arbitration, whichever is applicable.

When such a reference is made, it is not in the form of an appeal against the decision of the adjudicator, but involves consideration of the dispute or difference as if no decision had been made by an adjudicator.

## What happens if a party does not comply with the adjudicator's decision – how is a adjudicator's decision enforced?

In the vast majority of cases, parties do comply with the adjudicator's decision.

However, where this compliance does not occur, it must be remembered that adjudication is a contractual provision. Therefore, if an adjudicator's decision is not complied with, the non-compliance is a breach of contract and any action in respect of enforcement has to be under the contract.

First of all, the adjudicator's decision may not need to be enforced. For example, if an adjudicator decides that work that has been carried out complies with the contract, or if an adjudicator decides that a sub-contractor is entitled to an extension of time, then why do these decisions need to be enforced?

However, certain matters, normally those involving some form of payment, may need to be enforced if the party that is due to make payment (or to repay monies) refuses to make payment.

When a party is ordered by an adjudicator to make a payment and fails to do so and the other party starts proceedings for enforcement then the party that was due to make payment will be liable for the other party's reasonable costs in making the application even if the sum is paid in full prior to the court hearing.[38]

The Scheme has its own mechanism for the enforcement of adjudicator's decisions, through the use of an amended version of section 42 of the Arbitration Act 1996 (refer to paragraph 24 of the Scheme).

However, there are some drafting errors which exist when section 42 of the Arbitration Act 1996 is amended by paragraph 24 of the Scheme, and this makes

---

[38] *Outwing Construction Limited v. H. Randell & Son Limited* [1999] BLR 156.

the mechanism unwieldy and impractical. The problem with the resultant device was explored in *Macob* v. *Morrison*.[39]

Consequently, from *Macob* v. *Morrison* the application of (an amended) section 42 of the Arbitration Act 1996 is defunct, but a pattern for the enforcement of adjudicators' decisions in the English courts was set as being by way of a claim form followed by summary judgment.

This is now covered by part 24 of the Civil Procedure Rules (CPR) which enables the court to decide a claim or a particular issue without trial.

The courts have recognized the need to deal with enforcement speedily. This was first explored in *Outwing* v. *Randell*.[40] The general rule now is that when starting an application for summary judgment the matter should be marked 'concerning the enforcement of an adjudicator's decision', or similar. At the same time as issuing the CPR part 7 claim form, the claimant may issue an application without notice to abridge time for the defendant to acknowledge service of the claim form. That application can be heard on the same day as issuing the claim form. The Technology and Construction Court (TCC) will treat these matters as requiring an early hearing.

CPR part 8 is normally used where there is no substantial dispute of fact. It is possible to run a CPR part 24 proceedings seeking summary judgment, and CPR part 8 proceedings seeking a declaration, in tandem. The court may deal with both issues at once.

There is a school of thought that where the decision involves payment of money or makes a declaration that monies are due, enforcement ought to be a much simpler proposition, on the basis that there is no reason why the money should not be pursued through insolvency proceedings. A statutory demand for payment could be made under section 123 or 268 of the Insolvency Act 1986. The party on whom the demand is made has 21 days within which to pay the amount demanded. If payment is not made within this time a petition for bankruptcy or winding up of the defendant may be made to the court.

However, if there is a defence at all, or the claim is contested in some way, the court will not make a bankruptcy or winding-up order, and the party pursuing the claim may then be directed to pay the other party's costs associated with the court action. It is for this reason that insolvency proceedings are seldom used when enforcing adjudicators' decisions.

## Can monies be set-off against an adjudicator's decision?

The premise now is that set-off against an adjudicator's decision will not apply except in extremely limited circumstances.

Paragraph 23.2 of the Scheme requires the parties to comply with the adjudicator's decision.

---

[39] *Macob Civil Engineering Ltd* v. *Morrison Construction Ltd* (1999) 1 BLR 93.
[40] *Outwing Construction Limited* v. *H. Randell & Son Limited* [1999] BLR 156.

**If an adjudicator's decision is enforced and then the recipient of the money goes into liquidation, receivership or bankruptcy, where does this leave the payer in the adjudication who has a legitimate claim in further proceedings to get the money back or pursue a cross claim?**

In *Rainford House* v. *Cadogan*[41] a stay of enforcement was granted on the basis of strong *prima facie* evidence that Rainford House was insolvent. There was a similar result in *Isovel* v. *ABB*.[42]

In *Lovell* v. *Legg*[43] the court rejected an argument on the potential insolvency of the contractor as the reason for not enforcing an adjudicator's decision on the grounds that evidence of insolvency was insubstantial.

The Court of Appeal in the case of *Levolux* v. *Ferson*[44] voided any clause that would defeat the intentions of Parliament, as follows:

'But to my mind the answer to this appeal is the straightforward one provided by Judge Wilcox. The intended purpose of section 108 is plain. It is explained in those cases to which I have referred in an earlier part of this judgment . . . The contract must be construed so as to give effect to the intention of Parliament rather than to defeat it. If that cannot be achieved by way of construction, then the offending clause must be struck down. I would suggest that it can be done without the need to strike out any particular clause and that is by the means adopted by Judge Wilcox. Clause 29.8 and 29.9 (of GC/Works Sub-Contract, in this case) must be read as not applying to monies due by reason of an adjudicator's decision.'

In *A. Straume* v. *Bradlor*[45] it was held that adjudication fell within the meaning of 'other proceedings' in section 11 of the Insolvency Act 1986 so that adjudication proceedings could not be commenced against the insolvent party without the leave of the court.

**What are the principal reasons why an adjudicator's decision may not be enforced?**

The principal reasons why an adjudicator's decision may not be enforced are:

(1) Procedural unfairness or breach of natural justice.[46]
Procedural unfairness or error may cover the breach of the rules of natural justice but it cannot solely be classified on that basis.

There are no hard and fast rules of natural justice. The adjudication process must in essence be fair and allow each side a reasonable opportunity of presenting its case.

[41] *Rainford House Ltd* v. *Cadogan Ltd* [2001] BLR 416.
[42] *Isovel Contracts Ltd* (in administration) v. *ABB Building Technologies Ltd* (formerly ABB Steward Ltd) [2002] 1 BCLC 390.
[43] *Lovell Projects Ltd* v. *Legg and Carver* [2003] CILL 2019.
[44] *Levolux A.T. Ltd* v. *Ferson Contractors* [2003] BLR 118, CA.
[45] *A. Straume (UK) Ltd* v. *Bradlor Developments Ltd* [1999] CILL 1520.
[46] *Discain Project Services* v. *Opecprime Developments Limited* (No.1) (2000) 8 BLR 402; *Woods Hardwick Ltd* v. *Chiltern Air Conditioning* [2001] BLR 23.

It is now generally accepted that the adjudicator has to conduct the proceedings in accordance with the rules of natural justice or as fairly as the limitations imposed by Parliament permit.[47]

There is no reason in law why the adjudicator should not have individual telephone conversations with the parties.[48] Whilst the adjudicator may, due to time constraints, speak to the parties separately he should advise the other party himself of what he has learnt that may be relevant to his decision and not leave it to the party to inform the other party of what he told the adjudicator.[49]

The adjudicator may not take evidence from third parties without giving the parties an opportunity to comment on it,[50] nor should he give a witness statement for use in enforcement proceedings[51] because he would be seen not to be acting impartially.

However, it seems that the adjudicator may ascertain points of law and decide them without reference to the parties, even if the point had not been raised by either of them.[52]

There appears to be a difference between investigating fact and law.

Acting inquisitorially may mean making telephone calls but does not mean acting unfairly. There is a difference between telephone calls of a purely administrative nature and calls that convey or elicit 'relevant information'. The former would be sensible for the adjudicator to have made by a secretary, the latter should be made with great caution.[53]

The current position is that the Human Rights Act 1998 does not apply to adjudication.

In *Elanay* v. *The Vestry*[54] the court stated that:

'In my judgment, Article 6 of the European Convention of Human Rights does not apply to an adjudicator's award or to proceedings before an adjudicator and that is because, although they are the decision or determination of a question of civil rights, they are not in any sense a final determination. When I say that, I am not talking about first instance or appeals, but merely that the determination is itself provisional in the sense that the matter can be re-opened.'

(2) No clear identification as to the status of the decision or of the adjudicator himself.

(3) Unfitness of the adjudicator.

(4) Bias.

This does not have to be actual bias, favouring one party over another; it only needs to be a perception or danger of bias having occurred.[55] There is a simple

---

[47] *Glencot Development & Design Ltd* v. *Ben Barrett & Son Contractors Ltd* [2001] BLR 207.
[48] *Discain Project Services* v. *Opecprime Developments Limited* (No. 2) [2001] BLR 285.
[49] *Discain Project Services* v. *Opecprime Developments Limited* (No.1) (2000) 8 BLR 402.
[50] *Woods Hardwick Ltd* v. *Chiltern Air Conditioning* [2001] BLR 23.
[51] *Discain Project Services* v. *Opecprime Developments Limited* (No.1) (2000) 8 BLR 402.
[52] *Karl Construction (Scotland) Ltd* v. *Sweeney Civil Engineering (Scotland) Ltd* (2000) SCLR 766.
[53] *Discain Project Services* v. *Opecprime Developments Limited* (No. 2) [2001] BLR 285.
[54] *Elanay Contracts Ltd* v. *The Vestry* [2001] BLR 33.
[55] *Discain Project Services* v. *Opecprime Developments Limited* (No.1) (2000) 8 BLR 402; *Woods Hardwick Ltd* v. *Chiltern Air Conditioning* [2001] BLR 23; *Glencot Development & Design Ltd* v. *Ben Barrett & Son Contractors Ltd* [2001] BLR 207.

rule that where the adjudicator discovers something or has a view of his own it must be put to the parties and they must be allowed to address the point or points.

(5) Dealing with matters which are not part of the adjudication, that is exceeding jurisdiction.

(6) Failing to deal with matters which are part of the adjudication.

(7) No clear decision, for example failing to state an identifiable sum, and therefore being incapable of performance.

(8) Illegality.

# Arbitration

## What is arbitration?

Arbitration may be defined as a private procedure for settling disputes whereby a dispute between parties is decided judicially by an impartial individual (or a panel of individuals) either selected by the parties or appointed for that purpose.

An individual so appointed is referred to as an arbitrator, whilst his decision is referred to as an award. An arbitrator's award is legally binding on all the parties to the arbitration proceedings to whom it is addressed.

## How is arbitration incorporated into the JCT Standard Building Sub-Contract?

Arbitration is incorporated into the JCT Standard Building Sub-Contract by way of article 5 (of SBCSub/A or SBCSub/D/A) and clauses 8.3 to 8.8 (arbitration) (of SBCSub/C or SBCSub/D/C).

## Is arbitration automatically incorporated into the JCT Standard Building Sub-Contract?

Arbitration is not automatically incorporated into the JCT Standard Building Sub-Contract.

When the contract is executed, under item 2 of the sub-contract particulars (SBCSub/A or SBCSub/D/A) article 5 should be marked either to apply or not to apply.

If article 5 is not marked either to apply or not to apply, article 5 and clauses 8.3 to 8.8 will be deemed not to apply, and disputes and differences will be determined by legal proceedings.

Therefore, if disputes and differences are to be determined by arbitration and not by legal proceedings, a positive act must be taken if it is required that arbitration is used to finally determine disputes and differences.

## If arbitration is incorporated into the JCT Standard Building Sub-Contract, can the parties ignore this and refer a dispute to court?

If all of the parties to the sub-contract agree not to refer a dispute to arbitration but agree to refer the dispute to litigation, then the dispute can be referred to court.

However, if one party attempts to refer a dispute to court in the case where an arbitration agreement exists, section 9(1) of the Arbitration Act 1996 makes it clear that the other party may apply to the court in which the proceedings have been brought to stay the proceedings in favour of arbitration.

Section 9(4) of the Arbitration Act 1996 states that upon such an application being made, 'the court shall grant a stay unless it is satisfied that the arbitration agreement is null and void, inoperative, or incapable of being performed'.

Despite the above, the courts can intervene in removing arbitrators (as section 24 of the Arbitration Act 1996), in granting relief to an arbitrator who resigns (as section 25 (3) of the Arbitration Act 1996), in enforcing peremptory orders by the arbitrator (as section 42 of the Arbitration Act 1996), in determining the recoverable costs of the arbitration (as section 63(4) of the Arbitration Act 1996) and in determining preliminary points of law (as section 45 of the Arbitration Act 1996).

## What is the advantage of arbitration over legal proceedings?

The principal advantage of arbitration over legal proceedings is that disputes or differences are decided in private by a person whom the parties by agreement can select and who has the relevant experience of building contracts and knowledge of how the construction industry works, and who can set the timetable and procedure to suit the convenience of the parties and from whose award the courts have only limited powers to hear an appeal.

## What can be referred to arbitration?

Apart from the exceptions below, any dispute or difference between the parties of any kind arising out of or in connection with the sub-contract (which could imply that matters such as misrepresentation and tort may be referred to arbitration), whether before, during the progress, or after the completion of the abandonment of the sub-contract works or after the termination of the sub-contractor's employment (refer to article 5 of SBCSub/A and SBCSub/D/A), can be referred to arbitration.

The only exceptions to this is:

(1) any disputes or differences arising under or in respect of the Construction Industry Scheme or VAT, to the extent that legislation provides another method of resolving such disputes or differences; and
(2) any dispute or differences in connection with the enforcement of any decision of an adjudicator.

It should be noted that although the arbitration agreement is incorporated within the JCT Standard Building Sub-Contract it is separable from and survives the termination of the sub-contract. In this regard, section 7 of the Arbitration Act 1996, states that:

> 'Unless otherwise agreed by the parties, an arbitration agreement which forms or was intended to form part of another agreement (whether or not in writing) shall not be regarded as invalid, non-existent or ineffective because that other agreement is invalid, or did not come into existence or has become ineffective, and it shall for that purpose be treated as a distinct agreement.'

## What is the procedure that the arbitration must operate under?

Arbitration must operate under the Arbitration Act 1996 (as noted in clause 8.8 of SBCSub/C and SBCSub/D/C); clauses 8.3 to 8.8 of SBCSub/C or SBCSub/D/C; and the Construction Industry Model Arbitration Rules (CIMAR) current at the main contract base date (as article 5 of SBCSub/A and SBCSub/D/A, and clause 8.3 of SBCSub/C or SBCSub/D/C).

If any amendments to CIMAR have been issued by the JCT after the main contract base date, the parties may, by a joint notice in writing to the arbitrator, state that they wish the arbitration to be conducted in accordance with those amended rules.

The Arbitration Act 1996 confers wide powers on the arbitrator unless the parties have agreed otherwise, but leaves detailed procedural matters to be agreed between the parties or, if not so agreed, to be decided by the arbitrator.

To avoid problems arising, it is advisable to agree as much as possible of the procedural matters in advance, and the Standard Building Sub-Contract does this by incorporating CIMAR which are clearly written and self-explanatory.

CIMAR is a set of rules for the conduct of arbitrations that was originally initiated by the Society of Construction Arbitrators and which has now been adopted by most major construction institutions and bodies.

The objective of CIMAR is to 'provide for the fair, impartial, speedy, cost-effective and binding resolution of construction disputes, with each party having a reasonable opportunity to put his case and to deal with that of his opponent' (as rule 1.2 of CIMAR).

Where CIMAR is used in respect of JCT Contracts, the JCT Supplementary and Advisory Procedures must be read in conjunction with the (unamended) CIMAR Rules.

The arbitrator has the right and duty to decide all procedural matters, subject to the parties' right to agree any matter (CIMAR rule 5.1).

Within 14 days of the arbitrator being appointed the parties must each send the arbitrator and each other a note indicating the nature of the dispute and amounts in issue, the estimated length for the hearing (if a hearing is necessary at all) and the procedures to be followed (in line with CIMAR rule 6.2).

Within 21 days of the arbitrator's acceptance of appointment the arbitrator is to hold a preliminary meeting with the parties to discuss the above matters[56] (as

---

[56] A meeting does not need to be held if both parties agree, and if the arbitrator considers that a meeting is unnecessary (CIMAR rule 6.6).

CIMAR rule 6.3). A meeting does not need to be held if both parties agree, and if the arbitrator considers that a meeting is unnecessary (CIMAR rule 6.6).

The first decision for the arbitrator and the parties to make is whether CIMAR rule 7 (short hearing), rule 8 (documents only), or rule 9 (full procedure) is to apply.

The decision will, of course, depend on the scale and type of dispute.

Under all three rules referred to above, the parties exchange statements of claim and statements of defence, together with copies of documents and witness statements on which they intend to rely.

Under rule 8, the arbitrator makes his or her award based on the documentary evidence only.

Under rule 9, the arbitrator will hold a hearing at which the parties or their representatives can put forward further arguments and evidence. There may also be a site visit.

Under rule 7, a hearing of not more than one day (unless extended by agreement between the parties) is to be held, and the parties must exchange documents either at the same time or in sequence as the arbitrator may direct. In line with the JCT Supplementary and Advisory Procedures (rule 7.2) the parties should submit the written statements of case not later than seven days before the date of the hearing. The arbitrator should publish the award within one month of the hearing.

Under CIMAR rule 4, the arbitrator is given a wide range of powers including the power to obtain advice (rule 4.2); the powers as set out in section 38 of the Arbitration Act 1996 (i.e. general powers exercisable by the tribunal) (rule 4.3); the power to order the preservation of work, goods and materials even though they are part of work that is continuing (rule 4.4); the power to request the parties to carry out tests (rule 4.5); and the power to award costs (rule 13).

In respect of costs, the general principle is that costs should be borne by the losing party.

Subject to any agreement between the parties, the arbitrator has the widest discretion in awarding which party should bear what proportion of the costs of the arbitration.

In arbitration, costs may be awarded on an indemnity basis (i.e. reasonable costs reasonably incurred, with any doubt as to any costs that were reasonable or were reasonably incurred being resolved in favour of the receiving party) or on a standard basis (i.e. reasonable costs reasonably incurred proportionate to the matters in issue, with any doubt as to any costs that were reasonable or were reasonably incurred being resolved in favour of the paying party).

The standard basis is used in the majority of cases, and under the standard basis a successful party may (on average) only recover about 70 to 75% of his total costs incurred in the arbitration.

The 25 to 30% balance of the successful party's total costs is known as the non-recoverable costs, and this can be a significant factor when, on a large arbitration, a party's costs in the arbitration may run in to hundreds of thousands of pounds.

The question of costs in arbitration is normally greatly influenced by 'Calderbank' offers made by the defending party.

A 'Calderbank' offer is the arbitral equivalent of making a payment into court. When considering costs, the question for the arbitrator is whether the claimant has achieved more by rejecting the offer and going on with the arbitration than he would have achieved if he had accepted the offer. If the claimant fails to achieve more than he would have by accepting the offer, then he is likely to have an award of costs made against him.

The arbitrator will be entitled to charge fees and expenses and will apportion those fees between the parties as he sees fit. Irrespective of how the arbitrator apportions his fees and expenses, the parties are jointly and severally liable to the arbitrator for the fees and expenses incurred.

## Who will be the arbitrator?

Clause 8.4.1 of SBCSub/C or SBCSub/D/C notes that:

> 'the Arbitrator shall be an individual agreed by the Parties or, failing such agreement within 14 days (or any agreed extension to that period) after the notice of arbitration is served, appointed on the application of either Party in accordance with Rule 2.3 by the person named in the Sub-Contract Particulars'.

The arbitrator selected by the parties should, of course, be somebody with the necessary expertise and experience to deal with the dispute or difference being referred to arbitration.

The SBCSub/G (paragraph 68) notes that an appointer under item 14 of the sub-contract particulars SBCSub/A (or under item 15 of the sub-contract particulars SBCSub/D/A) should be made in case the parties are unable to agree as to who should be the arbitrator.

The appointer cannot name an arbitrator until at least 14 days after the notice of arbitration is served (refer to clause 8.4.1 of SBCSub/C and SBCSub/D/C).

The parties should agree upon one of a choice of three appointers under item 14 of the sub-contract particulars SBCSub/A (item 15 of the sub-contract particulars SBCSub/D/A), namely the President or a Vice President of:

- the Royal Institute of British Architects;
- the Royal Institution of Chartered Surveyors;
- the Chartered Institute of Arbitrators.

They should signify this agreement by deleting all but one of the three bodies listed above.

If the parties fail to select an appointer when executing the contract, then the default appointer will be the President or a Vice-President of the Royal Institution of Chartered Surveyors (as item 14 of the sub-contract particulars SBCSub/A; item 15 of the sub-contract particulars SBCSub/D/A).

## Is there any restriction on the appointment of an arbitrator?

There may be a restriction on the appointment of an arbitrator.

Clause 8.4.2 of SBCSub/C or SBCSub/D/C notes that where two or more related arbitral proceedings in respect of the main contract works or the sub-contract works falls under separate arbitration agreements, CIMAR rules 2.6, 2.7 and 2.8 must apply.

CIMAR rules 2.6, 2.7 and 2.8 impose duties on those who have the function of appointing arbitrators to give consideration as to whether the same or a different arbitrator should be appointed.

CIMAR rule 2.6 states:

'Where two or more related arbitral proceedings on the same project fall under separate arbitration agreements (whether or not between the same parties) any person who is required to appoint an arbitrator must give due consideration as to whether:

(i) the same arbitrator; or
(ii) a different arbitrator;

should be appointed in respect of those arbitral proceedings and should appoint the same arbitrator unless sufficient grounds are shown for not doing so.'

CIMAR rule 2.7 states:

'Where different persons are required to appoint an arbitrator in relation to arbitral proceedings covered by Rule 2.6, due consideration includes consulting with every other such person. Where an arbitrator has already been appointed in relation to one such arbitral proceeding, due consideration includes considering the appointment of that arbitrator.'

CIMAR rule 2.8 states:

'As between any two or more persons who are required to appoint, the obligation to give due consideration under Rules 2.6 or 2.7 may be discharged by making arrangements for some other person or body to make the appointment in relation to disputes covered by Rule 206.'

## How is an arbitration action commenced?

Arbitration is commenced by one party serving on the other party a written notice of arbitration.

## What is a notice of arbitration?

A notice of arbitration identifies (briefly) the dispute between the parties, and asks the other party to agree to the appointment of an arbitrator (refer to clause 8.4.1 of SBCSub/C and SBCSub/D/C).

Normally, a list of three names of prospective arbitrators should be included in the notice of arbitration.

## What are the powers of the arbitrator?

The powers of the arbitrator are derived from the Arbitration Act 1996 and the Construction Industry Model Arbitration Rules (CIMAR) current at the main contract base date (or any later amendments to those rules that the parties have, by a joint notice in writing to the arbitrator, stated that they wish the arbitration to be conducted in accordance with) (refer to clause 8.3 of SBCSub/C and SBCSub/D/C).

These powers are as detailed under clause 8.5 of SBCSub/C and SBCSub/D/C, and more specifically the arbitrator may:

- rectify the sub-contract so that it accurately reflects the true agreement made by the parties;
- direct such measurements and/or valuations as may in his opinion be desirable in order to determine the rights of the parties;
- ascertain and award any sum which ought to have been the subject of or included in any certificate and to open up, review and revise any certificate, opinion, decision, requirement or notice;
- determine all matters in dispute which must be submitted to him in the same manner as if no such certificate, opinion, decision, requirement or notice had been given.

## Are there joinder provisions within the arbitration rules?

There are joinder provisions within the arbitration rules.

First, clause 8.4.2 of SBCSub/C and SBCSub/D/C states that where two or more related arbitral proceedings in respect of the main contract works or the sub-contract works fall under separate arbitration agreements CIMAR rules 2.6, 2.7 and 2.8 must apply.

CIMAR rule 2.6 says:

'Where two or more related arbitral proceedings on the same project fall under separate arbitration agreements (whether or not between the same parties) any person who is required to appoint an Arbitrator must give due consideration as to whether:

(i) the same Arbitrator; or
(ii) a different arbitrator;

should be appointed in respect of those arbitral proceedings and should appoint the same Arbitrator unless sufficient grounds are shown for not doing so.'

Second, clause 8.4.3 of SBCSub/C and SBCSub/D/C states that after an arbitrator has been appointed either party may give a further notice of arbitration to the other party and to the arbitrator referring any other dispute which falls under article 5 of SBCSub/A and SBCSub/D/A to be decided in the arbitral proceedings and CIMAR rule 3.3 must apply.

CIMAR rule 3.3 states:

'After an Arbitrator has been appointed, either party may give a further Notice of Arbitration to the other and to the Arbitrator referring any other dispute which falls under the same arbitration agreement to those arbitral proceedings. If the other party does not consent to the other dispute being so referred, the Arbitrator may, as he considers appropriate, order either:

(i) that the other dispute should be referred to and consolidated with the same arbitral proceedings; or
(ii) that the other dispute should not be so referred.'

It should be noted that the courts will generally give a purposeful interpretation to joinder provisions that allow for the same arbitrator to be appointed in related disputes to avoid a multiplicity of proceedings, which might lead to excessive costs and inconsistent judgments.[57]

## What is the effect of the arbitrator's award?

Other than for the exceptions listed below, the arbitrator's award will be final and binding on the parties (refer to clause 8.6 of SBCSub/C and SBCSub/D/C).

The only exceptions being that the parties agree (pursuant to section 45(2)(a) and section 69(2)(a) of the Arbitration Act 1996) that either party may (upon notice to the other party and to the arbitrator):

(1) apply to the courts to determine any question of law arising in the course of the reference; and
(2) appeal to the courts on any question of law arising out of an award made in arbitration under the arbitration agreement.

Challenges to the award may also be made on the grounds of lack of substantive jurisdiction (section 67 of the Arbitration Act 1996), and on the grounds of serious irregularity (section 68 of the Arbitration Act 1996).

In respect of all of the challenges or appeals, section 70 of the Arbitration Act 1996 makes it clear that an application to the court cannot be made until any available arbitral process of appeal or review, or any available recourse under section 57 (correction or award or additional award) of the Arbitration Act 1996 has been exhausted.

If a party takes part, or continues to take part, in arbitral proceedings without making, either immediately or within such time as is allowed by the arbitration agreement, or the tribunal, or by the relevant provision of the Arbitration Act 1996, an objection that the tribunal lacks substantive jurisdiction, that the proceedings have been improperly conducted, that there has been a failure to comply with the arbitration agreement, or that there has been any other irregularity affecting the tribunal or the proceedings, he may not raise that objection later, before the tribunal or the court, unless he shows that, at the time he took part in the proceedings, he did not know and could not with reasonable diligence have discovered the grounds for the objection (as section 73 of the Arbitration Act 1996).

## *Legal proceedings*

Legal proceedings are often referred to as litigation and involve taking a dispute for resolution through the civil courts.

Procedure in the civil courts is governed by statutory rules called the Civil Procedure Rules (usually, simply referred to as CPR).

---

[57] *City & General (Holborn) Ltd* v. *AYH plc* [2005] EWHC 2494 (TCC).

The Civil Procedure Rules are contained in separate parts which set out the procedural rules to be followed in specific situations (e.g. disclosure, evidence, costs, offers of settlement, etc.).

The CPR implements recommendations of the Woolf Report (*Access to Justice*, 1994), and seeks to improve the speed, efficiency and accessibility of the civil court procedure.

Legal proceedings are incorporated into the JCT Standard Building Sub-Contract by way of article 6 (of SBCSub/A or SBCSub/D/A).

When the contract is executed, under item 2 of the sub-contract particulars (SBCSub/A or SBCSub/D/A) article 5 (relating to arbitration) should be marked either to apply or not to apply.

If article 5 is marked to apply then arbitration will apply and legal proceedings will not apply.

If article 5 is not marked either to apply or not to apply, article 5 will be deemed *not* to apply, and disputes and differences will be determined by legal proceedings.

Therefore, the default position is that disputes and differences are to be finally determined by legal proceedings.

The basic steps involved in a civil action in the Queen's Bench Division (where most disputes in the construction industry would be dealt with) are:

- the action is begun by the claimant issuing and serving a claim form (previously known as a writ);
- the defendant must then serve a defence or an acknowledgement of service;
- there may be a counterclaim from the defendant and reply from the claimant, in which the cases are defined;
- there is then a procedure of disclosure of documents and inspection of documents;
- witness statements are usually prepared;
- eventually, a trial is held which culminates in a judgment.

The process leading up to the trial is known as the interlocutory proceedings.

Most construction industry cases are dealt with in the TCC (the Technology and Construction Court), and in the TCC the same judge will normally make interlocutory orders as well as being the judge at trial. This makes them familiar with the issues before trial. The TCC procedures are incorporated in a practice direction under CPR.

A particular advantage of legal proceedings is the ease with which other parties may be joined in an action, unlike arbitration proceedings which are ordinarily limited to being between the two parties to the arbitration agreement.

At the commencement of any legal proceedings, the court will allocate the action to one of the three tracks on the following basis:

- small claims track – appropriate for claims not exceeding £5000. The small claims track is largely a documents only process.
- The fast track – appropriate for most cases where the amount claimed is over £5000 but does not exceed £15,000. The fast track is used where the trial is not expected to exceed one day.
- the multi-track – appropriate for all other cases.

In certain situations, a claimant may apply to a court for judgment on his claim on the ground that there is no (or no sufficient) defence. This is known as summary judgment, and if the defendant is unable to satisfy the court that there is an issue which ought to be tried, the claimant will be entitled to immediate judgment (i.e. summary judgment) on the claim or the part of the claim in question.

A successful party in legal proceedings is entitled to an order for payment of his legal costs by the loser (who must also pay his own legal costs).

However, the award of costs is at the discretion of the courts, and, because of this, a successful party may only recover somewhere in the region of 70% of its actual legal costs.

Because of the very high costs of legal proceedings, a defendant that considers he is likely to be found liable in some degree may obtain protection against liability for costs, both his own costs and those of the claimant, by making an offer of settlement. CPR part 36 allows the defendant to make a payment into court of the sum offered, and (in simple terms) if the claimant does not exceed the amount of that payment at the end of the trial then he is liable to pay both his own legal costs and the defendant's legal costs from the date of the notification of the payment into account (even though the claimant may have won the action).

In respect of litigation, the overriding objective of the Civil Procedure Rules (CPR) is to enable the courts to deal with cases justly and encourages the use of ADR (CPR rule 1.3) in helping to resolve disputes both quickly and cheaply.

If a party does not use ADR, then the consequence could be that a cost sanction could be imposed, because in deciding what order (if any) to make about costs, the court must have regard to all of the circumstances, including the conduct of the parties.

## Pre-action protocol for the construction and engineering disputes

In respect of disputes in construction, the CPR incorporates a pre-action protocol for construction and engineering disputes.

The objectives of the protocol are:

(1) to encourage the exchange of early and full information about the prospective legal claim;
(2) to enable parties to avoid litigation by agreeing a settlement of the claim before commencement of proceedings; and
(3) to support the efficient management of proceedings where litigation cannot be avoided.

The process of the protocol is:

- the claimant issues a letter of claim to the defendant;
- the defendant must acknowledge the letter of claim within 14 days (if the defendant does not acknowledge the letter of claim within 14 days, the claimant is entitled to commence proceedings without further compliance with the protocol;
- the defendant must issue a response (and, if appropriate, a counterclaim) within 28 days from the date of receipt of the letter of claim (the period of 28 days can be extended by agreement between the parties up to four months);

- the claimant is to issue a response to any counterclaim within the equivalent period allowed to the defendant to respond to the letter of claim;
- after the exchange of the above submissions, the parties should meet, in an attempt to narrow the issues in dispute, and to agree whether and what form of alternative dispute resolution procedure would be more suitable to settle the outstanding disputes, rather than litigation.

# Appendix A
# Schedule of Notices and Instructions

There are several obligations on both the contractor and the sub-contractor to issue notices in respect of matters, and the main items are summarized below:

## *Notices*

| Clause | Person to issue notice | Matter |
|---|---|---|
| 2.3 | Contractor | Notice to commence work |
| 2.10 | Sub-contractor and contractor | Discrepancy or divergence with or between documents |
| 2.11 | Sub-contractor and contractor | Discrepancies in SCDP documents |
| 2.12 | Sub-contractor and contractor | Divergencies from statutory requirements |
| 2.17.1 | Sub-contractor | Delay to progress or completion of sub-contract works |
| 2.17.3 | Sub-contractor | Further estimates of delay to progress |
| 2.18.2 | Contractor | No granting of extension of time |
| 2.20.1 | Sub-contractor | Date when sub-contract works are practically complete |
| 2.20.2 | Contractor | Dissent to date of practical completion |
| 2.21 | Contractor | Sub-contractor's failure to complete the sub-contract works |
| 3.2 | Contractor | Consent to sub-let any portion of sub-contract works |
| 3.5.1 | Sub-contractor | Disagreement to application of direction under clause 3.4 |
| 3.5.3 | Sub-contractor | Direction injuriously affects the sub-contractor's design portion |
| 3.6 | Contractor | Request to sub-contractor to comply with direction |
| 3.21 | Contractor | Suspension of works |
| 3.7 | Sub-contractor | Confirmation of instruction other than in writing |
| 3.20.1 | Contractor | New appointee as the planning supervisor |
| 3.25 | Contractor (upon receipt of a written request from the sub-contractor) | Completion date, date of possession, practical completion certificate, section completion certificate, certificate of making good, final certificate |
| 4.7 | Contractor and sub-contractor | VAT |
| 4.10.2 | Contractor | Specify payment due to sub-contractor |
| 4.10.3 | Contractor | Specify any amount to be withheld |

| 4.11 | Sub-contractor | Sub-contractor to suspend performance of sub-contract |
|------|----------------|-------------------------------------------------------|
| 4.12.2 | Contractor | Amount of final payment |
| 4.12.3 | Contractor | Amount to be withheld from final payment |
| 4.15.3 | Contractor | List of defects |
| 4.19 | Sub-contractor | Loss and expense due to matters affecting progress of the sub-contract works |
| 4.21.1 | Contractor | Loss and expense caused by sub-contractor's failure to progress sub-contract works |
| 5.3.1 | Sub-contractor | Information not sufficient to provide quotation |
| 5.3.3 | Contractor | Acceptance of quotation |
| 6.5.3 | Sub-contractor | Insurance against personal injury and property damage |
| 6.6.2 | Contractor | Loss or damage to work and site materials |
| 6.7.3 | Sub-contractor and contractor | Loss or damage to the sub-contract works or sub-contract site materials |
| 6.10.3 | Sub-contractor | SCDP professional indemnity insurance |
| 6.14.1 | Contractor | Remedial measures in accordance with joint fire code |
| 7.4.1, 7.4.2, 7.5.1, 7.6 | Contractor | Termination of sub-contract by contractor |
| 7.8.1, 7.8.2, 7.10.2.1 | Sub-contractor | Termination of sub-contract by sub-contractor |
| 7.10.1 | Contractor | Termination of sub-contract by sub-contractor |
| 8.2 | Contractor and sub-contractor | Intention to refer matter to adjudication |
| 8.4.1 | Contractor and sub-contractor | Notice of arbitration |

## Instructions

| Clause | Type of instruction (brief description) |
|--------|------------------------------------------|
| 2.10 | Resolution of discrepancy and divergence between documents |
| 3.4 | Direction in writing issued by contractor to the sub-contractor, which includes a variation |
| 3.4 | Written instruction of the architect under the main contract issued by the contractor to the sub-contractor |
| 3.22 | Suspension of the sub-contract works |
| 5.3.1 | Direction in writing issued by the contractor to the sub-contractor to submit a quotation |
| 3.10 | Opening up for inspection any work covered up, or arrange for test of any materials or goods. |
| 3.11.1 & 3.12.2 | Removal from site or rectification of such non-compliant work |
| 3.13.1 | Failure to comply with clause 2.1 (workmanlike manner and in accordance with health and safety plan) |

# Appendix B
# Standard Letter Precedents

## Notice to commence works – from the contractor

Dear Sirs

Pursuant to clause 2.3 of the sub-contract conditions, we now provide you with our written notice for you to commence your sub-contract works.

Yours faithfully

## Discrepancy or divergence with or between documents – from the sub-contractor

Dear Sirs

Upon inspection of the sub-contract documents we have noted that drawing no. CB/007A (which is a sub-contract document) is not included in the list of drawings attached to the main contract.
    We therefore give notice of this apparent omission to you, in line with clause 2.10 of the sub-contract conditions, and we now await your directions.

Yours faithfully

## Discrepancies in SCDP documents – from the sub-contractor

Dear Sirs

In accordance with clause 2.11.1 of the sub-contract conditions, we notify you that in the sub-contractor's proposals we have stated that Welsh slates will be used in respect of the roofing works, whilst in the SCDP analysis we have stated that Spanish slates will be used.
    We propose resolving this discrepancy by amending the wording in the sub-contractor's proposals to read Spanish slates rather than Welsh slates.
    We now await your directions.

Yours faithfully

## Divergencies from statutory requirements – from the sub-contractor

Dear Sirs

Upon a review of the sub-contractor's proposals it has become apparent that the proposed blocks do not meet the requirements of part L of the Building Regulations.

In line with clause 2.12.1 of the sub-contract conditions, we advise that to rectify this matter we propose upgrading the blocks in the sub-contractor's proposals from type 'SM' to type 'SH'.

We now await your directions.

Yours faithfully

## Delay to progress – from the sub-contractor

Dear Sirs

We give you notice, pursuant to clause 2.17.1 of the sub-contract conditions, that because of instruction number 008 that you issued to us, which added a further layer of plasterboard to the stud partitions, it is apparent that both the progress and the completion of our sub-contract works will be delayed.

We consider that instruction number 008 is a relevant sub-contract event in line with clause 2.19.1 of the sub-contract conditions, and we consider that this event will delay the completion of our sub-contract works by three weeks.

We would be grateful if you could now grant us an extension of time to the completion date of our sub-contract works in line with the above details.

Yours faithfully

## No granting of extension of time – from the contractor

Dear Sirs

We acknowledge receipt of your letter dated 1 April 2006, in which you request an extension of time to the completion date of your sub-contract works.

We have carefully considered your request, but must advise you that we are of the opinion that, if you increase your resources to suit the additional work now available to you, there is no reason why your period on site should be extended.

Therefore, as required by clause 2.18.2 of the sub-contract conditions, we notify you that we do not consider that you are entitled to any extension to the completion date of your sub-contract works.

Yours faithfully

## Date of practical completion – from the sub-contractor

Dear Sirs

Pursuant to clause 2.20.1 of the sub-contract conditions, we notify you by way of this letter that in our opinion our sub-contract works were practically complete on 1 April 2006.

We confirm that all as-built drawings were provided to you on 28 March 2006, and we attach herewith all of the information that you have requested from us in respect of the health and safety file.

The documents attached are:

• •
• •
• •

Yours faithfully

## Dissension from date of practical completion – from the contractor

Letter dated 5 April 2006

Dear Sirs

We acknowledge receipt of your letter dated 3 April 2006.

We do not agree that your sub-contract works are practically complete, and we would remind you that apart from the incomplete defect rectification works as listed within our letter dated 28 March 2006, you have not yet commenced in rooms G004, F005 and S006, at all.

We request that you deal with all outstanding works without further delay.

Yours faithfully

## Failure of sub-contractor to complete on time – from the contractor

Letter dated 3 April 2006

Dear Sirs

Despite the fact that you were due to complete your sub-contract works on 28 March 2006, we note that your works on site remain incomplete.

We therefore, must record this fact in line with clause 2.21 of the sub-contract conditions, and we now give you notice that we will seek to recover from your company the amount of any direct loss and/or expense that we suffer or incur as a result of your failure to complete your sub-contract works on time.

Yours faithfully

## Direction injuriously affecting the sub-contractor's design portion – from the sub-contractor

Letter dated 3 March 2006

Dear Sirs

We acknowledge receipt of your instruction number 005 dated 1 March 2006 which instructs us to add further tank supports to the structural roof members.

We must notify you that because of the proprietary nature of our roof design, the structural roof members are designed to take a minimal load only. The roof as designed by us will not carry the weight of the additional tank supports and tanks proposed, and therefore your instruction injuriously affects the efficacy of our design.

We issue this notice to you pursuant to clause 3.5.3 of the sub-contract conditions.

We now await your further directions.

Yours faithfully

## Confirmation of instruction other than in writing – from the sub-contractor

Letter dated 3 March 2006

Dear Sirs

We confirm that on 1 March 2006, your site agent, Mr Jones, instructed our site foreman, Mr Smith, to add an additional layer of 12.5 mm plasterboard to both faces of the studwork walls to walls A, B and C in Room S004.

Yours faithfully

## Payment notice – from the contractor

Letter dated 24 February 2006

Dear Sirs

With reference to your interim payment no. 4 which has a payment due date of 22 February 2006, we now attach to this letter full details of the gross valuation that we propose to pay.

The said gross valuation is in the sum of £12,524.56 (excluding VAT).

Please accept this letter as being the written notice required under clause 4.10.2 of the sub-contract conditions.

Yours faithfully

## Withholding notice – from the contractor

Letter dated 2 March 2006

Dear Sirs

Further to our letter dated 24 February 2006 which provided full details of the gross valuation that we propose paying in respect of your interim payment no. 4, we now advise you that we propose withholding from the above amount, and from your interim payment no. 4, the sum of £563.00 (excluding VAT).

We propose withholding the said sum of £563.00 as this relates to the repairs to a brick wall that was damaged by your operatives whilst moving your materials around site. The said damage was notified to you on site incident form number 003A dated 1 March 2006. The amount of £563.00 is in line with invoice number AB005263 submitted by A. Hod & Co. Ltd, the bricklaying sub-contractor that repaired the damage.

Yours faithfully

## Sub-contractor to suspend performance – from the sub-contractor

2 April 2006

Dear Sirs

We draw to your attention that you have failed to make any payment to us in respect of our interim payment no. 4 by the final date for payment.

In line with clause 4.11 of the sub-contract conditions, we hereby give you notice that if you fail to make payment to us of the amount that is due within the next seven days, we intend to suspend the performance of our obligations under the sub-contract until payment is made by you of the amount that is due.

Yours faithfully

## Sub-contractor suspends performance – from the sub-contractor

10 April 2006

Dear Sirs

Further to our letter dated 2 April 2006 we record that we still have not received any further payment from you in respect of our interim payment no. 4.

We therefore must reluctantly advise you that, pursuant to clause 4.11 of the sub-contract conditions, we intend to suspend the performance of our obligations under the sub-contract as from 11 April 2006.

Yours faithfully

## Final payment notice – from the contractor

3 May 2006

By special delivery

Dear Sirs

With reference to your final payment which has a payment due date of 2 May 2006, we now attach to this letter full details of the gross valuation that we propose to pay.

The said gross valuation is in the sum of £26,356.25 (excluding VAT).

Please accept this letter as being the final payment notice as required under clause 4.12.2 of the sub-contract conditions.

Yours faithfully

## Withholding notice from final payment – from the contractor

Letter dated 15 May 2006

Dear Sirs

Further to our letter dated 3 May 2006 which provided full details of the gross valuation that we propose paying in respect of your final payment, we now advise you that we propose withholding from the above amount, and from your final payment, the sum of £2663.00 (excluding VAT).

We propose withholding the said sum of £2663.00 as this relates to:

(1) The repairs to a brick wall that was damaged by your operatives whilst moving your materials around site. The said damage was notified to you on site incident form number 003A dated 1 March 2006. The amount of £563.00 is in line with invoice number AB005263 submitted by A. Hod & Co. Ltd, the bricklaying sub-contractor that repaired the damage.
(2) The cost of materials that you asked us to provide to you as fully detailed in your letter dated 20 April 2007, all in line with the attached invoice number DD030DE submitted by Ace Materials Co. Ltd, in the sum of £2100.00.

Yours faithfully

## Application for loss and expense – from the sub-contractor

Dear Sirs

Pursuant to clause 4.19 of the sub-contract conditions, we hereby make application to you for the direct loss and/or expense that we are incurring as a result of the disruption to the regular progress of the sub-contract works that we are suffering.

This said disruption is as a direct result of the direction that you issued to us which passed on the architect's instruction that no work is to be carried out between the hours of 10.00 a.m. and 2.00 p.m., even though the site working hours stipulated in the sub-contract documents are from 8.00 a.m. to 5.00 p.m.

The issue of your direction referred to above is a relevant sub-contract matter under clause 4.20.1 of the sub-contract conditions.

We attach to this letter full details of the direct loss and/or expense that we are incurring, and would ask that you advise us by return if you require any further details or information from us at this stage.

Finally, and as noted under clause 4.22 of the sub-contract conditions, we wish to record that this application is being made without prejudice to any other rights or remedies that we may possess.

Yours faithfully

## Application for contractor's reimbursement – from the contractor

Dear Sirs

We must notify you that the regular progress of the main contract works is being disrupted as a result of your failure to complete your works to Zone A of Building B.

The other sub-contractors in that area are being disrupted because of the need to re-sequence their works as a result of your non-completion.

We have received loss and/or expense applications from three of the sub-contractors affected, which we have been obliged to pay, and we attach those applications to this letter.

The attachments are:

XX Co. Ltd's letter dated 4 March 2006
YY Co. Ltd's letter dated 6 March 2006
ZZ Co. Ltd's letter dated 3 March 2006

Accordingly, in line with clause 4.21 of the sub-contract conditions, we give you notice of the above and we advise you that we will seek to recover the said direct loss and/or expense from your company.

We would ask that you advise us by return if you require any further details or information from us at this stage.

Finally, and as noted under clause 4.22 of the sub-contract conditions, we wish to record that this notification is being made without prejudice to any other rights or remedies that we may possess.

Yours faithfully

## Default by sub-contractor – from the contractor

By special delivery (optional)

Letter dated 12 February 2006

Dear Sirs

We give you this notice pursuant to clause 7.4.1 of the sub-contract conditions.

We record that, for no good reason, you have failed to attend site for the last four working days. It is therefore clear that you are failing to proceed regularly and diligently with the sub-contract works. This a specified default covered by clause 7.4.1.2 of the sub-contract conditions.

We must now put you on notice that if you continue this default for ten days from receipt of this notice we will terminate your employment under the sub-contract.

We trust that it will not be necessary for us to take this action, and hope that you will now immediately proceed regularly and diligently with the sub-contract works.

Yours faithfully

## Default by sub-contractor – from the contractor

By special delivery (optional)

Letter dated 28 February 2006

Dear Sirs

We would refer you to our letter dated 12 February 2006.

We are extremely disappointed to record that you have failed to return to site and have failed to proceed regularly and diligently with the sub-contract works.

Because of this continuing default by your company, we have no option other than to terminate your employment under the sub-contract; pursuant to clause 7.4.2 of the sub-contract conditions.

We refer you to clause 7.7 of the sub-contract conditions which sets out the consequences of termination under clause 7.4, and we will write to you with further details of these consequences under separate cover.

Yours faithfully

## Default by contractor – from the sub-contractor

By special delivery (optional)

Letter dated 12 February 2006

Dear Sirs

We give you this notice pursuant to clause 7.8.1 of the sub-contract conditions.

We record that, for no good reason, you have failed to proceed with the main contract works and this has seriously affected the progress of our sub-contract works. This is a specified default covered by clause 7.8.1.2 of the sub-contract conditions.

We must now put you on notice that if you continue this default for ten days from receipt of this notice we will terminate our employment under the sub-contract.

We trust that it will not be necessary for us to take this action, and hope that you will now immediately proceed with the main contract works.

Yours faithfully

## Default by contractor – from the sub-contractor

By special delivery (optional)

Letter dated 28 February 2006

Dear Sirs

We would refer you to our letter dated 12 February 2006.

We are extremely disappointed to record that, despite our notice, you have failed to proceed with the main contract works.

Because of this continuing default by your company, we have no option other than to terminate our employment under the sub-contract; pursuant to clause 7.8.2 of the sub-contract conditions.

We refer you to clause 7.11 of the sub-contract conditions which sets out the consequences of termination under clause 7.8, and we will write to you with further details of these consequences under separate cover.

Yours faithfully

# Table of Cases

# Table of Statutes and Regulations

# Clause, Recitals, Articles and Sub-Contract Particulars Index

# Subject Index